C000230920

Mind changing! It is relatively rare to read a book that actually changes your mind about something that really matters to you. Refreshingly, this is such a book. As one who has consistently and confidently prophesised the decline, if not demise, of the professions of the built environment, I am now renewed with a fresh hope for their future. Simon Foxell has, somewhat inspiringly, given hindsight, insight and foresight to this engaging subject. Hindsight, in the way he scholarly describes their historical foundations and subsequent flourishing. Insight, by skilfully examining what some see as their present malfunctioning, misdirection and marginalisation. And, most of all, foresight, for his strategic exploration of their future prospects, promise and potential with his outline 'six-point plan' and collaborative sense of purpose. The challenge of reforming the professions of the built environment remains daunting, but I am now imbued with a greater feeling of optimism, expectancy and direction.

John Ratcliffe
Professor Emeritus DIT & Founder, The Futures Academy

Professionalism for the Built Environment raises our awareness of the foundations of professional society and sets the agenda for a radical reappraisal of the educational and institutional structures we may require in the future. This book stimulates debate on the professions' responsibilities to civil society, company and self. It provides an authoritative briefing for practitioners, educators, clients and all concerned with creating a built environment that is equitable, fosters wellbeing and improves livelihoods.

John Worthington
Co-founder DEGW. Formerly Commissioner of Independent Transport
Commission and Director The Academy of Urbanism

This comprehensive discourse explores the origins, development, rise and uncomfortable transformation of our Built Environment Professional Institutions, all fascinating. For more than 25 years Simon Foxell has been exploring these ideas and has assembled a great story. This book concludes with a collective challenge that is beautifully articulated by considering the many pressures facing a new young chartered professional. Yes, professionalism clearly has great value but, in a rapidly changing world with automation threatening, can the institutions pool their efforts for the greater good of humankind and our planet? This is essential reading for aspirants, practitioners and academics.

Robin Nicholson
Senior Partner Cullinan Studio

A real gem for any professional interested in improving collaboration amongst built environment professionals. *Professionalism for the Built Environment* touches on the past to inform an effective plan for the future of our industry. As a Head of Department for Architecture and the Built Environment, I am delighted to see a book that will bring the professions closer together and will become the 'must have' for all built environment professionals.

Elena Marco
Head of Department Architecture and the Built Environment,
University of the West of England

This book looks at the professions' need to reinforce the actual and perceived value which ultimately justifies their existence: highly topical and of genuine importance in the UK but also globally. The author treads a careful path between dispassionate and survivalist analysis, with wide-ranging examples from earlier times and other sectors. The thorny issue of whether professionals can reclaim the benevolent arbiter role without having any (or much) "skin in the game" is addressed as is the role of ethics, supported by a comparison of institute code of conduct topics.

John Field
Past President CIBSE (2016–17)

Anyone entering the built environment professions should ground themselves in an understanding of how today's society arrived at its current relationship with the professions. As the labour market evolves, the dominant status of professions within our class structure is being challenged. Professional practices are definitive in governing how we build, but must be clearer in guiding the professional to reconcile competing priorities, balancing public good with private gain. In the future, built environment professions will have to reckon with the planet's most serious challenges. We should therefore be both critical and concerned with the fitness of built environment professions. This book equips the professionals themselves to make that assessment of their profession. It is accessible to all who work building our future, yet stands on the shoulders of the intellectual giants who have analysed the emergence of professions over three centuries – a history intertwined with that of the RSA.

Matthew Taylor
Chief Executive, Royal Society for the Encouragement of Arts,
Manufactures and Commerce (RSA)

This is an energetic and engaged survey of ethical issues raised in the evolution and current state of the professions most responsible for the built environment in England. It offers a stimulating and valuable contribution to debates over the ethical responsibilities of professionals more broadly.

Melissa Lane
Class of 1943 Professor of Politics, Princeton University

Simon Foxell continues the important work begun by RIBA Past President Frank Duffy. For the profession of architecture to survive and prosper we must build a body of knowledge that is of more value and accessible to society and to our clients. In an era when most service delivery is advertised as being 'professional' we must set new standards for the definition of ethical professionalism. We must play our part in the huge challenge to diversify and rejuvenate the building industry, improving how we develop and retain talented people from all walks of life. Deploying artificial intelligence alongside these strategies, we can increase the perceived value of our service, advice and design solutions. We can thereby expect in due course to receive a more valuable return for that contribution. Simon Foxell's book provides us with vital and insightful analysis towards achieving this goal.

Benjamin Derbyshire
President, Royal Institute of British Architects (RIBA)

This is a hugely interesting and important book examining professionalism and the built world in which we live. Ranging from the occupationally created social order of Florence in the middle ages to Nazi Germany's use of architecture, and ending up at today, Foxell has produced a magnificent portrait of how society, economic structures and the built environment intermesh. A compelling analysis is presented of how engineering, surveying and architecture impact on us today. A truly impressive book, which I wholeheartedly recommend to anyone interested in understanding why their lived experience is as it is.

Professor Gerard Hanlon
Queen Mary University of London

In an era of abundant information, rapidly developing machine intelligence and an erosion of public confidence in 'experts', questions of what it means to be professional have never seemed more pressing. This excellent book is a timely reminder of the importance of professional values, not only for all those involved in construction and the built environment, but for society more generally. Simon Foxell has provided an honest, insightful and compelling analysis of contemporary challenges facing built environment professionals and their institutions. In doing this he has positioned the role of professionalism where it should be: in supporting the creation and development of a built environment that is fit for everyone. A very worthy achievement indeed. Highly recommended.

Professor John Connaughton
Head of Construction Management and Engineering,
School of the Built Environment, University of Reading

This is a must have book for professionals working in the built environment. It is absolutely bursting with useful information and insight. For the new entrant there is a clear explanation of what it means and how to be a true professional. For the student of history there is a comprehensive and intriguing account of how we got here and why we are the way that we are. For practitioners and business leaders there is a compelling analysis of the current state of the professional services sector with many clues about how to build a value proposition based on professionalism and avoid the commodified race to the bottom. For leaders of the professions there are both challenges and supportive suggestions about how we need to deliver social value in a highly competitive and fast changing market economy. I always like a book that helps organise my, often disparate, thoughts and this was definitely one of those!

Ed McCann
Director of Expedition Engineering and Vice President of
the Institution of Civil Engineers.

This book provides a fascinating insight into the history of the professions of architects, surveyors and civil engineers, which puts into perspective many of the challenges resulting from the fragmentation of the construction industry we are currently witnessing. By drawing out the parallels between the industries in the UK, Germany, France and America the book explores the organisation of professional bodies further. It lays out the bare facts of the challenges to our industry, which we need to learn from and use to push our profession forward.

Caroline Buckingham
Director caroline buckingham architects and
RIBA Vice President Practice & Profession

Given the rise in public awareness and expectations of the 'professional' in any sector of the economy, there has never been a more important time to explore the key aspects of professionalism in the built environment arena. Given the Grenfell Tower disaster and similar catastrophic events around the world, the need for a professional approach when addressing construction methods, building regulations, health and safety guidelines, environmental statute and land use policy is a prerequisite to ensure public confidence. Otherwise, fundamental ethical behaviour and the financial transparency of the property sector are put at risk. This is a superbly written review of a wide range of professional areas all supporting the built environment and I endorse it wholeheartedly.

Dr Louise Brooke-Smith
Partner Arcadis LLP – UK Head of Town Planning and Head of Social Value,
RICS Past Global President (2014–15)

The professions need to move with the times and what Simon Foxell's book points out is that professional organisations **do** have the ability to change with society's changing demands, even though often it involves a lot of kicking and screaming. His historical analysis of the development of professions globally and across the principle disciplines of the built environment is both comprehensive and eminently readable. From the incongruous detail of the 1887 RIBA entrance requirements, through the turbulent history that led to the development in the 1930s of the separation of the profession from its regulating authority, to Egan, Morrell and the rise and fall of CABE, this is a definitive study that will stand the test of time. Perhaps for longer than the professional climate it defines.

Peter Clegg
Senior Partner Feilden Clegg Bradley Studios

How best can the professions, for whom certainty is a touchstone, navigate such very uncertain times? Simon Foxell finds a compelling path that starts with a meticulous and fascinating examination of the development of construction professions and proceeds to an illuminating comparison of their cultures across continents. Running through these stories are masterful analyses of both foundational and contemporary questions of ethics and morals and their intersection with practicalities. Through these Foxell suggests the way ahead. Although the focus is on construction professions, all will find much food for thought here. A culmination of two decades of engagement with professional dilemmas, this book goes further and deeper than any other.

Sunand Prasad
Senior Partner Penoyre & Prasad LLP and RIBA Past President (2007–09)

There has never been a more appropriate time to examine the relationship between our built environment professions and the society that they serve. Recent events such as the Grenfell fire and the collapse of Carillion have brought into question the level of trust in which our professionals are held. Simon Foxell has produced a comprehensive review of how the professions in the built environment have developed over the last 200 years and, importantly, addressed what it means to be a professional in this industry today. I commend this book to current practitioners, students and those operating in the policy arena.

Professor Peter Hansford
Chair of Construction and Infrastructure Policy at
University College London and former Chief Construction Adviser
to the UK Government

This is a book on a subject that should be dull but which is actually engaging and full of vital lessons for society. Tracing the history of three powerful institutes, Foxell shows how dilemmas professionals struggle with today – public interest vs. private gain, occult knowledge vs. openness, established order vs. rapid change – are rooted in how the professions formed and how they guarded their privileges. Proving that this can't go on, the author offers solutions to keep the professions relevant and valuable. All professionals and policy-makers (as well as clients and the public who rely on them) should act on these solutions.

Dr Richard Simmons
Chief Executive the Commission for Architecture and
the Built Environment 2004–11

This is a fascinating and important book, which by tracing the historical constitution of professions, leads us to a critical point in the contemporary context, where the service provision aspect of professional life in a financialised world has overwhelmed the professions' wider responsibilities to society. Foxell ends with a compelling set of suggestions as to how the professions might regain the trust from clients and the public, a trust that has been rapidly eroded in recent years.

Professor Jeremy Till
Head of Central Saint Martins and Pro Vice-Chancellor Research,
University of the Arts London

Britain is belatedly waking up to the full damage that will have been done, once all facets of society are conceived in terms of efficiency maximisation and return on investment, especially in the built environment. In this urgent and timely study, Simon Foxell looks at the ethos of professionalism, and how it carves out a space of responsibility between market and state. As Foxell convincingly argues, a revival of long-standing professional ideals will be crucial, if we're to achieve a better balance between social and economic responsibilities in future.

William Davies
Reader in Political Economy, Goldsmiths, University of London

Simon Foxell captures so well the broad sweep of the history, the current state and the potential futures and dilemmas of the built environment professions – especially the threats to their ethical codes and integrity and how their public interest mandate etched into their origins is endangered. This book is enriching. It makes the topic feel urgent. It forces us to ask what kind of cities we want and how the professions need to collaborate across boundaries so cities become the best they can be while addressing the global issues that really matter to all of us.

Charles Landry
Author of *The Creative City: A Toolkit for Urban Innovators*

Foxell's comprehensive reader on ethics, morals and the public interest is a rich mine of information, explanations and discussions providing the reader with many entry points to the subject. You won't agree with everything, but the important thing is that these subjects become part of our everyday discussions and arguments about how to balance the survival of our professions and the planet.

Stephen Hill
Director C$_2$O futureplanners

This book is both timely and important, coming as it does in the wake of the Grenfell fire in London, at a time of acute housing shortage in the UK and in the context of global climate change. By charting the evolution of the built environment professions, the author encourages us to consider the changes now required if they are to remain relevant to both society and their members in today's world.

Peter Oborn, Director Peter Oborn Associates and former RIBA Vice President International (2011–17)

Now that sociologists have largely abandoned serious study of the professions, it falls to those on the inside to do this necessary work. In *Professionalism for the Built Environment*, architect Simon Foxell embraces this challenge. He combines a much-needed snapshot of the current state of the UK's building professions with an in-depth historical survey explaining how we got here. This presentation of a coherent set of ideas for moving forward could well reinvigorate the entire field of study.

Professor Tom Spector
Oklahoma State University

This book should be read for three fundamental attributes: (1) its description of the building professions together; (2) its description of professionalism as a concept and practice that is constantly under construction; and (3) its in-depth, cross-disciplinary, and cross-national histories of the construction of our building professions. The book's analysis of the building/design professions shows how changes in the concept of a "profession" are always linked to permutations in the economic regime (e.g. capitalism). The story told here is not of an oligopoly of socially-motivated citizens operating outside the market but, rather, of an entrepreneurial class trying to establish a brand. Foxell is much too judicious to use these terms in his precise historical narrative, but that doesn't diminish the fact of its message nor its call for the need for change.

Professor Peggy Deamer
Yale University

This is an amazingly thorough analysis of how the professions of the built environment have developed. Focussing on architecture, civil engineering and surveying, Simon Foxell tracks how the notion of what it means to be a professional has been influenced by changes in the political and social climate, both in the UK and abroad. He also shows how the relationships between education, the professional institutions and practice have continuously evolved. In this fast changing world, with the increasing impact of technology, globalisation, global warming and events such as the Grenfell Tower disaster, he challenges the professionals through their institutions to re-evaluate their modus operandi and to put the public interest, including concern for the wider environment, first.

Jane Wernick
engineersHRW

Foxell presents an optimistic overview of current professionalism. This is unbelievably timely in the post-Grenfell era when questions arise about the trust with which construction professionals are held. We have known that a cavernous performance gap exists, but with the highly public exposure of the failing eco-system which created a perfect storm for Grenfell, this has now been extended into a credibility gap. This book provides a critical approach to the past in order to create a view of a better future. A strong case is made for the key Institutes collaborating and rebuilding trust. All professionals need to read this excellent book and re-invigorate our performance, ethics and attitudes, working together across the professions to earn and rebuild public trust.

Jane Duncan
RIBA Immediate Past President

PROFESSIONALISM FOR THE BUILT ENVIRONMENT

In the aftermath of the Grenfell Tower tragedy, this new book provides thought provoking commentary on the nature of the relationship between society, the prevailing economic system and professionalism in the built environment. It is both an introduction to and an examination of professionalism and professional bodies in the sector, including a view of the future of professionalism and the organizations serving it.

Simon Foxell outlines the history of professionalism in the sector, comparing and contrasting the development of the three major historic professions working in the construction industry: civil engineering, architecture and surveying. He examines how their systems have developed over time, where they are currently and some options for the future, whilst asking difficult questions about ethics, training, education, public trust and expectation from within and outside the industry.

The book concludes with a six-point plan to help, if not ensure, that the professions remain an effective and essential part of both society and the economy; a part that allows the system to operate smoothly and easily, but also fairly and to the benefit of all.

Essential reading for built environment professionals and students doing the professional studies elements of their training or in the process of applying for chartership or registration. The issues and lessons are applicable across all building professions.

Simon Foxell is Principal of The Architects Practice. He has designed buildings across the UK including houses, schools and offices and in 1996 was awarded the prestigious Benedictus Prize by the Union of International Architects. In recent years he has worked with schools and local authorities, designing new facilities and advising on development programmes. He was lead design adviser to Birmingham City Council's Transforming Education programme from 2007 to 2010.

He is a core member and co-ordinator of the Edge, the built environment think tank, and is on the Green Construction Board's Routemap group. He is a former member of the RIBA Council and Board and was Chair of both Policy and Strategy and RIBA London region as well as Vice chair of Practice and Building Futures.

BRI Book Series

New interdisciplinary and transdisciplinary approaches need a forum for information and discussion.

This book series shares similar aims and scope to the journal, *Building Research & Information*, but allows for a deeper discussion, together with more practical material.

SCOPE: This book series explores the linkages between the built, natural, social and economic environments, with an emphasis on the interactions between theory, policy and practice. Emphasis is on the performance, impacts, assessment, contributions, improvement and value of buildings, building stocks and related systems: i.e. ecologies, resources (water, energy, air, materials, building stocks, etc.), sustainable development (social, economic, environmental and natural capitals) and climate change (mitigation and adaptation).

If you wish to contribute to the series then contact the series editor Richard Lorch at richard@rlorch.net with a short note about your ideas.

The Rebound Effect in Home Heating
A Guide for Policymakers and Practitioners
Ray Galvin

The Recovery of Natural Environments in Architecture
Air, Comfort and Climate
C. Alan Short

PROFESSIONALISM FOR THE BUILT ENVIRONMENT

Simon Foxell

LONDON AND NEW YORK

First published 2019
by Routledge
2 Park Square, Milton Park, Abingdon, Oxon OX14 4RN

and by Routledge
711 Third Avenue, New York, NY 10017

Routledge is an imprint of the Taylor & Francis Group, an informa business

British Library Cataloguing-in-Publication Data
A catalogue record for this book is available from the British Library

Library of Congress Cataloging-in-Publication Data
Names: Foxell, Simon, author.
Title: Professionalism for the built environment / Simon Foxell.
Description: Abingdon, Oxon ; New York, NY :
Routledge/Taylor & Francis, 2018. | Includes bibliographical references.
Identifiers: LCCN 2018009988 | ISBN 9781138900202 (hardback : alk. paper) |
ISBN 9781138900219 (pbk. : alk. paper) | ISBN 9781315707402 (ebook)
Subjects: LCSH: Civil engineers–Professional ethics. | Architects and builders–Professional ethics. | Construction worksrs–Professional ethics. | Social responsibility of business. | Building failures–Prevention. | Building–Moral and ethical aspects.
Classification: LCC TA157 .F69 2018 | DDC 174/.9624–dc23
LC record available at https://lccn.loc.gov/2018009988

ISBN: 978-1-138-90020-2 (hbk)
ISBN: 978-1-138-90021-9 (pbk)
ISBN: 978-1-315-70740-2 (ebk)

Typeset in Univers
by Out of House Publishing

Printed and bound by CPI Group (UK) Ltd, Croydon, CR0 4YY

CONTENTS

PREFACE

Professions are unique social configurations, which flourished with the emergence of the liberal state and the rise of the bourgeoisie. In their classical forms, independent producers served individual consumers while acting collectively to improve the ethics and ensure the competence of their peers. Professions persuade the state to protect them from market forces by arguing that commercialism is inconsistent with their market calling. At the same time, professions invoke market imperatives to resist state control, insisting that they must preserve their 'independence' in order to serve their clients loyally.

(Abel 1989, 'Between market and state', p285)

During the writing of this book there have been many points at which questions of professionalism have arisen as matters of public debate:

- Surgeons have been found guilty of abusing their professional positions and the trust patients placed with them to carry out unnecessary procedures and in some cases cause grievous bodily harm
- A public relations firm was found to have run a secret campaign intended to foment ethnic and racial tensions in South Africa and was expelled from its trade association for unethical and unprofessional behaviour. At the start of 2018 the relevant professional body had yet to rule on the case
- Large numbers of cases of sexual harassment have been reported to have occurred in the workplace
- The tragedy at Grenfell Tower in west London, which caught fire on 14th June 2017 and resulted in the deaths of 71 residents.

The last of these has been the most devastating with its rapid spread up the 27 storeys of the residential tower and the fury of the inferno. Many contributing factors have been identified for the spread of the fire, with several of them connecting directly to questions about the responsibilities and the professionalism of the many parties involved with the design, installation and management of the works to the building. All of these incidents, but especially Grenfell, have resulted in an intense discussion about the nature of the relationship between society, the prevailing economic system and professionalism. These debates have yet to provoke any significant change – but it will almost certainly come.

Other tragedies as well as incidents of corruption and misconduct in public office are related in this book; including, in the latter part of the 19th century, the collapse of railway bridges as a result of substandard design or construction with the loss of many lives, and the discovery that a number of well-respected public sector construction professionals were taking bribes and kickbacks. Such events and the media pressure that followed them led directly to changes in the overall system of professionalism. Measures were implemented to tackle problems of trust; regulations introduced or upgraded; and action taken to ensure that professionals were better qualified, more effectively informed or were obliged to follow official procedures and protocols. The professions reacted, made modifications to their *modus operandi* and carried on. However the longer-term outcomes weren't always predictable and at times huge unintentional shifts occurred over a relatively short period in response to relatively minor incidents or policy announcements. What appeared to be stable and time-honoured has frequently been revealed to be highly adaptable under pressure. Such malleability over two centuries has been the key reason for the impressive survival of the professional system – even while its superficial trappings have appeared little changed.

This book is primarily a comparative study of the three major historic professions working in the construction industry – civil engineering, architecture and surveying – only drawing in the other professions in the industry in its later stages. It examines how their systems have developed over time, where they are currently and some options for the future. It makes no attempt to be all encompassing, but seeks to place the changes in the context of their time. The narrative tends to expand where participants and witnesses have left a solid trail of evidence and there has been a degree of controversy and resistance to change. Less emphasis is given to periods when little has been recorded or progress has been smooth. At times a greater amount of space is inevitably devoted to the activities of the more voluble architectural profession with its extensive media coverage, while the engineers and surveyors have often kept their own counsel and more quietly dealt with the same issues. Apologies are due where developments are illustrated mainly by architectural examples alone and the other professions have been afforded less attention.

The juxtaposition of the three professions provides a fuller perspective on the development of professional services in the construction industry. It makes it possible to observe the progress of the separate but closely interlinked strands, from their foundations in the early 19th century to the present, when the individual disciplines are beginning to show cautious signs of coalescing. Achieving a wide-angle view has meant stepping back from some of the controversies that, at the time, obsessed individual professions. Once established, the separate professions have tended to proceed single-mindedly; each focusing on its own issues as if they were unique to itself. Yet, often enough, the issues they were tackling had already been addressed and decided by another organization. Observing them moving, however haphazardly, across their shared territory it is easier, with all the benefits of hindsight, to see where they were heading, even if they couldn't see it themselves at the time.

Over the decades the principal opposing poles attracting the professions have been public service on one side and self-interest on the other and high levels of institutional energy have been devoted to maintaining a degree of balance between them. This book makes it apparent that the professions have been moving away from the public service pole for some considerable time, under pressure from the market and also from supposedly cash-poor governments that consistently, and with little regard for quality or social equity, want to buy more 'service' for less 'reward'. This approach has shredded many of the core values of professionalism and a course corrective is, as a result, long overdue. The experience of the Grenfell Tower fire, if nothing else, teaches us that. Professionalism is under threat even while it offers potential solutions to urgent social challenges.

Predictions of the death of the professions have been commonplace for a long time now, at least since the collapse of Keynesianism and triumph of the market in the 1970s. Books such as *The Future of Work* (Handy 1984), *The End of the Professions* (Broadbent *et al.* 1997) and *The Future of the Professions* (Susskind and Susskind 2015) have detailed the coming destruction of the sector as a source of stable employment for almost as long. Despite these forecasts, there is no suggestion that professionalism itself will become redundant. If anything, the need for the honest and impartial intermediary role will increase as society grapples with how to control ever more powerful market players and protect the interests of both individuals and society. Whether society can work without arbiters, equipped with expertise and experience, is a question rarely asked or answered.

It is this position: 'between market and state' (to adopt part of the title of Richard Abel's 1989 paper quoted above) that the professions have forgotten to properly maintain or interest themselves in. The lure of (as well as the neoliberal injunction towards) helping themselves has pulled them gradually away from public service and their duty to the wider good. This has in turn contributed to the crisis of professional relevance and the generalized existential threat. The professions as collective entities and in particular their representative institutions, urgently need to focus on their position and role in society again. For a while, at least, there needs to be a corrective bias towards the public interest in order to establish the role of trusted intermediary.[1]

This, and much more, is discussed in the later chapters of this book, but first it is necessary to set out some of the extensive background to the development of the three professions. The earlier chapters describe how they formed and reached their present state alongside some of the ever evolving thinking that has attempted to analyse and interpret the elusive idea of professionalism.

The current study follows on from, and is influenced by, a number of other pieces of work. These, in particular, include: *The professionals' choice: the future of the built environment professions* instigated by the author in the early 2000s for the Building Futures think tank and published by them in 2003; the special issue of the *Building Research & Information* journal on 'New Professionalism' guest edited by Bill Bordass and Adrian Leaman, published in 2013;[2] and The Edge Commission on the Future of Professionalism and the subsequent report, *Collaboration for Change*, authored by Paul Morrell and edited by the current author, published in 2015, together with numerous sessions and discussions before and after publication of each one of them. All three of these publications have shaped the thinking evident in the main text – although I take full responsibility for the contents that follow and cannot place any of the blame on these foundational works.

The sources for this study are as wide as I have been able to make them, but they are also all, intentionally, in the public domain. Where the institutions are criticized for not publishing relevant information, it is for not making it publicly available, even when it may be circulated internally.

In sketching out the development of professionalism in the industry I have been very dependent on and drawn extensively from histories covering each of the three professions/institutions: Barrington Kaye's *The Development of the Architectural Profession in Britain: a sociological study* (1960), F.M.L. Thompson's *Chartered Surveyors: the growth of a profession* (1968) and Garth Watson's *The Civils: The story of the Institution of Civil Engineers* (1988). These vary in their focus (and degree of institutional pride), however all three run out of steam or interest during the first half of the 20th century and each of these professions are in need of serious studies that bring the histories of their institutions up to date. This work makes no attempt at this onerous task.

I am a practising architect rather than an academic. I hope, but doubt, that a degree of familiarity with the subject matter atones for any deficiency in critical apparatus and method. As a former Council member of the RIBA I know most about the internal workings of that organization, but over the years I have also taken part in many discussions and sessions with senior members and staff from the other institutions in the industry addressing issues of professionalism and institutional organization. I hope, in so far as it is possible to overcome one's own internal biases, that I have been able to be even-handed between them.

Thanks are due to the many who have contributed knowingly and unknowingly to this book, but in particular to my family and wife Anne, my editor at *Building Research & Information*, Richard Lorch, who has had the unenviable task of reading through and commenting insightfully on early drafts, as well as fellow members of the Edge who have added their many thoughts to the process and have helped to shape some of the ideas contained in these pages.

Five hundred years ago, when Martin Luther published his 95 *Theses* in Wittenberg and triggered the Reformation, he can have had little idea that he was going to change the world of work as well as that of religion. By challenging the prevailing concept of holy poverty and labour-free salvation, he endowed ordinary work with both moral and spiritual value and made the concept of a 'calling' or 'vocation' (*Beruf*) applicable to all forms of useful endeavour. What mattered according to the Reformation historian, Euan Cameron, was 'the "life of service" and consideration for the needs of society' (2012, p419). To achieve this the clergy was re-imagined as a task-focused cadre rather than as representatives of divine grace. They professed their vows and became part of a community based around a shared education, knowledge and occupation. This established itself as the default model that other professional groupings would follow from then on. They have been considering and professing ever since.

Simon Foxell, London, January 2018

Notes

1 It is tempting, of course, to write 're-establish' but there is little evidence that the professions have every truly held the position of independent and impartial 'intermediate agent', first described by John Soane in 1788 (see Chapter 3) and by many others since, however much they may have aspired to it.

2 See www.tandfonline.com/toc/rbri20/41/1

FOREWORD

This book considers, with the benefit of both experience and insight, the future of the professions at a time of challenge and change.

It is probably the conceit of every generation to imagine that it stands at a crossroads; at a point where major decisions must be made about how the future might differ from the past. It also happens that, at such a time, the loudest voices are too often those who either profess that all the fundamentals are fine, but if we just weather the storm then we will emerge into calmer waters, or argue that we've reached a dead-end and that we have to start again from scratch. Whilst these extreme positions tend to capture the headlines, the reality is that the institutions that are in it for the long term need to twist and turn and trim their sails to suit prevailing winds.

Few organizations have shown themselves as capable of adapting as the professions. It is not, however, a mere conceit to see the threats currently lining up against those who hold to a 'steady as she goes' course. These include a general breakdown in trust in almost all institutions, and particularly in the opinion of 'experts', and a suspicion of anything posing as altruism. There is an increasing sense that lowest cost is the clearest measure of 'best value'; a growing belief (by everybody but themselves) that much of the work of, and therefore need for, professions will be rendered redundant by technology and the internet; a parallel confusion of 'knowledge' with 'judgement'; and an accumulation of grand challenges (economic, social and environmental). At present the professions seem incapable not just of answering these challenges, but of even finding an agreed route to an answer.

Yet each of these threats presents a related opportunity. For example, anybody who has made the mistake of googling their medical symptoms will find that it rarely dispenses with the need for further professional advice. Professionals need to harness the benefits of technology to bolster the quality of their own performance, but must also make the case for providing expert guidance and judgement.

Responding to these threats and opportunities requires much thinking and discussion. This must involve individuals and practices offering professional services as well as the institutions whose validation reinforces the promises of individual practitioners.

In choosing a path, we need a guide (or a series of guides), and this book is one. Foxell starts with a well-researched history of how the modern professions came into being and ends by considering the current challenges faced by the professions, their clients and by wider society. A key issue is how these needs can be made to converge to the benefit of all. Although the past isn't always a reliable guide to the future, knowing how we got to where we are is a necessary foundation to thinking about where we go next. It is also a fascinating read.

It does not, of course, provide all of the answers. (If it did, then at least some of them would be suspect, given the amount of change that is occurring.) However, in a quiet voice expressing a depth of thought, it frames the questions and points the way to some of the answers. Foxell leaves us with the most critical question of all, as we move deeper into the 21st century: just what is it that the professions actually profess?

That the question has to be asked, and even more that it is not easy to find a comprehensive answer, are evidence for a loss of the certainties that previous generations of professionals lived by. This is no bad thing. The immediate neighbour of certainty is complacency, and that is not a strategy either to survive or prosper in a time of change.

As professional businesses seek to commercialize, and commercial businesses seek to professionalize, it becomes increasingly difficult to isolate just what it is that differentiates the established professions from those simply seeking to act 'professionally'. Professional businesses have a choice. They can retreat back into history, erect barricades, lobby for some form of protectionism and seek to deny market access to those who do not carry their badge. Alternatively, they can recognize the winds of change and set their course accordingly. They can search for a genuinely differentiated level of skill, knowledge and judgement, all anchored within a strong ethical framework, which both their clients and society depend upon. Nowadays this is called a brand, a promise that will be kept.

Paul Morrell, Chief Construction Adviser to the UK Government 2009–12

1

Profession

Wir haben nicht nur einen Beruf, sondern der Beruf hat auch uns.
(Not only do we have a profession, but a profession has us.)
(*Eduard Spranger,* Psychologie des Jugendalters, *1927, p251*)

New arrivals have come to knock on the door of their chosen institution to request admittance. It has been a long journey to reach this place and time; requiring years of study, effort and determination; and now – bearing skills, knowledge and hard-won experience – turning back is not an option. They primarily want to be welcomed in, and, if that is allowed and agreed, it is likely to be for the duration of their lives. It will require submission to a set of rules – some relatively opaque, others almost invisible – that currently appear to have an uncertain future. Do these eager and deserving postulants know what to expect behind the impressive doors? Their chosen profession appears the very essence of stability, but it too has been on a journey to meet them at this particular moment and the doors, along with the pillars that frame them, may just be for show. What do the new recruits really know of the profession they want to join and the commitment they are about to make?

The long-established construction discipline they want to join, despite its appearance and talk, is itself not a stand-alone entity, but an element in a much greater system of professions that exists across the UK and the world; a system they will also be signing up to become a part of. The system of professions comes with its own extended history and with an essential, if problematic and ever dynamic, relationship to society. It is a system, which will adopt them as much as they adopt it. For they are about to enter a compact, not so much with their institution as with wider society, their clients and employers; to behave in a certain way, to maintain expertise and provide a range of skills and services; and, in return, win status, opportunities and, to an extent, a living. The nature of the compact or deal isn't necessarily very clear, but it has been arrived at after centuries of trial, error, push and shove; a process that will carry on shifting the underlying terms of their working lives for the foreseeable future and well beyond.

The professional system has developed over both time and geography in response to the need to foster and provide insight, expertise and knowhow in specialist types of work that rely on intellect, judgement and, critically, good faith. Such services are difficult to describe clearly, certainly in advance, and hence difficult to price, especially when the knowledge is in an area that is beyond a commissioning party's own comprehension. Society requires solutions that provide effective answers to immediate challenges while ensuring that wider and longer-term views are taken, the level of risk is minimized and those with privileged and unchallengeable knowledge do not take advantage of their position. The best if not the only way to deliver such solutions has proved over time to be through a system of trust. The challenge has been to build and maintain that trust while being able to nurture and guarantee the proficiency to tackle fresh problems.

The solution reached in many parts of the world has been for society and government to accept and recognize self-regulating associations of experts who have an interest in maintaining quality and who will peer review and police one another's activities to ensure high standards are maintained. To achieve this those associations need to control access to their group and maintain the authority to eject members who do not meet, or continue to meet, their collective standards of both work and behaviour. In return they expect a monopoly over the provision of their particular service in a geographic area and can be expected to fight off any competitors they see as infringing it.

This solution forms the foundation for the professional system and, because it is effectively a compact between those with expertise and the ruling and administrative system of a society, it can be prone to political influence, interference and expediency. For example, during the US Presidency of Andrew Jackson (1829–37), the recognition of professional standing was abruptly cancelled, and anyone was allowed to practice, resulting in huge damage and mistrust (Sarfatti Larson 2013 [1977]). Alternatively, if professionals are given too much licence, there is a real danger of self-aggrandisement and abdication of responsibility. It is easy for professionals to become a self-serving elite. Those who get too close to power assume a ruling position in society is theirs by right, or alternatively, they lose their essential independence and ability to self-regulate. The compact needs to be frequently reviewed and kept in good shape with each side respecting the role of the other as it is renegotiated for fresh circumstances and times. Professionalism needs to be externally based, maintaining separation from government, clients and individual practice and as a result gaining perspective.

The form of the compact varies significantly from jurisdiction to jurisdiction but in Britain it originally and deliberately placed the professions just outside the competitive capitalist system, in so far as that was possible. This was grounded in the belief that the temptation of additional earnings should not have an influence on critical judgements and decisions on behalf of society. The independent application of expertise and knowledge should not be, or perceived to be, compromised by financial factors. In the construction sector a degree of separation was sought between commerce and professionalism, leading to fixed fee scales to avoid competition by price and any financial involvement with manufacture, building or development in order to protect professional independence and save it from moral jeopardy. This approach, of 'people with power based on their expertise, neither knights nor peasants but able to front the middle to tell both what to do' (Pye 2014, p154), has been excised in recent decades, both in the UK and elsewhere, and today built environment professionals are exhorted to be as entrepreneurial and business-minded as possible. This has unquestionably redefined the relationship between the professional and society, but the consequences have been little discussed, other than to bemoan the downgraded influence of the professions, and there has been scarcely any attempt to consciously recast or re-invent the professional compact for modern times and circumstances.

Roots of professionalism in Britain

The origins of professionalism in Europe lie in the three 'learned professions': the church (including scholars), medicine and the law. The separate social and legal status of these occupations provided the necessary licence for highly trained individuals to work relatively freely outside the restrictive employment norms of the feudal system and ensured a means of control over who could and couldn't practise a discipline. A parallel system of trade guilds controlled a much wider range of crafts and occupations, including all the building trades and, initially at least, the design and construction oversight roles. The guilds were generally more restrictive and rule-bound than the professions and although they largely faded from sight during the early modern period, only reappearing much later and in a new guise as trade unions, they left behind a substantial legacy of occupational traditions and demarcations for new professional organizations eventually adopt. While the guilds declined the learned professions thrived and gradually integrated their activities with those of the rapidly developing universities that opened across Europe during the Renaissance and Enlightenment periods.

In early modern continental Europe the universities supported a wide range of disciplines including mathematics, science and technology, and provided a home and training for a great breadth of specialisms. In contrast, English universities focused narrowly on theology and classical studies, leaving the higher levels of medical education to teaching hospitals and legal training to the Inns of Court. Other disciplines relied almost entirely on apprenticeships with most skills and knowledge either handed down from a master or learnt on the job. This separation, with the resultant lack of both an academic base and state imposed administrative support for professional disciplines during their formative periods, created a distinct professional ethos in England and Wales – Scotland adopted a more European approach. The new professions, as each emerged, needed to be largely self-sufficient and independent, both academically and politically. Perversely, and for both good and bad, it also gave them a more amateur, do-it-yourself quality than their continental equivalents, making them reliant on an active and participatory membership in order to thrive. The relative self-sufficiency of English and Commonwealth professionalism continues to be distinctively different from the state-sanctioned system prevalent elsewhere in Europe.

In the 18th century, with ever growing technical expertise, an expanding workforce and an increasingly connected world, specialist workers along with many enthusiastic amateurs intent on increasing their knowledge and engaging with their peers and colleagues, met in newly founded clubs and societies. This was as true for the well-connected founders of the Royal Society of Arts (RSA), established in London in 1725 as the Society for the Encouragement of Arts, Manufacture and Commerce; the members of the Bath and West

Society (1777), founded to promote and improve agriculture; and the group of Midlands' industrialists, natural philosophers and inveterate inventors who met monthly as the Lunar Society of Birmingham (1775–1813). Generally such organizations would establish themselves long before seeking official or royal approval. Only in rare cases, usually when national security was at stake, would bodies be instigated directly by the State (e.g. the Royal Observatory in 1647 or the Ordnance Survey in 1747).

Professional formation

With less prominence, similar meetings began to be organized across Britain amongst various proto-professional groups and, often separately, between their apprentices and trainees. In March 1771 a group of 13 senior engineers, including John Smeaton (1724–92) and Robert Mylne (1734–1811), started meeting in the King's Head Tavern in Holborn, London following the regular sessions of Parliamentary Committees dealing with legislation for the new roads, navigations and canals they were involved with.[1] Architects, having first gathered as sub-groups within a number of artist societies and fraternities as well as the Royal Academy of Arts (1768), struck out on their own with the 'Architects' Club' in October 1791 and, from then on, met weekly at the Thatched House Tavern in St James, London. Members included James Wyatt (1746–1813), Henry Holland (1745–1806), George Dance junior (1741–1825) and Samuel Pepys Cockerell (1753–1827) (Kaye 1960, pp57–8). The founders of the City Company of Surveyors having been excluded from the Architects' Club for their unwelcome commercial links, established the 'Surveyors' Club' only a few months later in 1792 (Thompson 1968 pp71–2).

Initially these societies were dining clubs, but by early in the next century the expectations and needs of their members had expanded. A strong demand had developed for ways to represent and protect the interests of the various different groups, for forums in which technical information could be shared and prepared for publication and for a more structured and improved system of education for new recruits. Above all there was a demand for means of establishing social respectability and helping the public to distinguish the skilled and educated from their lesser competitors and the interlopers who were believed to be damaging their social standing and threatening their livelihoods. It was a call for something we would recognize as a professional body. A period of institution building had begun.

For the groups of friends, colleagues and rivals who met in coffee houses and rented meeting rooms in the rough and ready period prior to establishing more formal associations, there was little doubt that they were already professionals or that they had a discipline to call their own. They wanted to meet to exchange ideas and share their enthusiasms, but not necessarily to found worthy institutions. No longer were they the long bearded seniors familiar with the corridors of Parliament, but impatient young men who wanted to make a difference and build themselves the professional home that they saw as essential to support and facilitate their future career. The Institution of Civil Engineers (ICE) was formed by a group of young engineers, one still an articled pupil, with an average age of 25, but they were also realistic and worldly-wise enough to know that they needed a figurehead of much greater reputation and years than themselves, and in 1820 persuaded the revered, 63 year old Thomas Telford to accept the post of President. It was a post he occupied for another 14 years, until his death in 1834 (Watson 1988, p9).

The founding of the Institute of British Architects was messier, comprising the merging and fracturing of several rival architectural groups, including the Society of British Architects. A new journal, the *Architectural Magazine* reported on their first meeting in The Freemason's Tavern in Covent Garden, London in January 1834 (Loudon 1834, p89). Under the heading '*Domestic Notices*' it noted:

> *Architectural Societies.* – Endeavours seem to be making among the architects and surveyors of London and its vicinity to establish Architectural Societies. A number of architects have met occasionally, for mutual improvement in Exeter Hall, since 1831, and they contemplate the establishment of an Architecture College, after the manner of those in Germany. Some other architects and surveyors have had meetings, for the purpose of establishing a society for the study of architecture and architectural topography. ... There is also a third body of architects, who meet occasionally, but chiefly, as our correspondent Scrutator informs us, for the purpose of dining together.

Following the meeting architects delegated by the new Society wrote to their own local grandee, Sir John Soane (1753–1837), asking him to become their first president. He declined, conveniently citing the rules of the Royal Academy, where he had the post of professor (Kaye 1960, p77). Lacking Soane's eminence and pulling power, the Society never established itself fully enough and became one of the many precursor groups that acted as tributaries feeding the Institute of British Architects, founded a year later. The founders of the Institute chose their first president more astutely, picking the First Lord of the Admiralty, Earl de Grey. He wasn't an architect, even if he had designed his own country house (with assistance); but he was supremely well connected.

The young civil engineers of the fledgling ICE saw with a high degree of clarity that they needed a legal and organizational structure that would provide them with public recognition and social status but, above all, would project stability and engender confidence. The medical colleges and various legal associations, as well as familiar institutions including the RSA and The Royal Institution (1799), all used the mechanism of a Royal Charter to achieve this, but on consideration the engineers preferred the easier to manage form of a legal trust.

One problem with a Royal Charter was that although this ancient system of granting royal recognition and incorporation had a long track record; having originally been used to grant rights and duties to cities and towns and then been adapted for use by worshipful companies, hospitals, colleges, schools and learned societies; by the end of the 18th century it was being employed by almost any venture looking to achieve an advantageous market position. Recent charters included the Incorporated British, Irish and Colonial Silk Company (1825) and the London Portable Gas Company (1827) as well as overseas trading firms such as the Van Diemen's Land Company (1825). Nor, at the time, did the King, George IV, have a very savoury or popular reputation.[2] The standing of the Royal Charter system was at its nadir in the early 19th century

Table 1.1 The process of professional formation as defined by Geoffrey Millerson (1964, pp10–13)

Professional formation, or to use the ugly academic term, professionalisation, is now a recognised process that disciplines and occupations go through to emerge as a 'profession' – an exclusive body of practitioners, conforming to rules and social norms and in return being given a protected designation and a formal status. The four steps in the process are:

1. The ability to define and recognise an area of expertise and activity that is the particular focus of the profession including a knowledge base and a distinct group of practitioners that separates them from other adjacent groups and disciplines. For some this needs to be accompanied by the development of a theoretical apparatus that covers the discipline. Critically the realisation that a discipline has a clear identity of its own needs to be accepted both internally and externally.

2. The opportunity to acquire and share knowledge and best practice, often combined with the intention to improve standards and make them consistent and uniform across the discipline. Amongst the early professions this was rudimentary but developed into an education system with a curriculum, training facilities, testing and certification. In recent times the concept of lifelong learning and continual professional development (CPD) has extended this principle.

3. The demand from emerging professionals for a separate domain to be recognised. This is often expressed as dissatisfaction with current practice, training, standards and opportunities for improvements and recognition. Sometimes it may relate to a strong difference of opinion in an existing profession or the sense that one party is an overlooked junior discipline.

4. Recognition from outside, including official acknowledgement.

and sensibly the ICE was looking for a more robust alternative. Nonetheless, their solicitor insisted that establishing a trust presented 'insuperable difficulties' (Watson 1988, p19) and that, despite everything, applying for a Royal Charter was still the best option.

The engineers' capitulation to legal advice and the cluster of professional institutions that followed in their wake rescued the status of the Royal Charter system, but only after a modest innovation. The beginning of the 19th century was a period of intense philhellenism and a dedicated revival of the culture of Pericleian Athens, as exemplified by the life (and death in 1824 at Missolonghi) of Lord Byron, was in progress. The values of classical Greece were being, briefly as it turned out, promoted over the sturdy Roman virtues that had inspired the French Revolution and that would later enthuse the Victorians. The Greek belief in putting the welfare of the city-state and the community above that of the individual, the idea of civic virtue, first found its way first into the charter of the Royal Society of Literature (RSL), granted in 1825, via the enthusiasm of its classically minded first president. The ICE consciously followed the RSL's example in 1828 and the RIBA did so again in 1837. As a result the idea that professions should incorporate an ethos of service to the wider public benefit, rather than simply profit or even individual charitable good works, became an essential characteristic of all the professions. The Privy Council is now absolutely clear that 'incorporation by Charter should be in the public interest' (Privy Council 2017).

With public interest a defining quality of the professional institutions these new organizations found themselves, possibly unexpectedly, with a clear and powerful narrative that explained their place in and benefit to society, a narrative that justified their exclusiveness as well as the special privileges they sought and were frequently granted. It succeeded in turning a diverse group of professional individuals into a force, not only within society, but also in politics and the economy: a profession. Collectively such groups rapidly became 'the professions' and by the second half of the 19th century this new social order had established itself as the basis of the *arriviste* middle class and became a force radically different from anything that had gone before. An indication of this can be seen in the following, from *The Choice of Profession* by Henry Byerley Thomson, an 1857 career guide (p5):

> The importance of the professions and the professional classes can hardly be overrated, they form the head of the great English middle class, maintain its tone of independence, keep up to the mark its standard of morality, and direct its intelligence.

Although the ICE and RIBA had been operating for several decades by the middle of the century the question of who was and wasn't a professional in the construction sector was still a very live issue for society as well as the institutions. An explanatory note in the Official Registrar's report accompanying the 1851 Census for England and Wales commented that 'many of the 2,971 architects are undoubtedly builders: and here the want of a better nomenclature is felt' (cited in Wheeler 2014, p.60). The institutions had moved almost straight to the final stage of professional formation (see Table 1.1) without stopping to clarify who was properly qualified to join them. The education and accreditation process only became mature enough at the end of the 19th century to resolve the issue and minimum entry standards were only raised to degree level over the course of the 20th century.

The 1851 census listed both civil engineers and architects in Class IV (Persons engaged in Literature, the Fine Arts, and the Sciences) and certainly not in the more prestigious Class III (Learned Professions). Surveyors were categorized further still down the schedule and were linked with builders in Class XI (Person engaged in Art and Mechanic Productions) (University of Leicester 2003). In 1861 architects were moved to Class V Industrial (Persons engaged in Art and Mechanic Productions) (Wheeler 2014, pp60–1), but by the time of the 1881 census all three professions had been upgraded to Class I PROFESSIONAL CLASS (subsection 3 – Persons engaged in Professional Occupations) (Woollard 1999). Achieving respectability had been a 60 year struggle, but the new institutions had eventually secured themselves a position in society, even though their numbers remained small – see Figure 1.1.

Defining professionalism

The hard won collective status of 'profession' and with it the qualification of professional, enabling a wide range of individuals to use their institutional membership to enhance their standing, caused a social revolution, even if of the mildest mannered and most bloodless kind. The establishment of so many new 'professions', including veterinarians, accountants and actuaries, marked the arrival of what the social historian Harold Perkin (1989) has labelled 'professional society', a period lasting from approximately the 1880s to the 1980s,

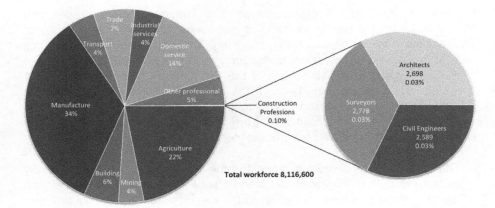

Figure 1.1
Occupations in England and Wales, 1851

Source: 1851 Census data – Corfield (1995, p32); Armstrong (1972, pp191–310).

when professionals were in the ascendant. The rise of professionalism may not have attracted anything like the same attention from historians as, say, that of the welfare state, but the question of what constitutes 'professionalism', has nonetheless been the subject of a continuous stream of academic discourse and social science analysis, with attempt after attempt to identify the essence of its character and to pin down the elusive nature of its success. This work has, over the decades, produced a series of insightful studies that have lit up different aspects of professionalism and provided a language for discussing and understanding it. But they have also left something else: a tangible record of the way professionalism has changed over the last two centuries; the demands and expectations placed on it; and the way society has reacted to a system that has come bearing gifts, but has not always entirely succeeded in producing benefits. As Perkin (1989, p117) says:

> The professional is offering a service that is … esoteric, evanescent and fiduciary – beyond a layman's knowledge or judgement, impossible to pin down or fault even when it fails, and which must therefore be taken on trust – he is dependant on persuading the client to accept his valuation of the service rather than allowing it to find its own value in the marketplace. His interest therefore is to persuade society to set aside a secure income, or a monopolistic level of fees, to enable him to perform the service rather than jeopardize it by subjecting it to the rigours of capitalist competition in the conventional free market. It is the success of such persuasion which raises him (when he succeeds) above the economic battle, and gives him a stake in creating a society which plays down class conflict (in the long if not in the short term) and plays up mutual service and responsibility and the efficient use of human resources.

The theorizing and attempts to produce definitions started with Emile Durkheim in 1893 and his doctoral thesis *The Division of Labour in Society*, in which he addressed the origins and early workings of professional formation. While it is important to recognize that French has a wider occupational usage of the term 'profession' than English, he identified, in a spirit of optimism, that:

What we particularly see from professional groupings is a moral force capable of curbing individual egoism, of nurturing among workers a greater feeling of common solidarity, and preventing the rule of the strongest from being applied too brutally in industrial and commercial relationships.[3]

The historian and social critic R.H. Tawney picked up the thread of Durkheim's argument in his book, *The Acquisitive Society* (1920), an excoriating attack on modern society in England and the selfishness that was driving it. He believed that contemporary mores '[made] the individual the center of his own universe, and dissolve[d] moral principles into a choice of expediencies' (p31) and saw professionalism as a force that could potentially counteract and control the dangerous instincts of both industry and capital and become the principle that would allow nationalization of the country's assets to function for the wider benefit of all (p92):

A Profession may be defined most simply as a trade which is organized, incompletely, no doubt, but genuinely, for the performance of function. It is not simply a collection of individuals who get a living for themselves by the same kind of work. Nor is it merely a group which is organized exclusively for the economic protection of its members, though that is normally among its purposes. It is a body of men who carry on their work in accordance with rules designed to enforce certain standards both for the better protection of its members and for the better service of the public.

Tawney was followed by the team of Alexander Carr-Saunders and Paul Wilson, who in their classic 1933 study, *The Professions*, equally saw the professional system as a force that could resist the incursions of both industrialism and official bureaucracy, but with the fundamental difference that they believed that professionals encouraged the stability of society rather than providing a means for overturning the ownership of the means of production. In their view professional associations (cited in Macdonald 1995, p2):

inherit, preserve and pass on a tradition ... they engender modes of life, habits of thought and standards of judgement which render them centres of resistance to crude forces which threaten steady and peaceful evolution.

Carr-Saunders and Wilson's work was developed by the American sociologist Talcott Parsons – an early proponent of systems analysis. In his essay 'The professions and social structure' (1939, p34) and with reference to *The Acquisitive Society*, he wrote:

It seems evident that many of the most important features of our society are to a considerable extent dependent on the smooth functioning of the professions. Both the pursuit and the application of science and liberal learning are predominantly carried out in a professional context. Their results have become so closely interwoven in the fabric of modern society that it is difficult to imagine how it could get along without basic structural changes if they were seriously impaired.

Table 1.2 William Goode's traits of professions

A 'community of profession' has the following traits:

1. Its members are bound by a sense of identity
2. Once in it, few leave, so that it is a terminal or continuing status for the most part
3. Its members share values in common
4. Its role definitions vis-à-vis both members and non-members are agreed upon and are all the same for all members
5. Within the area of commercial action there is a common language, which is understood only partially by outsiders
6. The community has power over its members
7. Its limits are reasonably clear, though they are not physical and geographical, but social
8. Although it does not produce the next generation biologically, it does so socially, through its control over the selection of professional trainees, and through its training processes it sends these recruits through an adult socialization process.

(1957, p194)

Parsons' later, more comprehensive study, *The Social System*, attempted to describe how modern society worked. Specifically, it illustrated how the various parts (the capitalist economy, the legal framework, the social order and the professions) were interconnected and had adapted to provide a finely balanced social order (1951, p25):

> It is the structure of the relations between the actors as involved in the interactive process which is essentially the structure of the social system. The system is network of such relationships.

This relatively stable model was refined by another American sociologist, William Goode, into a broader definition of the special characteristics of professional communities (see Table 1.2).

Goode placed his faith in both the professional and wider social communities to police behaviour (1957, p156) – a view very much of its time:

> The very great prestige of the professions is a response of the society to their apparent self-denial, i.e., they can, but typically do not exploit. This is not to say that professionals are nobler than lay citizens. Instead, the professional community holds that exploitation would inevitably lower the prestige of the professional community and subject it to stricter lay controls.

It was a view that was taken apart in the decades that followed, during which academics largely reverted to Adam Smith's presumption that professions were 'a conspiracy against the public' (1776, Chapter X). Sociologists no longer saw society as a well functioning machine and preferred to examine why professional groupings were so effective at consolidating their social position and economic power. Margali Sarfatti Larson in the introduction to her groundbreaking study, *The Rise of Professionalism* (2013 [1977], p.xvi), confronted the issue:

> My intention is to examine here how the occupations that we call professions organized themselves to attain market power. I see professionalization as

the process by which producers of special services sought to constitute and control a market for their expertise. Because marketable expertise is a crucial element in the structure of modern inequality, professionalization appears also as a collective assertion of special social status and as a collective process of upward social mobility.

Larson drew particularly on the work of Karl Polanyi, who in *The Great Transformation* (1957 [1944]) saw the rise of professionalism in the context of the destruction of the old world order effected by the 1914–18 war and the rise of the market system in the 20th century that took its place. She identified the ease with which the professions, with some minor adaptation, fitted into the new dispensation; noting 'the professional project is part of a basic structural transformation – namely, the extension of exchange relations under capitalism to all areas of human activity' (p209). She also worried about the inward-facing nature of professional groups defining them, more often than not, as 'colleague-orientated, rather than client orientated' (p226) and 'collectively separated from the laity by inaccessible or "tacit" knowledge[4] as much as by testable and explicit expertise' (p206).

Andrew Abbott (1988) revisited many of Larson's themes, if not her politics, in his authoritative study, *The System of Professions*. He offered and then tested a systems theory to explain the formation and development of the professions, proposing that they were in a state of continual dynamic competition both with each other and with the wider world of work (p297):

> Professions evolve together. Each shapes the others. By understanding where work comes from, who does it, and how they keep it to themselves, we can understand why professions evolve as they do.

Abbott's theory was more comprehensive than previous attempts to explain how the professions interlocked with society and it encompassed not only the formation of institutions, with their accompanying and contested jurisdictional boundaries, but also the challenges of mature and declining professional groupings. It is a theory that allows for a great variety of professional motivations, without the need for complex control mechanisms (p7):

> For some professionalism was a means to control a difficult social relation; for others, a species of corporate extortion. For still others its importance lay in building individual achievement channels, while a fourth group emphasised how it helped or hindered general social functions like health and justice.

Abbott's work gave professionalism a general hypothesis that described its internal workings with reasonably clarity: a hypothesis that, despite challenge, has largely stood the test of time. It has also, of course, encouraged new work and theorizing, either expanding on and challenging Abbott's work or using it to cover areas that were outside his original theoretical construct. In particular a growing body of studies has provided a platform for sociologists to better describe and examine the impact of changing social, economic and technological pressures on the system of professions.

Political pressure on professionalism has increased relentlessly since the 1980s, changing large swathes of both the context and social structure in which practitioners work. Ulrich Beck in *Risk Society: Towards a new modernity* (1992, pp209–11) argued that a professional group can deal with (and profit from) the 'hazardous conditions' that result from the current state of society, but it:

> First … must succeed in protecting its access to *research* institutionally and thus opening for itself the sources of innovation. Second it must succeed in essentially (co)determining the standards and contents of *training* and assuring in that way the transmission of professional norms and standards to the next generation. Third, the most essential and least often surmounted hurdle is taken where even the *practical application* of the knowledge worked out and the trained abilities occurs in professionally controlled organizations.

Another side of the struggle between society and the professions has been depicted by the English sociologist, Gerard Hanlon, who has described the way 'the state is engaged in trying to redefine professionalism so that it becomes more commercially aware, budget-focused, managerial, entrepreneurial and so forth' (1999, p121). This directly challenged many of the long-standing principles of professionalism and continues to threaten the traditional proposition that placed the professions outside the market as a separate, third force driven, at least in part, by non-market-based rewards.

Eliot Freidson (2001), combining these perspectives in *The Third Logic*, has described how 'the properties of professionalism fit together to form a whole that differs systematically from the free market on the one hand and the firm, or bureaucracy, on the other' (pp3–4). He argued for the importance of a professional occupation maintaining control of its own work, leading to a central concern that (p213):

> What is likely to be most at risk for the professions is their freedom to set their own agenda for the development of their discipline and to assume responsibility for its use. Thus the most important problem for the future of professionalism is neither economic nor structural but cultural and ideological. The most important problem is its soul.

Freidson's analysis of the outsider status of the professions is key to understanding what he terms the 'assault on the professions' that has been

Table 1.3 Eliot Freidson's interdependent elements of the 'ideal type' of professionalism

1.	specialized work in the officially recognized economy that is believed to be grounded in a body of theoretically based discretionary knowledge and skill and that is accordingly given special status in the labor force;
2.	exclusive jurisdiction in a particular division of labor created and controlled by occupational negotiation;
3.	a sheltered position in both external and internal labor markets that is based on qualifying credential, which is controlled by the occupation and associated with higher education; and
4.	an ideology that asserts greater commitment to doing good work than to economic gain and to the quality rather than the economic efficiency of work.

Source: Freidson (2001, p127).

underway since the 1980s with a series of governments attacking the value of professional work followed by non-professional businesses appropriating it. The assault has been a pincer-movement between the private and public sectors, with the aim on the one hand of the conversion of professional work into entrepreneurial activity that competes in the marketplace and on the other for it to be turned into a service provision that is principally delivered for profit. At present the professions appear to have little answer to this process and no significant new narrative to offer defining and explaining their worth to society. Freidson (2001, p190) despairs in the case of the medical profession that 'the economically self-interested actions of the profession and its failure to undertake responsibility for assuring the quality of its members' work weakened its claims and appeared to confirm the truth of the assumptions of consumerism and managerialism'. His prescription runs wider and in particular he writes 'if professionalism is to be reasserted and regain some of its influence, it must not only elaborate and refine its codes of ethics but also strengthen its methods of adjudicating and correcting their violation' (p216).

This transition to a new version of professional work has concerned a number of observers including Julia Evetts at the University of Nottingham, who has analysed the shift from the occupational professionalism championed by Freidson to a new form where, 'the effects are not the occupational control by the worker practitioners but rather control by the organizational managers and supervisors' (2012a, p6). This, in her view, has created the situation where 'professional work is defined as service products to be marketed, price-tagged and individually evaluated and remunerated; it is, in that sense, commodified' (Svensson and Evetts 2003) – see also Table 1.4.

Table 1.4 Aspects of professionalism from Julia Evetts, 'Professionalism in turbulent times'

The image of the ideology of professionalism as an occupational value that is so appealing involves a number of different aspects. Some might never have been operational; some might have been operational for short periods in a limited number of occupational groups. Aspects include:

- Control of the work systems, processes, procedures, priorities to be determined primarily by the practitioner/s;
- Professional institutions/associations as the main providers of codes of ethics, constructors of the discourse of professionalism, providers of licensing and admission procedures, controllers of competences and their acquisition and maintenance, overseeing discipline, due investigation of complaints and appropriate sanctions in cases of professional incompetence;
- Collegial authority, legitimacy, mutual support and cooperation;
- Common and lengthy (perhaps expensive) periods of shared education, training, apprenticeship;
- Development of strong occupational identities and work cultures;
- Strong sense of purpose and of the importance, function, contribution and significance of the work;
- Discretionary judgement, assessment, evaluation and decision-making, often in highly complex cases, and of confidential advice-giving, treatment, and means of taking forward;
- Trust and confidence characterize the relations between practitioner/client, practitioner/employer and fellow practitioners.

Source: Evetts (2012b, pp12–13).

Table 1.5 The definition of a profession

A profession:

- Is a full-time liberal (non-manual) occupation;
- Establishes a monopoly in the labor market for expert services;
- Attains self-governance or autonomy, that is, freedom from control by outsiders, whether the state, clients, laymen or others;
- Training is specialized and yet also systematic and scholarly;
- Examinations, diplomas and titles control entry to occupational practice and also sanction the monopoly;
- Member rewards, both material and symbolic, are tied not only to occupational competence and workplace ethics but also to contemporaries' belief that their expert services are 'of special importance for society and the common weal'.

Source: Burrage, Jarausch and Siegrist (1990, pp. 203–25).

The professional

The development of professionalism has been on a rollercoaster for two centuries, but the definition of what it is to be a profession has taken a smoother journey, incrementally evolving as it passes from the hands of one academic to another. There is an approximate list of characteristics, rehearsed by the various sociologists quoted above, which roughly includes association, a knowledge base, institutionalized training, accreditation, peer control and a code of ethics. For agreeing to work within this regime the professional is rewarded by a less well defined set of privileges including functional recognition, social respect and status, work autonomy, and sometimes control and monopoly over an area of work. A British–German/American–Swiss trio, Michael Burrage, Konrad Jarausch and Hannah Siegrist, developed the current reigning definition of a profession in 1990 (see Table 1.5).

Current state of play

One hundred and sixty years of development have taken the totality of the professions (as defined in national statistics) from 5% of the UK workforce in 1851 to over 20% in 2016, with the construction professions, in particular, increasing their headcount by nearly 45 times. Yet, as a system, it feels unstable, undervalued and under threat. What is the real state of play?

The professions face a series of universal threats that they are expected, and, indeed, expect themselves, to help manage and conceivably resolve. These include 'wicked' problems such as climate change – both mitigation and adaptation, long-term demographic trends, deepening inequality, maintaining safety and social cohesion and dealing with constant disruption and destruction (creative or otherwise) in an ever changing world. At the same time the professions' degree of influence has diminished as: 'possible solutions to client problems and difficulties are defined by the organization (rather than the ethical codes of the professional institution) and limited by financial constraints' (Evetts 2012b, p22).

The construction industry professions have already accepted the challenge of dealing with environmental change, although they seem to be struggling to know how to effectively respond. Taking responsibility could hardly be shirked

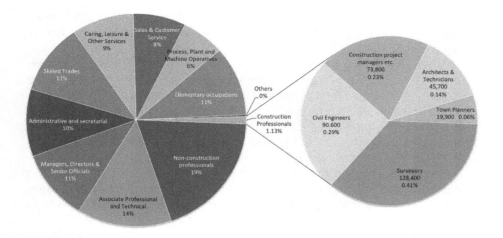

Figure 1.2
UK employment data
2nd quarter 2016 (Total
workforce 31 661 000)

Source: Office for
National Statistics.

given the clear role buildings and built infrastructure play as a major part of the problem, but an explicit role has also been thrust on the industry professions as governments and international bodies have legislated for future industry action, thereby embedding tasks and targets for the sector into their forward planning and treaty commitments. To cite one example: the UK's commitment to the 2016 Paris Agreement on Climate Change envisages reducing greenhouse gas (including CO_2) emissions from buildings (and their power sources) to zero by 2050, or shortly after, through extensive deployment of 'heat pumps, district heating and low-carbon hydrogen' (Committee on Climate Change 2016, p40).

Reducing greenhouse gas emissions to zero is a clear and numerically precise challenge, if an intransigent one. As it has received much discussion and deliberation, it is also reasonably well understood and accepted. But climate change also presents many other issues for resolution, including: designing and adapting existing buildings for hot and less temperate weather, creating and managing urban and social environments that are resilient in changing conditions, providing greater security and infection control, and doing so as cities and towns become denser, more crowded and have to cope with the potential influx of climate refugees from elsewhere.

The need to build more sustainably, to pollute less and use resources more wisely, is likely to dominate the work of the construction professionals in the decades to come. But climate change is just one of the nine planetary boundaries defined by the Stockholm Resilience Centre and one of the four, also including biosphere loss, land-system change and nitrogen and phosphorus pollution of the atmosphere and oceans, that is causing increasing or high risk to the planet (Steffen *et al.* 2015). All these four are being driven by the impact of human actions and their extent is directly related to the increase in overall global population.

Current (2017) UN population forecasts, according to their medium-variant projection, see the world's population growing from its current 7.55 billion to 8.55 in 2030, 9.77 in 2050 and 11.18 billion by the end of the century (UN 2017, p1). While Europe has a roughly stable population, the UK has an annual growth rate similar to the Americas of 0.8% with a population projected to grow from

65.6 million in 2016 to 70 million in 2026 and over 76 million by 2046 – subject to any changes in global population patterns caused by, for example, climate-induced migration (ONS 2017a).

The headline figures however disguise dramatic differences in age profiles across the world. In the West with improvements in lifestyle and health the proportion of the population in retirement is rapidly increasing. In the UK 18% of the population was aged over 65 in 2016. This will rise to 25% by 2050 (ONS 2017a). In the Middle East and South Asia populations are currently overwhelmingly young, but are projected to age significantly as the century progresses. But as other regions age, Africa will stay young throughout the 21st century, countering the demographic trends elsewhere, but resulting in a massive global imbalance. As the UN's 2017 population report dryly notes, 'Africa will play a central role in shaping the size and distribution of the world's population over the next few decades' (UN 2017, p4).

Professionals are aging along with the rest of the population – see Figure 1.3 and also Chapter 9 – but they are also coping with even more rapid change to their working lives. Technology is altering the way professionals function, communicate and compete and is endlessly updating the tools they need to work with to stay ahead. Together with increasingly powerful information handling it is providing the capacity to undertake new and unprecedented tasks, but it is also threatening jobs for professionals and non-professionals alike. Predictions appear on a regular basis that much skilled and expert labour will soon be replaced by intelligent computer based services. In a paper on the future of employment (2013, p18), Carl Frey and Michael Osborne noted:

> Occupations that require subtle judgement are also increasingly susceptible to computerisation. To many such tasks, the unbiased decision making of an algorithm represents a comparative advantage over human operators.

Their paper includes a table of 702 occupations ranked according to the likelihood of their computerization in the near future (from least to most). Architectural and engineering managers are ranked 78th, environmental engineers 81st,

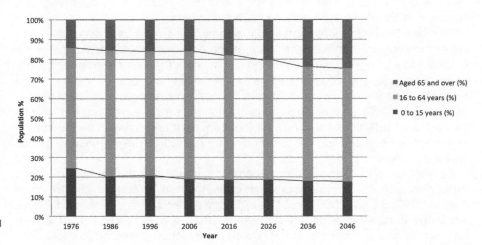

Figure 1.3
Age profiles (UK)
Source: Office for National Statistics.

architects 82nd, civil engineers 84th, landscape architects 133rd, surveyors 262nd and construction and building inspectors 350th.

Others, including Richard and Daniel Susskind (2015, p3), are gloomy about whether there is a future for professionals at all:

> Professionals play such a central role in our lives that we can barely imagine different ways of tackling the problems that they sort out for us. But the professions are not immutable. They are an artefact that we have built to meet a particular set of needs in a print-based industrial society. As we progress into a technology-based Internet society, however we claim that the professions in their current form will no longer be the best answer to those needs.

The message has been heard often before: from John Maynard Keynes in a 1930 essay 'Economic Possibilities for our Grandchildren', predicting that their generation would only work a 15 hour week; to Jeremy Rifkin in 1995, announcing that IT would remove millions of jobs in sectors including agriculture, manufacturing and the service sector. But the voices appear to be getting more insistent and the technology ever more competent and competitive.

> 'We are approaching a time when machines will be able to outperform humans at almost any task' – Professor Moshe Vardi speaking at the American Association for the Advancement of Science, Washington 2016.[5]

Massive systemic disorder has become a norm in an age that wants little more than a veneer of novelty over a base of prosperity and stability. The world is more international, with both workers and capital footloose and attracted to wherever offers the best deal, while simultaneously borders are closing and nationalism is loudly asserting itself. The professional system will adapt to these new circumstances, but it needs to do so extraordinarily fast in order to move with the pace of change. The professional system's response to these, and multiple other, issues has, so far, been to seek safety in numbers and agglomerate into ever fewer and ever larger multi-national and multi-disciplinary professional service firms (PSFs), as detailed in Chapter 8. In a fragile trading environment for professional work, size allows for bets on a particular sector or part of the world to be hedged and for staff and technology to be readily deployed to wherever the market dictates.

In 2016 the UK's 20 largest construction PSFs employed 26 670 chartered construction professionals in the UK (Building 2016)[6] accounting for approximately 12.8% of the 207,564 strong professional construction workforce reported by the professional umbrella group, the Construction Industry Council (CIC 2016, p5). The 20 firms had a total UK workforce of 65 600, combining chartered, non-chartered and technical staff representing 18% of their international workforce. Their combined UK-earned fees were reported as £6.3 billion. In comparison with equivalent survey ten years earlier fees had risen by 75%, although chartered staff had increased by only 10% (Building 2006).

The transition to large shareholder[7] and overseas-owned companies, while an effective bulwark against an uncertain world, has resulted in the threat to professional structures identified by Evetts, as the staff of international firms move beyond the reach of national institutions and any ability to effectively enforce standards or require collegiate behaviour. Individual institutions have less appeal when the staff numbers employed by the bigger companies are comparable to or even greater than the membership of the largest of them, they are able to run their own show and their links to politicians and other decision-makers are often far closer and more frequent.

The institutions nonetheless still remain an essential part of the professional landscape. They continue to provide key services, especially relating to education and accreditation, for the whole sector, although they are fully relevant only to medium and smaller-sized firms. The picture for individuals is moving to a similar position, as, increasingly, a choice of radically divergent careers is presented to graduating professionals; they can choose to work and gain support either in the large professional service firm sector or work for SMEs and gain support from their institution and local professional groups. It is still possible to move between the two partway through a career, but in practice it is becoming ever more difficult.

As large professional companies become more corporate and managerial, they are also less likely to be appointed directly by public bodies or other commissioning clients and more likely to be working for a contractor or an intermediate development organization. This has resulted from the increased use of contracts such as design and build and its numerous variants, and also procurement methodologies such as public–private partnerships or sale and leaseback, where the design function is contracted out and consultants become second tier suppliers with no direct line of contact to the body commissioning a building or other facilities. Such methodologies are seen as providing greater certainty of cost and programme but also a much simpler, 'one-stop shop' approach for clients. In achieving this they place building professionals in a more junior position than they traditionally held, far less able to enforce standards or communicate directly with building users or decision-makers. Professionals have lost their position as brokers and arbiters in this version of the design and construction system as well as a great deal of their independence and autonomy, generally leaving professionalism weaker and less able to intervene.

The relative weakness of both professions and institutions at present is illustrated by the attitude of governments and large corporations, which generally prefer to engage management consultancies or large PSFs when seeking advice rather than ask professional representatives or institutions, as they have in the past. This begs the question as to whether the institutions should continue as they are, surviving reasonably well despite difficult times and making themselves useful where they can, or should they collectively seek to reinvigorate the professional system and attempt a bold move to make professionalism an indispensable part of the construction sector well into the future? There have been a number of attempts to answer this question over recent years and to produce possible route maps for navigating professionalism in the construction industry through the 21st century.

In 2003 the joint RIBA-CABE sponsored think tank, Building Futures, published a study, co-ordinated and edited by the present author. *The professionals' choice: the future of the built environment professions*, concluded: 'the challenge for the construction professions is clear. They need to take ownership

Table 1.6 Elements of a new professionalism. BR&l

1.	Be a steward of the community, its resources, and the planet. Take a broad view.
2.	Do the right thing, beyond your obligation to whoever pays your fee.
3.	Develop trusting relationships, with open and honest collaboration.
4.	Bridge between design, project implementation, and use. Concentrate on the outcomes.
5.	Do not walk away. Provide follow-through and aftercare.
6.	Evaluate and reflect upon the performance in use of your work. Feed back the findings.
7.	Learn from your actions and admit your mistakes. Share your understanding openly.
8.	Bring together practice, industry, education, research and policy making.
9.	Challenge assumptions and standards. Be honest about what you do not know.
10.	Understand contexts and constraints. Create lasting value. Keep options open for the future.

Source: Bordass and Leaman (2013, p6).

of the managerial and financial techniques that they have traditionally viewed as threatening, so that they restore their members to positions of authority in their industry' (Davies and Knell 2003, p14). Building Futures published follow-up studies in 2010 (*Practice Futures: Risk, entrepreneurialism and the professional institute*) and 2011 (*The Future for Architects?*), which were no less certain that change had to come.

In 2013 the international journal *Building Research & Information* published a special issue on 'New Professionalism', guest edited by Bill Bordass and Adrian Leaman, in which they outlined the elements of a new professionalism (see Table 1.6) which they believed could unite 'all built environment professionals, their institutions and their educational systems – [in] the equivalent of the Hippocratic Oath'.[8]

In 2014, the discussion on 'New Professionalism' was taken up by the multi-disciplinary construction industry think tank, the Edge, through a series of public debates and then a commission of inquiry on the future of professionalism, chaired by the former government chief construction adviser, Paul Morrell. Morrell's Commission report *Collaboration for Change*, published in May 2015, concluded 'it is critical that the professions maintain (or, on some interpretations, regain) their legitimacy by engaging in some of the great challenges facing society, as they relate to the built environment' (Morrell 2015, p87).[9]

No one, and certainly not any of the institutions, their executives or their presidents, envisage that the professions can stay as they currently are and rely on past glories. What should they be doing to improve the output and working lives of their members and most critically the society and world they claim to serve?

Opportunities

The political system and with it society at large has been painfully remoulding itself in recent years. There is a strong awareness that the neoliberal consensus of the last 30 plus years is not delivering an equitable outcome, resulting in society being harshly divided into the have and have-nots. Over that time inequality has risen steadily and the gains in social mobility made in the mid-20th century have largely been washed away. Thomas Piketty (2014, pp23–4), shows that:

> The top decile claimed as much as 45–50 percent of national income in the 1910s–1920s before dropping to 30–35 percent by the end of the 1940s. Inequality then stabilised at that level from 1950 to 1970. We subsequently see a rapid rise in inequality in the 1980s, until by 2000 we have returned to a level on the order of 45–50 percent of national income. The magnitude of the change is impressive. It is natural to ask how far such a trend might continue. ... This phenomenon is seen mainly in the United States and to a lesser degree in Britain.

Entrenched social inequality has led to a surge in populist remedies in recent years and in widespread support for political upheaval, often simply for its own sake and frequently careless of the unknown and unpredictable costs involved. UK politicians, having declared that 'society is broken' (Cameron 2008) and mocked the electorate by claiming 'we're all in this together' (Osborne 2009), eventually saw the *status quo* disintegrate in 2016. At the time of writing the dust is still settling, but it is clear that almost all sides see the necessity for a rewriting of the social contract and a wholesale realignment aimed at radically improving equity and cohesion and that places quality of life for the great majority over enormous financial gain for a very few. Such sentiments are, of course, standard political boilerplate that have justified any number of policy initiatives in the past, but they also appear to represent an insistent public demand for change at a moment when one system is patently running out of steam but its replacement has yet to emerge.

If, and it's an uncertain if, economics and politics are on the cusp of a change, as a broad political consensus appears to suggest or wants to believe is possible, then professionalism has an essential role to play in helping to rebuild a more equitable society. Professionalism possibly has a stronger position of influence than the politicians, who have relatively few tools left at their disposal. Alternatively, if the system continues to bump along ineffectively, as it has been doing for some time, then an effort to redefine professionalism and make it a more effective force for change will hardly have been wasted. In either event there is an important job to do that can only improve the credibility of construction sector professions, which have for some time been vulnerable to accusations of being overly self-interested and insufficiently committed to the wider public interest.

The earliest building practitioners worked primarily for the rich and powerful, emperors, kings and dictators, even when constructing many of the great religious buildings of history. Over the last two centuries, a period encompassing the entire professional era, they found other clients: democratically elected governments, community organizations and locally based businesses, together with a great many relatively ordinary individuals. The opportunity to serve the wider community and to work to improve the public welfare fitted well with the aspirations of professionalism, described at length above. Political pressure in the 1980s led to the demolition of unfashionable measures used to protect professionalism from market forces and dismantled a majority of public sector professional employment. Since then the profit motive has dominated construction professionalism, breaking down the barriers between professional and entrepreneurial work, leaving many of today's highly respected professionals once again reliant on the world's most powerful individuals, dynasties and dictatorships for a substantial portion of their workload. It can be work that has dubious political purposes, is built using a modern version of slave labour and is paid for with dirty money. This was not in the original script for the development of the professions, but it is where we are now.

Standards have slipped, even while projects have become more formally innovative and ambitious, and professionals have become party, however unwillingly, to cost cutting and value engineering that have cheapened the built (and lived) results. As with the disparity in wealth a gulf has opened up between the standards of public and private projects. J.K. Galbraith's warning of private opulence and public squalor (1970 [1958], p207) is more relevant now than when he made it in 1958. Governments have felt obliged to step in to improve standards; by introducing and upgrading building and health and safety regulations, implementing architectural policies and establishing advisory and reviewing bodies such the Commission for Architecture and the Built Environment (CABE). Equally, and at other times, governments have sought to reduce standards and guidance, to cut regulations and aggressively reduce costs. Professional organizations have rarely been conspicuous in their attempts to raise standards or promote best practice.

It has become apparent in recent decades, following any number of post-occupancy evaluations (POE), that buildings rarely match up to the performance standards claimed for them at the time of their design. This has been particularly apparent for energy use in buildings but is also true for most other functional and social outcomes. The ubiquity of the situation has led to the familiar phrase, 'the performance gap'. Knowing that outcomes will not match up to the promises made at the outset cannot be anything other than unprofessional, yet it is, and continues to be, the norm. Remedies for the situation are available and some individuals and companies have embraced them, yet the only effective solutions will be those that are implemented widely across the industry. The UK government has begun to address the issue by making the Soft Landings (Bateson 2015) initiative obligatory for public buildings, but the professional institutions have yet to make their move.[10]

These are very limited examples of where professionalism in the UK construction sector appears to be falling short, and appearances matter when trust is the greatest commodity that the professions have to work with. The professions need to be in the business of building and maintaining trust. To achieve this they must ensure that they have an effective and inclusive working relationship with society, based around the creation of a body of knowledge and expertise that is openly and transparently shared, continuously maintained and improved and that underpins the professional offer to all sectors of that society and all parts of the economy (see the work of Frank Duffy on architectural knowledge in Duffy and Hutton 1998 or Duffy and Rabeneck 2013).

The professions need to regain the initiative. They must ensure that their centuries-old values, skills and objectivity continue to be relevant in the modern world and are an essential component of a society that delivers its rewards more broadly and fairly than the current dispensation allows for. Professionalism as a proposition, despite its apparent recent decline and in the face of many substantial threats, offers the opportunity to make a sub-stantial contribution to the economy, society and environment. But to do so it needs to be revitalized.

Renewing the bargain

For the modern world to function it is necessary to place trust in both people and systems, whether it is with money, health, reputation or vulnerable dependents. Expertise is required, but it can be impossible to judge whether it is up to the specific task or even to know what the task is. Trust is required, but when expertise has to be paid for it is again impossible to know how much it is worth, especially when under the current rules of the market the expert is encouraged to do the least possible for the greatest reward. The concept of professionalism was developed to navigate these impasses and the resolution takes the shape of a complex asymmetric bargain between the supply of and demand for expert services.

The ability to supply expert services means that there are substantial sunk costs that have been used to pay for gaining proficiency through extensive training, practice and experience. These costs can only be repaid by selling, at a fairly high rate, skill and time to those prepared to buy. However, the years dedicated to developing expertise have frequently convinced the prac-titioner that their craft has a value, both of itself and to the wider world, pos-sibly beyond monetary calculation. Indeed, the practitioner's commitment to their craft may overrule other concerns, resulting in exploitation and a failure to adequately recoup their costs. But it means that if there is a com-pact or bargain to be struck it will have to include potentially abstract issues associated with the quality and purpose of the work as well as purely mon-etary reward.

The compact to be agreed is delicate and complex, but both sides can obtain from it most of what they need. It is the genie at the heart of professionalism, what the writers Richard and Daniel Susskind have called 'The Grand Bargain' in which they envisage a fair, if not excessive, payment, combined with a reason-able mix of trust, respect, status and autonomy, should be exchanged for a set of agreed services performed by a well and appropriately trained, adequately resourced and up-to-date individual or group who agree to work to specific standards, put their clients' interests before their own and to do so with both honesty and integrity (2015, pp9–45) (See Appendix A for the full version of the Susskinds' rendition of the bargain.)

In practice, of course, any compact will be mediated by multiple other players; including educational providers, professional institutions, previous employers, regulators and insurers, as well as by society as a whole. Each of these may add to or amend the terms of the bargain, gradually making the deal itself more intricate and less clear. The hope, nonetheless, is that both sides emerge happy and adequately rewarded by the exchange.

In the modern world the idea that professionals have a special role in both business and society, but at least partially outside the marketplace, has lost much of its force. Essential items such as trust, autonomy and the interests of others have been stripped away in the interests of enabling rational choice theory to run unhindered. If professionals are to develop a new role, or even a new version of the old one, the compact needs renewing in an explicit way that spells out very clearly the nature of the exchange and the benefits that should accrue to those not directly involved in its striking; including the workforce, wider society and the environment.

The nature of such an enhanced version of the professional compact will be explored at greater length in the final chapter of this book. The intervening chapters look at the roots and development of professionalism in the UK construction industry and the journey that has taken the industry to this possible point of inflection.

Notes

1 The Society of Civil Engineers, later renamed The Smeatonian Society (Watson 1988, p7).
2 'There never was an individual less regretted by his fellow-creatures than this deceased king.' *The Times*, Obituary for George IV, 15th July 1830 cited in Hibbert (1988, p342).
3 *Ce que nous voyons avant tout dans le groupe professionnel, c'est un pouvoir moral capable de contenir les égoïsmes individuels, d'entretenir dans le cœur des travailleurs un plus vif sentiment de leur solidarité commune, d'empêcher la loi du plus fort de s'appliquer aussi brutalement aux relations industrielles et commerciales.* (Durkheim 1893, preface to second edition, p.xxxix).
4 Tacit knowledge refers to the ideas of Karl Polanyi's brother Michael, expounded in books such as *The Tacit Dimension* (1966).
5 Quoted by Sarah Knapton in *The Telegraph* (2016).
6 The consultant firms in *Building* magazine's survey are ranked according to total UK chartered staff (inc. QSs, architects, engineers, planners and project managers).
7 Companies in the UK are obliged by the Companies Act 2006 to behave in the way their directors consider:

> would be most likely to promote the success of the company for the benefit of its members [shareholders] as a whole, and in doing so have regard (amongst other matters) to –
>
> a) the likely consequences of any decision in the long term,
> b) the interests of the company's employees,
> c) the need to foster the company's business relationships with suppliers, customers and others,
> d) the impact of the company's operations on the community and the environment,
> e) the desirability of the company maintaining a reputation for high standards of business conduct, and
> f) the need to act fairly as between members of the company.

> This is in potential conflict with the professional ethos where the interests of society need to be put before that of company owners, e.g. the RICS standard that requires all members to act 'consistently in the public interest when it comes to making decisions or providing advice' (RICS 2007a).

8 The proposal for a Hippocratic Oath for scientists, put forward by the Junior Pugwash Group, was promoted by the physicist, Joseph Rotblat, in his 1995 Nobel acceptance speech and could easily be adapted by the construction professions:

> I promise to work for a better world, where science and technology are used in socially responsible ways. I will not use my education for any purpose intended to harm human beings or the environment. Throughout my career, I will consider the ethical implications of my work before I take action. While the demands placed upon me may be great, I sign this declaration because I recognize that individual responsibility is the first step on the path to peace.

9   *Collaboration for Change: the Edge Commission Report on the Future of Professionalism* (Morrell 2015) can be downloaded from www.edgedebate.com. Simon Foxell acted as instigator and co-ordinator of the Commission of Inquiry and editor of the final report.

10 The Soft Landings methodology was developed by the Usable Buildings Trust (Way and Bordass 2005) as a process that would encourage greater collaboration between the various stakeholders in design and construction projects and assist in the delivery of better buildings. Further collaboration with BSRIA led to the first edition of the Soft Landings Framework in 2009.

2

Beginnings

Although the divisions into the now familiar professional sectors in the UK construction industry have been apparent for many centuries, the transformation of informal occupational groupings into professional organizations is a comparatively recent, 19th century development. The 19th century was good at developing new traditions: several were expressed through buildings such as the 'gothic' Church of England or highland, baronial Scotland with its tartan embellishment; others through ceremony such as the Royal Family or the celebration of Christmas; and a wide range of institutions with their regalia and full sets of whiskers (Hobsbawm and Ranger 1983). Such myth creation has obscured the origins and motivations behind the modern professional disciplines. This chapter considers the journey from early organization to fully established professional bodies, starting with one of the best-documented and fully functioning examples of the mediaeval craft guild system in Renaissance Florence.

Florence

In 1236 the Republic of Florence published its first schedule of '*Arti*' or guilds in the city.[1] The 21 authorized guilds covered a range of occupations including merchants, manufacturers, bankers, craftsmen and innkeepers, as well as various occupations we would recognize as 'the professions', and provided the basic civic building blocks of the *popolo*, the great majority of the people; those outside the ranks of the elite and powerful ruling families.

Entrance into a guild was not easy and required a direct and legitimate family connection, usually a father in the same guild; proofs of competence; and an expensive joining fee. But, once in, membership provided a degree of security, political influence, a position in society and some influence on the city government. Guild members were required to periodically renew a set of sworn promises including, in the case of the Arte di Calimala, the cloth merchants and the oldest of the Florentine guilds (*Statuti dell'Arte dei Rigatti e Linaioli di Firenze*; see Najemy 2006, p40), to:

> 'acknowledge, observe and implement everything that the consuls of the merchants of Calimala shall require of me within the terms of their office', never to defraud creditors, to observe and conduct themselves in accordance

with the Guild's statutes, and to advise the consuls 'as best as I know how' whenever requested. .

In return for maintaining work and behaviour of a high standard, both rigorously policed; keeping to the stipulated hours and days of work; and providing various public services, the guild members were provided with a protected market for their work within the city, benefited from a system of trade and financial guarantees and had a degree of welfare protection for themselves and their families.

Florentine guilds were self-governing bodies, if state-regulated ones, and had many of the characteristics we recognize as defining professions:

- Members were required to have an appropriate education and to show the relevant competence and skill to practice
- Members signed up to a code of ethics that they were required to follow
- The organization maintained a system of discipline and punished or expelled those members deemed to have transgressed its rules
- The guilds provided services to its members
- In a few cases, notably that of the Guild of Doctors and Apothecaries with its library and museum of medicines, the guilds took care to develop a repository of knowledge to inform their members' practice.
- The guilds operated as monopoly providers for their range of goods and services and took measures against those encroaching on their territory.

The 21 guilds were strictly ranked by the city authorities and sub-divided into seven major guilds (*Arti maggiori*), including lawyers, bankers, doctors and various merchants, and 14 minor (*Arti minori*) ones ranging from the butchers (number 8 in precedence) to the bakers (number 21). Architects and engineers such as Michelozzo (1396–1472) were expected to join the Guild of Builders and Stonemasons (*Arte dei Maestri di Pietra e Legname*), ranked at number 12, at least until the tanners and curriers were demoted beneath them in 1280, when they went up one notch on the social scale. Some architects got around this lowly status by practising another, higher-ranked discipline. The architect Brunelleschi (1377–1446) was a member of the *Arte della Seta*, the silk merchant's guild, ranked at number 5, owing to his early career as a goldsmith. Attempts at establishing a separate Association, if not a full guild, the *Maestri a Architetti*, in the early 13th century, appear to have flourished briefly, foundered and then disappeared (Staley 1905).

The Florentine guild system was one of the most organized examples of the occupation-based social order that existed throughout mediaeval Europe; a system that has left its mark on many modern day organizations and activities. It provided both monopoly and political power; set standards for a wide range of trades, including those we would recognize as the fore-runners of the traditional professions, led by the doctors and the lawyers; and was instrumental in shaping the European university system both as an extension of guild organizations and a means of providing the necessary education and qualifications for entry into the professions. As an aside the Florentine records also show the reputational damage suffered by the doctors as a result of their failure to cope with the outbreak of the Black Death in 1348. Following the epidemic members of elite families abandoned the doctors' guild and without their political authority the guild lost much of its status and influence (Park 1985).

Although the professional classes established positions of power and often wealth as members of guilds, they largely failed, despite their best efforts, to ensure that occupations were only to be passed down within families and to consolidate a caste-based inheritance as a distinct stratum of society. They remained commoners; part of the Third Estate, comprising about 97% of the population, in the tripartite social system recognized across Europe. This kept them permanently outside both the First (the clergy) and Second (the nobility) Estates and without any special privileges or exemptions in law. Nonetheless, along with the other powerful Third Estate groupings of merchants and bankers in the upper echelons of the Estate, those with professional qualifications and occupations gradually began to be recognized as part of an identifiable *bour-geois* or *burgher* caste; literally 'those who live in the borough'; an urban class possessing substantial social and financial capital.

The professions in question were principally legal and medical: advocates, *procureurs* (prosecutors), notaries, doctors, barber-surgeons and apothecaries (Braudel 1982, p483). Not all practitioners were trained in universities or medical schools, and many who did were reputed to only have attended a few lectures for appearance's sake. Nonetheless the standing of the great schools, such as the University of Bologna, founded by legal scholars in the 12th century, or the University of Paris with its four faculties of Arts, Medicine, Law and Theology, kept the professional ideal alive. Both groups drew their authority from revered foundational texts, with medicine looking to the 5th century bce Hippocratic oath and the teachings of Galen, and the lawyers studying the series of Roman legal codes, known as the *Digesta seu Pandectae*, rediscovered in the 12th century. Both disciplines focused on book learning and drew on the considerable aura of authority around scholarly activity. The growth and strength of the universities in turn created a further sector of professionals, the scholars themselves or the professoriate. Perhaps in tacit recognition of this, it would be the scholars that the professions would later rely on to provide their disciplines with the necessary knowledge, validity and accreditation to define themselves in the modern world.

In contrast with law and medicine the concept of distinct built environment professions barely travelled beyond Renaissance Italy, although the idea remained alive in the Islamic world, especially in the Ottoman Empire and in *al-Andalus* (modern Spain and Portugal). In the majority of the western world, until approximately the mid-17th century, although architects, engineers and surveyors were recognized as loosely belonging to 'the professions', there were few organizations, formal qualifications or guilds that defined standards or codes of behaviour. Such training as there was, was a combination of apprenticeship and applied experience gained on the job.

Guilds

If the idea of the professions remained restricted to a narrow set of scholarly occupations the idea of the craft guilds and their control of a wide range of different trades spread across Europe between the 12th and 15th centuries, forming in towns and cities initially as a means of controlling production and maintaining standards as well as for mutual support and protection and later as a mark of citizenship and then governance of urban areas and lives. Guilds came with an abundance of rules including, in 1258 in the case of the masons of Paris, a limit on the number of apprentices (but not helpers and servants), the hours of work (unless he is finishing off a stair or the arch of doorway), rules for measuring quantities, the oaths to be sworn about the quality of mortar, the extent of civic guard duty required and the loyalty owed to the master of the Guild (Frisch 1987, pp53–5). Other guilds heavily policed their collective honour, expelling anyone if he, his wife or any members of his household, infringed the Guild's standards of workmanship or of economic and sexual probity.

The guilds established the formal system of master, journeyman and apprentice that remained the default arrangement for the construction professions until the late 19th century, long after the guilds had disappeared across most of Europe. The system was controlled by the Guild itself. Starting at the level of new recruits apprentices were vetted by the Guild for moral suitability and family connections.[2] The Guild also laid down the number of years an apprenticeship would last before qualification as a journeyman. The journeymen of Europe were an itinerant workforce, trained and qualified in a craft but too badly paid to settle down and marry and thereby obtain full citizenry rights. Their lifestyle, based around their all-male hostels, developed a misogynistic culture at odds with the respectability of the masters that the guilds continually struggled to keep under control. Eventually and after many years on the road working for different masters they hoped to become one themselves, as one of their employers died or retired. Again the relevant Guild would make the decision and set the terms with a particular emphasis on restricting the numbers of masters allowed to settle in a town (Rowlands 2001, pp57–9; Braudel 1982, pp314–16; Mundy 1991, pp118–19; Krause 1996, pp3–4).

The guilds were generally craft or trade-based and didn't see themselves in the intermediary role of a traditional professional. One rare example to the contrary is the organization of Ordained Surveyors of Edinburgh that emerged around 1803 as a subgroup within St Mary's Chapel, the trade guild of carpenters and plasterers in the city. The Ordained Surveyors drew up a set of agreed rules and standards for measuring building works, a practice that was then copied elsewhere in Scotland. The long lasting Scottish system worked as a partnership between the guild system, 'the trades, skills and "mysteries" carried on within a city's limits' and municipal authority. Measuring was recognized as a distinctive 'mystery' and Ordained Surveyors were sworn in a ceremony officiated at by the Edinburgh Sheriff following an examination of competence conducted by two senior ordained measurers. Maintenance of standards was firmly enforced and in 1750 one sworn measurer was 'defrocked for attesting a false measurement'. The Scottish Surveyors kept their independence for over a century, only amalgamating with The Surveyor's Institution, the forerunner of the RICS, in 1937 (Thompson 1968, pp73–5).

Unlike Scotland, England's craft guilds had diminished to almost complete insignificance centuries earlier. In the early 16th century local government was using the sale of monopoly rights to raise money, destroying the power of trade associations in the process. In a similar way rights over various occupations were distributed to veterans of the Civil War following the Restoration in the

17th century and by the beginning of the 19th century the remaining guilds and guildhalls had become clubs and meeting places for wealthy merchants (Krause 1996, p7). Any links between the English mediaeval guilds and the modern construction professions are a much later confection intended to give them a veneer of historical authority.

If the guilds' privileges were sold from under them, the professional disciplines faced a similar, if more deliberate, fate as faculties in universities dealing with medicine and law were closed down or suppressed during the reign of Henry VIII (Krause 1996, p13). For centuries afterwards practical subjects only rarely made it onto the syllabus and during the 17th and 18th centuries graduates of Oxford and Cambridge universities, if they intended to practice a profession at all, were destined either for the civil service or the church. Doctors in England were trained outside the university system (Scotland had a different model and provided medical training in its universities) with both physicians and surgeons learning their skills as apprentices and through attending lectures and demonstrations at the London teaching hospitals. Barristers took their articles and qualified at the Inns of Court. The apprenticeship model, a system that survived largely intact until reforms in education in the early 20th century, became the *de facto* method for training professionals in England. It was a system exported across the globe to the various outposts of the British Empire as well as, initially, the United States. This distinctive system with its non-academic approach to learning and strong belief in self-sufficiency resulted in a divide developing between universities and professional practice in much of the Anglophone world that, despite a change of approach to training in the second half of the 20th century, largely continues to this day. If the lack of an academic base has affected issues such as well-integrated built environment research, it has also been responsible for the growth and proliferation of strong professional institutions in many countries when a university qualification was not, until relatively recently, necessary for gaining professional status (see Chapter 7).

Reformation

The beginnings of religious reform in the 16th century – Martin Luther's objection to the selling of indulgences in 1517 in Saxony and the English King Henry VIII's split with Rome over his marital arrangements between 1532 and 1534 – produced, within two generations, a clergy who were interpreters of religion with a direct relationship with secular government, rather than bestowers of absolution gaining their authority from god, via the papacy. This, in Euan Cameron's analysis (2012, p403), 'injected a large intellectual element, requiring knowledge and understanding' and:

> The clergy developed an intense self-consciousness as a profession, and subscribed to a particular collective set of values. Lay people might reasonably have felt that their ministers were actually less familiar and recognizable than their priests had been.

This had the effect of formalizing the idea of an appropriately (and expensively) educated group who were both qualified and accredited at a national level and who were intended to serve, but also set apart from, the wider community. Their needs led directly to the foundation of large numbers of universities across Europe in the 16th century. The development produced an occupational grouping with all the characteristics of a modern profession – a defined membership, a ruling group, accreditation and disciplinary processes, a knowledge base and a mission – and established the various national models for professional organization.

But Luther and his followers went further. The protestant theology he promoted saw work as the prime means of achieving spiritual worth. The idea of a vocation (*beruf*) was applied to a wide range of 'useful' occupations. There was no longer any encouragement for being poor and holy: getting an education and bettering yourself through work was now (almost) a religious duty. It was no surprise that the first 'new' professionals in England started to appear in the reigns of the protestant monarchs, Henry VIII and his daughter Elizabeth I.

Recognition of new professional groupings outside the quartet of church, law, medicine and scholasticism developed slowly in England from the 16th to the 19th century, with a series of new roles emerging in construction and elsewhere gradually gaining status and an acceptance within the professional system and society at large. These roles, if not always the personnel filling them, had a shared root in the building trades but as specialisms appeared and differentiated themselves, each discipline eventually took its own journey towards professionalization. The surprise is perhaps that, during the 19th century, what had become very distinct streams arrived at broadly similar professional and institutional solutions despite their earlier diversity.

The identifiably separate building professional can be first seen emerging in John Fitzherbert's *Book of Husbandry and Book of Surveying*, published in 1523, where he stated 'The name of a *Surveiour* is a French name and is as *moche* to *saye* in *Englysche* as an Overseer' (cited by Thompson 1968, p2). This generally vague definition came with the qualification that the overseeing he had in mind should mainly be of farmland, but also that knowledge of '*Castels* and other *buyldinges*' and 'what the *walles, tymber*, stone lead, *sclate, tyle* or other coverings is worth' was essential (p2). The breadth of the term usefully allowed it to cover almost everyone involved in estate management, cartography, measurement, valuation, urban development, building design and quality control. But almost as soon as it was adopted to differentiate the building and estate supervisor (a word derived from the same Latin root as surveyor) from the craftsman and labourer, individual disciplines began to calve from the larger body of surveyors to define their own particular areas of operation; what Andrew Abbott would later term, 'jurisdictions' (1988, p33). In time this process would leave behind the diverse body of modern professional surveyors.

Architects

Although buildings in England had long been designed and built by a range of master masons and carpenters working in the craft tradition, before the Elizabethan period there was little sense of deliberate authorship and almost no means of consolidating the expertise behind designs, enabling scholarly input on new styles in building or supervising either the works or the monies spent. The one significant exception to this ad hoc way of working was the King's Works or the Office of Works, a department of the Royal Household. The Office of Works, established in 1378, employed the best talent available to design and maintain the royal palaces, castles and national defences (Summerson 1993, p28).[3] By the reign of Henry VIII (1491–1547) the King's Works comprised a central office in Westminster and five relatively autonomous regional bases, along with whatever temporary bases were required for on-going projects. It was, to all intents and purposes, a professional organization in its own right; developing new ideas, maintaining standards and training the next generation.

Each branch office was run by a 'surveyor', using the term's literal meaning, with a supporting cast of comptroller, purveyors, clerks, masons and carpenters. The officers at the Westminster Office were granted the grander titles of Surveyor of the King's Works, Comptroller of the King's Works, Master Mason, Master Carpenter, King's Glazier, Serjeant Plumber etc. Up to the mid-16th century appointment to the position of King's Surveyor tended to be an internal promotion from a role such as that of Master Carpenter, but from 1578 when Thomas Blagrave (c1522–90), a courtier and member of the minor gentry, was made the Surveyor of the Queen's Works under Elizabeth I it became an appointment usually given to those who already had social status at court. This served not only to enhance the role of the Office but it also increased the status of subsequent office holders, including both Inigo Jones (1573–1652) and Christopher Wren (1632–1723), who made their names as Surveyors of the King's Works and executed a majority of their projects through the Office. After the hiatus of the English Civil War (1642–51) the Office also grew to encompass more disciplines, including the Surveyor of the King's Private Roads and the Surveyor of the King's Gardens, with the first Architect of the Works, Sir William Chambers (1723–96), being appointed in 1761. The Office of Works eventually became the Works Department within the Office of Woods, Forests, Land Revenues Works and Buildings in 1851, a government department, and was finally merged into the Ministry of Works in 1940.

The reputation of the Office of Works and its position within the Royal Court had the effect of making it acceptable for educated individuals from good families to involve themselves with the practice of architecture and construction, establishing a long tradition in Britain of the gentleman builder and architect. Sir John Thynne (1515–80), steward to the Duke of Somerset took charge of all the Duke's building projects and eventually propelled himself, through a tumultuous life and marriage to the daughter of Sir Richard Gresham, to become a powerful landowner in his own right, a member of Parliament and, in semi-political exile, the builder and owner of one of the first of the Elizabethan 'prodigy houses', Longleat in Wiltshire (Summerson 1993, p46). Thynne's contemporaries with a similar enthusiasm for hands-on building included his friend Sir Thomas Smyth (1513–77), a onetime ambassador to Paris, and most prodigiously of all Sir William Cecil (1520–98), Secretary of State and Lord High Treasurer to Queen Elizabeth, both of whom actively, and competitively, engaged in building grand houses for themselves on their English estates. The gentlemanly tradition of building became one of the founding traits of English architecture, a tradition maintained in the following century by aristocratic architects including Sir John Vanburgh (1664–1726) and Richard Boyle, Earl of Burlington (1694–1753). This idea of the gentleman architect would go on to have significant influence on the belief-system and establishment of the profession in England.

If Sir John Thynne was instrumental in instigating one version of the construction professional he was also responsible for the rebirth of another. In 1568 on the recommendation of the Master Mason in the Office of Works, Thynne hired a mason, Robert Smythson (1535–1614) to work on his second house at Longleat, the first having burnt down in 1567. Working under Thynne, Smythson flourished as a designer and within ten years was working for himself, claiming credit, along with the French mason Allen Maynard, for much of the success of Longleat. 'The ordenanse thereof', he wrote, 'came *frome* us' (Summerson 1993, p62), and he went on to design two of the other great masterpieces of the period, Hardwick Hall and Wollaton Hall. Following his death a memorial was erected in Wollaton Church with an epitaph reading '*Architector and Survayor unto the most worthy house of Wollaton with diverse others of great account*'.[4] If Smythson wasn't necessarily the first to be given the new English title of architect, as the term continued in occasional use throughout the mediaeval period (Pevsner 1942), he exemplified a new independent position in both business and society; one grown out of, but now very separate from, that of the builder. Maintaining that distinction came to be one of the great preoccupations of the profession until late in the 20th century.

Robert Smythson's decision to move from the position of master craftsman to the then relatively unknown role of 'architect', was certainly groundbreaking. But even through his route into the profession continued to be followed by others – including Robert Mylne (1734–1811), who trained as a mason in Edinburgh, Henry Holland (1745–1806), originally a master builder, and John Carr (1723–1807), the prominent Yorkshire architect, who initially worked in his father's, grandfather's and great grandfather's trade as a stonemason (Jenkins 1961, p92) – it remained an extremely difficult step to take and most masons and carpenters preferred to carry on contributing designs in support of their building work without ever claiming professional status. For a builder, becoming an architect meant making an awkward class transition to an arguably more elevated but much riskier position in society. Even if such an 'advancement' could usefully be passed on to the next generation – Smythson's son John (d.1634) also became an '*architecter*' as did his grandson, Huntingdon (d.1648) – in practice there was greater incentive for a master builder to retain his position and use his influence to promote subsequent generations to professional or quasi-professional status by apprenticing them in their teens to an architect, a route taken by many sons of masons and surveyor builders. As a result it was the new men in the generation who followed Smythson, a generation exemplified by Inigo Jones and who clearly saw themselves as gentlemen – part of the Royal Court and servants of the King, who adopted and secured the title 'architect', rather than those with artificer or craftsmen roots.

Jones may in fact have been apprenticed as a joiner and certainly had a humble background as the son of a Smithfield cloth worker (Colvin 1978), but by the time he started his career as an architect he had already been to Italy on his first study tour and was an integral part of the Court in his role as costume and stage designer for the elaborate masques staged, in collaboration with the playwright Ben Jonson, to entertain the King and his entourage. Even in this role he was described, in the preface to the masque written by Thomas

Campion and performed to celebrate the marriage of James I's daughter, Elizabeth, to the future king of Bohemia, as 'our *kingdome*'s most Artful and Ingenious Architect' (Summerson 1993, p108). Less than three months after this performance this King's favourite had been appointed to be the Surveyor of the King's Works, placing him in the single most important position in the world of building design. It was a role he took up in 1615 after a 19-month study trip to Italy with his influential and highly cultured patron, Thomas Howard, the second Earl of Arundel; a trip taken as a lengthy diversion from their prime task of delivering Elizabeth to her new husband in Heidelberg (Summerson 1966, p35).

It is within the Office of Works under Jones that the idea of the professional's office and in particular that of the architect, first took recognizable shape. It was still a multi-disciplinary office, handling, in John Summerson's words, an 'unremitting stream of crude business' (1966, p40), but overlaid with a new intellectual and artistic strand carried out at a remove from the administration of practical work on the various royal and government buildings. The architecture it produced was based on earnest scholarship and Vitruvian principles using the collection of books and drawings that Jones had brought back from his Italian travels. The library included copies of Palladio's *Quattro libri dell' architecttura* (1601), Alberti's *L'architettura* (1565), Philibert de L'Orme's *Le premier tome de l'architecture* (1567), Vignola's *Regola delli cinque ordini d'architettura* (1607) as well as volumes of Euclid, Herodotus and Aristotle (Anderson 1993). Ben Jonson (c1572–1637), the playwright and for a time a close colleague, satirized Jones's move in a long-running and bitter feud from trade to proto-professional respectability and used the very existence of his library to attack him (1631, ll1–10 and 101–4):

> Master Surveyor, you that first began
> From thirty pound in pipkins, to the man
> You are; from them leapt forth an architect,
> Able to talk of Euclid, and correct
> Both him and Archimede; damn Architas,
> The noblest engineer that ever was!
> Control Cstesibius: overbearing us
> With mistook names out of Vitruvius!
> Drawn Aristotle on us! And thence shown
> How much architectonic is your own!

He ends with a direct assault on Jones's professional probity:

> Live long the Feasting Room. And ere thou burn
> Again, thy architect to ashes turn!
> Whom not ten fires, nor a parliament can
> With all remonstrance make an honest man.

Clearly the transition from workman to gentleman professional achieved by both Smythson and Jones was not effortless or accomplished without treading on others' sensibilities.

The Office of Works under the 27 years of Inigo Jones's surveyorship was more of an administrative hub than a design office and its principal task was to keep the series of 'houses of access' used by the King and the Royal Court in a good state of repair and ready to be put to use at short notice. This work was largely managed by a staff of clerks-of-works based at the individual properties, who would use outside contractors to carry out all but the most routine tasks. It was Jones's job to keep the whole operation running smoothly and working to its annual allocated budget. Other regular business included enforcing the regulations relating to building in London and trying to control the endless illegal encroachments and breaking of the rules by individual Londoners, as well as acting as London's *de facto* chief planner, responsible for developments such as the design of the streets and *piazza* of Covent Garden. The regular stream of architectural projects, mainly for royal palaces around London and a series of gateway structures, as well as the various private projects that Jones designed for aristocratic clients, were carried out on top of the main business and were done during his first ten years at the helm, judging from the drawings, directly by Jones himself. Construction of such projects was organized in-house, using the long experience and well-honed skills of the Office's team including the King's Master Mason Nicholas Stone, the Master Carpenter, William Portington, and George Weale and Andrew Dunnant, Clerks of the Works – all of whom worked together on the building of the new Banqueting House in Whitehall (1619–22) (Summerson 1993, p111).

By the second period of Jones's leadership of the Office of Works he was entrusting his pupil John Webb (1611–72) to act as his amanuensis and proxy within the office. Webb produced many of the architectural drawings to Jones's instructions. This development of a clear career path within an office was an important step towards establishing and normalizing a pattern of advancement leading from pupillage to professional status. In time Webb was expected to take over as the King's Surveyor, but the English Civil War forcibly interrupted his succession to office. Following the Restoration he was forced to appeal to Charles II for a job that he considered was his, by both right and inheritance (see Jardine 2002, pp156–7):

> To the Kings most excellent Majesty
> The Humble petition of John Webb Architect
> Humbly Sheweth
> That your petition was by the especial command of yor Ma[jes]ties Royall father of ever blessed memory. Brought up by Inigo Jones esq. yor Ma[jes]ties late Surveyor of the Works in ye study of Architecture for enabling him to do yor Royall father and yor Ma[jes]ties service in ye said Office. In order whereunto the late unhappy warre appointed his deputy to overrule the said place in his absence wch yor petitioner did, until by a Comtee of parliament in ye yeare 1643 he was thrust out.

Study and experience were not sufficient qualification to secure Webb the job he craved and in any case Charles had significant favours to return. In his place the newly restored King appointed a Royalist supporter and poet, Sir John Denham (1615–69) to the post. Webb was dismayed that 'though Denham may have, as most gentry, some knowledge of the theory of architecture, he can have none of the practice' (Summerson 1993, p173). In the event Denham acted as an able administrator of the Office of Works and, although ever remaining the underling, Webb was given a series of major projects to undertake in recompense, including the new royal palace at Greenwich. From then on the system of royal patronage usually allowed the elite to take the prime posts, for example in 1660, another gentleman architect, Sir Hugh May (1622–84), was appointed Paymaster of the King's Works and designed a series of grand houses for court officials and nobility. The Office ran on an uneasy marriage of training, experience and social status that allowed the arriviste sons of tradesmen like Jones and Webb to gain sufficient prestige and position to work with and for aristocrats and landed gentry, even if they had little chance of ever joining their ranks.

The greatest challenge for the Office and a major disruption to their principal work of building and maintaining the network of royal palaces came with the Great Fire of London in 1666. Suddenly all its resources were focused on the rebuilding effort of the devastated city and further work on palaces became unthinkable until years later when May began remodelling work on Windsor Castle in 1675 (Summerson 1993, p221). The rebuilding work in the wake of the fire required a huge expansion of both means and personnel; experienced surveyors were initially and urgently required to measure, assess and confirm ownership of the ruins, new plans were promoted by a range of opportunists of various degrees of respectability and skill and new designs were needed for thousands of buildings including the destroyed old St Paul's Cathedral.

Within a week of the fire being brought under control a royal proclamation announced that the City would be rebuilt with broader streets and using fire-proof brick and stone in place of timber construction. Within a month Charles II established a Rebuilding Commission in collaboration with the Corporation of the City of London with the King appointing two architects as commissioners, Hugh May and Sir Roger Pratt (1620–84) – the only other gentleman architect of any standing, together with a newcomer, a mathematician, scientist and astronomer, Christopher Wren. The City reciprocated with two builder-surveyors, Peter Mills (c1600–70) and Edward Jerman (d1668–9), and another scientist, Wren's friend and colleague Robert Hooke (1635–1703). Webb did not get a look in.

The prestige of this appointment was reflected in a letter to Pratt from his cousin in October 1666 (Gunther 1928, p11):

> I am exceeding glad the King hath commissioned you (no question in the first place) with Mr. May and Dr. Renne: they will get more secrets of your art brought from Rome, & so from Athens, then you from them. If you have a hand in repairing St. Pauls, the most Royall peice in Christendom, it will be a Jacob's ladder to carry you to heaven beyond all profit & honour in the world.

The six commissioners rapidly became three following the retirement of Pratt and deaths of Jerman and Mills. Wren was directly employing Hooke, whom he had known since they were students at Oxford, as his right hand man, surveyor, co-designer and office manager. This left May, who was unable to compete with the unstoppable force of Wren's progress and was bitterly disappointed in 1669 when his younger rival was appointed to replace Denham as Surveyor of the King's Works with effective charge over the entire operation. Samuel Pepys (1633–1703) noted in his diary (Pepys 1669, 21st March):

Met with Mr May, who tells me the story of his being put by Sir John Denham's place, of Surveyor of the King's Works, who, it seems, is lately dead, by the unkindness of the Duke of Buckingham, who hath brought in Dr. Wren: though, he tells me, he has been his servant for twenty years together, in all his wants and dangers.

With this appointment Wren began a 49-year role as head of what was effectively the only professional construction office in England.

Wren, like Jones before him, had relatively little experience of building when he was appointed Royal Surveyor. He started his adult life as a philosopher, mathematician and astronomer but with a restless energy that threw him from discipline to discipline. As a young student in 1655–6 he was involved in assisting Sir William Petty (1623–87) in his cadastral survey of confiscated land in Ireland, the Down Survey, for Oliver Cromwell's son Henry (Jardine 2002, p69). His fortunes turned in 1660 with the restoration of the monarchy and the return of various friends of his father (the former Dean of Windsor) from exile alongside the new King and the release of his uncle, Matthew Wren, the Bishop of Ely, from prison in the Tower of London. Wren now not only moved in the well-connected and international circles of the scientific *virtuousi* who would form the core of the Royal Society but also had the royalist connections that would allow him to meet and present his ideas to Charles II.

Job opportunities started to flow in. In 1661 Wren was appointed Savilian Professor of Astronomy at Oxford, another offer came from the Royal College of Physicians and his modest surveying experience allowed him to be put forward by his cousin Matthew Wren (son of the Bishop and now secretary to the Lord Chancellor) as a suitable candidate to lead the surveying and mapping work for the fortified harbour planned for Tangier. The job offer came with significant temptations, as later explained in a note by Wren's son (Wren 1750, p260):

A Commission to survey and direct the Works of the Mole [breakwater], Harbour and Fortifications of the Citadel and Town of Tangier in Africa, was at this Time proposed for him, (being then esteemed one of the best Geometricians in Europe) with an ample salary, and Promise of other royal Favours, particularly a Dispensation for not attending the Business of his Professorship, during his Continuance in his Majesty's Service abroad; and a Reversionary Grant of the Office of Surveyor-General of the royal Works on the Decease of Sir John Denham: all of which was signified to him by Letter from Mr. Matthew Wren, Secretary to the Lord Chancellor Hyde.

It was an offer that Wren, wisely in the event, turned down, claiming ill health, but he had clearly been marked as a potential successor to Denham on the basis of his mathematical ability and social connections rather than his architectural expertise. Wren's subsequent pursuit of architecture, following a seven-month study tour of France, which included a meeting with Bernini in Paris, seems primarily driven by his family and friends and not least by the King's interest in his career. Wren's *entrée* to London and to the Royal Court, once established, was promoted through his involvement in science and as a founding member of The Royal Society, but whatever his other interests, he was continually manoeuvred by his circle towards architecture and it was perhaps inevitable that he was appointed Surveyor of the King's Works at the first opportunity. The King, ignoring the hopes of Webb, May and Pratt, chose his own man rather than an architect associated with the Commonwealth period.

The work that Wren directed at the Office of Works was extensive, with royal palaces, 51 new churches in the City of London and, above all, the rebuilding of St Paul's Cathedral, as well as many other buildings both in and out of London. To achieve this workload the office varied its involvement in individual projects from the production of a simple model for craftsmen to interpret to the continuous on-site presence that was provided at St Paul's. Wren acted as principal architect, assisted by Hooke who also took on surveying and administration duties with another experienced surveyor Edward Woodroffe (c1622–75), in charge of the practical building work. In time, Wren's trainees, Nicholas Hawksmoor (c1661–1736) and William Dickinson (c1671–1725), both rising from apprentice positions in the office, together with the surveyor John Oliver, took over from Hooke and Woodroffe.

If the Office of Works became a professional workplace under Inigo Jones, under Wren it matured into an extraordinarily productive machine, if not a remunerative one. For all of his accomplishments, work ethic, honours and achievements Wren was never very good at obtaining adequate recompense from his employers or even at getting paid those fees he had agreed. His son, also Christopher, in the family memoir, 'Parentalia', describes the business terms for the work on St Paul's (Wren 1750, p344):

> The Surveyor's Salary for building St. Paul's, from the Foundation to the Finishing thereof, (as appears from the publick Accounts) was not more than 200 l. per Annum. This, in Truth, was his own Choice, but what the rest of the Commissioners, on the Commencement of the Works, judged unreasonably small, considering the extensive Charge; the Pains and Skill in the Contrivance; in preparing Draughts, Models, and Instructions for the Artificers in their several Stations and Allotments; in almost daily overseeing and directing in Person; in making Estimates and Contracts; in examining and adjusting all Bills and Accounts etc. Nevertheless he was content with this small Allowance, nor coverted any additional Profit, always preferring the publick Service to any private Ends.

There is something very familiar to modern ears about Wren's lack of business acumen and the frequent financial embarrassments he ran into as a result. Professional practice and money in the building industry have maintained a difficult relationship ever since.

As Wren's son recorded, the Office of Works took responsibility for all aspects of building design and execution, including the more complex engineering solutions and calculations. In practice a great deal was left to individual trades to carry through to completion. The historian Howard Colvin notes (1978, p23) that the workmen Wren employed:

> were quite capable of designing and building a church without the supervision of the Royal Society. And to a great extent they actually did so: the

ceilings, altar-pieces, pulpits and other furnishings were usually designed by those who executed them.

However, if necessary Wren involved himself directly in detailed contract arrangements. In 1681 he advised the Bishop of Oxford on the best method of contracting for his building works, writing that (Wren Society 1943, cited by Colvin, 1978, p21):

> There are 3 ways of working: by the Day, by Measure, by Great; if by the day I can make a sure bargain neither to be overreached nor to hurt the undertaker: for in things they are not every day used to, they doe injure themselves, and when they begin to find it, they shuffle and slight the worke to save themselves. I think the best way in this business is to worke by measure: according to the prices in the Estimate or lower if you can, and measure the work at 3 or 4 measurements as it rises. But you must have an understanding trusty Measurer, there are few that are skilled in measuring stone worke, I have bred up 2 or 3.

The reference to 'breeding measurers', in addition to the training we know the Office provided for young men like Hawksmoor and Dickinson, indicates that the Office made serious efforts to educate the next generation of professionals. Colvin notes how 'the importance of the Office of Works in maintaining the tenuous thread of architectural experience is shown by the number of eighteenth century architects who were either "bred up in the King's Works" or who held office under the Surveyor General' (1978, p30). This route allowed a range of relatively humble artisans and clerks who worked in and progressed through the Office to emerge as professional architects. Isaac Ware (c1704–66), the son of a London cordwainer, was apprenticed at about 14 years old to Thomas Ripley (1683–1758), a man who had himself risen from being a carpenter to the position of Comptroller of the King's Works, and by the age of about 26 he had obtained his first independent design commission, the conversion of a London mansion into a hospital. In similar fashion John Hallam, described by John Vanbrugh as 'a poor mean Country Joyner', became Secretary to the Board of Works under the patronage of the Surveyor General, Thomas Hewitt (1656–1726), and eventually architect for a series of buildings in Nottinghamshire (Colvin 1978).

If the Office was a route to professional status for a few talented sons of workman, they rarely bridged the class divide between officers and gentlemen in the profession. The artisan route through the Office might lead to appointment as Secretary to the Board, as achieved by both Ware and Hallam, but membership of the Board itself was reserved for a separate stratum of gentlemen architects, whose experience of design was gained working on one of their own or their relatives' projects and their education involved a study trip to continental Europe. Many well-connected architects, including Colen Campbell (1676–1729), William Kent (?1685–1748), William Chambers (1723–96), James Wyatt (1746–1813), Robert (1728–92) and James Adam (1732–94), joined the Office at a senior level following their formative experience on the Grand Tour across Europe. At the most their training may have included a period studying at a French school, as Chambers did at J.F. Blondel's *Ecole des Arts*, but more frequently it only extended as far as preparing drawings of classical remains in Italy (Colvin 1978).

For many travellers, the Grand Tour alone may have been sufficient qualification for a part-time career in architecture. Jeremy Black in his study of the Tour (1992, pp302–3) records that:

> a lively interest in Architecture was one of the attributes of gentility in eighteenth-century Britain, and many members of the elite were knowledgeable enough to play a role in the construction or alteration of stately houses … Tourists paid great attention to Palladian buildings and, particularly in the first half of the century, a visit to Vicenza was considered an essential part of the Italian section of the Grand Tour.

The general approach appeared to reflect the advice of Sir Roger Pratt to would-be clients, in a treatise written in 1660 (cited in Gunther 1928, p60), to:

> get some ingenious gentleman who has seen much of that kind abroad and been somewhat versed in the best authors of architecture: viz. Palladio, Scamozzi, Serlio, etc. to do it for you, and to give you a design of it in paper, though but roughly drawn (which will fall out better than one which shall be given you by a home-bred architect for want of his better experience as is daily seen).

Similar advice was provided by Roger North (?1653–1734) (cited in Kaye 1960, p47), another contemporary of Wren's, who in the early 18th century recommended a do-it-yourself approach for the landed gentry: 'For a profest architect is proud, opiniative and troublesome, seldome at hand, and a head workman pretending to the designing part, is full of paultry vulgar contrivances; therefore be your own architect, or sitt still'.

The architectural profession at the time of Wren's death in 1723 was small and only growing gradually but it remained badly split between commoners and gentry. For the commoners entry was possible through a training in the Office of Works or one of a very few other offices or by promotion from the ranks of master mason or carpenter to architect, a route satirised in one poem written in 1786 that claimed that even 'The carpenter, turn'd architect, *designs*' (Busby 1786). One such was Robert Brettingham; who advertised in the Norwich Mercury in 1753 that 'leaving off his business of Mason, he intends to act in the character of an Architect, drawing plans and elevations, giving estimates, or putting out work, or measuring up any sort of building for any Gentleman in the Country' (Colvin 1978, p138).[5] Meanwhile the better-connected architect *manqué* could more simply trade on their experience on the Grand Tour, especially if they could also boast some training abroad. The most potent example of this is the career of Richard Boyle, 3rd Earl of Burlington, 'the architect Earl'. Initially he employed James Gibbs (1682–1754) and then Colen Campbell to remodel his London home, Burlington House. Then, following a series of tours to Europe between 1714 and 1719 and as an enthusiastic Palladian convert, he took over himself with the support of William Kent, at that moment a decorative painter, two young trainees, Henry Flitcroft (1697–1769) and Daniel Garrett

(?–1753), and a draughtsman, Samuel Savill; for its time, it was a substantial proto-professional office.

Burlington, wealthy and influential, designed extravagant homes for himself and his friends and built several public buildings including the York Assembly Rooms, but it was his influence over the rest of the profession that was the most marked. He relentlessly promoted his friends, especially Kent, secured appointments for his *protégés* in the Office of Works and was widely consulted by anyone contemplating building anything significant. Sir Thomas Robinson was one of the many who ensured that Burlington passed judgement on the proposed designs for his house at Wentworth Woodhouse before he started work, commenting that 'the whole finishing will be entirely submitted to Lord Burlington' (cited in Colvin 1978, pp128–30).

The tension within the profession between those who rose 'through the ranks' and the gentlemen back from the Grand Tour, although never entirely relieved, was eased as more formal structures were introduced to train architects through a system of articles of apprenticeship (pupillages) and assistantships, revisiting old guild practices. At the end of an apprenticeship students might even be encouraged to travel, with their master's blessing, not only to Rome and Renaissance Italy but possibly further afield to Sicily, Greece and the Levant. Although neither of Burlington's two trainees travelled abroad they were two of the earliest beneficiaries of pupillage outside the Office and both went on to have successful careers in their own right, with the undoubted advantage of his continued support and social contacts.

During the 18th century the pupillage system gradually became accepted as the preferred and most popular route into the profession. The career of a successful architect from the next generation, Robert Taylor (1714–88) illustrates the point of change. He was the son of a master mason monumental sculptor and was apprenticed at 18 to another sculptor. His father managed to let him have 'just enough money to travel on a plan of frugal study to Rome' which was curtailed by his bankruptcy and death. On his return, Taylor worked through an undistinguished period as a sculptor to become the preferred architect to bankers and merchants in the City of London as well as taking a series of public positions as surveyor to the Bank of England and several London hospitals. He ran an extensive practice, run on strictly business-like lines and trained a large number of the next generation of architects, including Charles Beazley (1760–1829), George Byfield (1756–1813), Samuel Pepys Cockerell, (1753–1827), Charles Craig (c1754–1833), John Nash (1752–1835) and William Pilkington (1758–1848). Cockerell later referred to Taylor in his will as 'The affectionate prop and support of my early steps … to whom … I was chiefly indebted for my first advancement in life' (Colvin 1978, p226).

As the system of apprenticeships developed, London pupils could also, if time and their duties permitted, attend the lectures delivered at the Royal Academy that started with the appointment in 1768 of the first Professor of Architecture, Thomas Sandby (1721–98). It was the first stirrings of a process of formal instruction for the profession outside practice, and the two approaches would run together in the English system from then on. (See Chapter 5 for more on pupils and their training.)

Clubbing together

Georgian (1714–1830) London saw the founding of endless new organizations, societies, clubs and academies. These ranged from the deeply serious to the highly licentious, and they provided plentiful opportunities for architects and others to come together. This was initially in wider circles of artists and intellectuals such as the 'Society for the improvement of knowledge in Arts and Sciences' at a meeting of which, its founder, the architectural writer Robert Morris (1701–54), read his Lectures on Architecture in about 1730 (Colvin 1978, p32); or the Society for the Encouragement of Arts, Manufactures, and Commerce (now known as the RSA) founded in 1754, which encouraged the 'Art of Drawing' and gave small monetary awards to promising student architects. But it was with the founding of the Royal Academy of the Arts in 1768 that architecture, as a profession, was first and deliberately given both recognition and a formal role, with five architects included among the original 36 Academicians.

The training at the Royal Academy, such as it was, comprised six annual public lectures delivered by the Professor of Architecture. These were to be (see Colvin 1978, p34):

> calculated to form the taste of the Students, to instruct them in the laws and principles of composition, to point out to them the beauties or faults of celebrated productions, to fit them for an unprejudiced study of books, and for a critical examination of structures.

It is not difficult to see that students would have been looking for more input, not least in the years 1798 to 1806 after Sandby's death when his successor George Dance the Younger (1741–1825) failed to deliver any lectures at all, followed by a three year wait until 1809 when Dance's successor John Soane (1753–1837) started to deliver his, now famous, set of 12 lectures. Such paucity encouraged new, specifically architectural, societies to appear.

In October 1791 four architects – James Wyatt, Henry Holland, George Dance and Samuel Pepys Cockerell – met at the Thatched House Tavern in St. James's, London to launch the Architects' Club, with, initially, 15 highly select architect members and a further four honorary members. Ostensibly the club was a study association, but as Barrington Kaye in his study of the founding of the profession in Britain notes, it 'rapidly degenerated into little more than an after-dinner gathering' (1960, p58). The Architects' Club was followed into existence by the London Architectural Society in 1806; The Architects and Antiquaries' Club in 1819; and the Architectural Society in 1831. Each of these organizations had at their founding a declared professional purpose, such as the London Architectural Society's 'for the encouragement of architecture', statements that covered many an unspoken motivation, but as the society's papers reported only two years later, the 'few clubs which have been formed by persons in the profession, are rather to enjoy the pleasures of good fellowship among men engaged in the same pursuit, than for the advancement of the art' (London Architectural Society 1808, cited by Colvin 1978, p36). Papers were presented at some of the dinners organized by the clubs and annual exhibitions of members' work held, but little of professional consequence was discussed or agreed. The various societies for all their fellowship and educational usefulness failed to engage with an existential concern facing architects at the start of the 18th century: did they have a future in the rapidly industrializing and newly capitalist world? A concern expressed almost exclusively as anxiety about their social and cultural standing.

The 'wholesale destruction of the traditional fabric of society' in Britain as described by Karl Polanyi (1886–1964) (1957 [1944], p77) among others, caused by the onset of the Industrial Revolution during the period from 1795 to 1834, shredded the certainties of the fledgling architectural community. Kaye celebrates the freedom that this brought as: 'in transferring his practice from the noble patron to the committees who were responsible for the erection of town halls and clubs the architect was able at last to shed the humiliating relationship that patronage entailed', but notes that this also involved losing 'its relatively uncritical, and, on the whole, beneficient protection. Henceforth the architect found he was obliged to sell his services in an open market' (1960, p57). In historian Eric Hobsbawm's words 'there was an order in the universe, but it was no longer the order of the past. There was only one God, whose name was steam and spoke in the voice of Malthus, McCulloch, and anyone who employed machinery' (1977a, p229).

One dominant concern of architects was that there should be a clear separation between them and builders. In part this was to ensure that they were completely free of any, well-founded, suspicion that if they were both supervisor and contractor they would use the opportunity to defraud their client, but it was also a clear attempt to avoid the taint of an association with trade. Architects were very eager to be seen as gentlemen, as arbiters of taste and discrimination and even beyond that, for the additional kudos it carried, as 'men of science'.

Colvin (1978, p39) notes one exchange reported in *The Builder* (XX), from around 1817, between a barrister and the architect, Daniel Asher Alexander (1768–1846), he was cross-examining:

'You are a builder, I believe?'
'No sir; I am not a builder; I am an architect.'
'Ah well, builder or architect, architect or builder – they are pretty much the same, I suppose?'
'I beg your pardon; they are totally different.'
'Oh indeed! Perhaps you will state wherein this difference consists.'
'An architect, sir, conceives the design, prepares the plan, draws out the specification – in short supplies the mind. The builder is merely the machine; the architect the power that puts the machine together to set it going.'
'Oh very well, Mr Architect, that will do. A very ingenious distinction without a difference. Do you happen to know who was the architect of the Tower of Babel?'
'There was no architect, sir. Hence the confusion.'

The opening essay that accompanied the establishment of the London Architectural Society in 1806 (collected in 1808 and cited in Jenkins 1961, p115) also tried to separate the two occupations, perhaps over candidly for its cause:

In England, architecture has been considered rather as trade than as a science. The public buildings in which it has had an opportunity to display itself, have been comparatively few; and it has perhaps been imagined that, in the usual practice of house-building, the accomplished architect has had little advantage over the mechanical builder. Yet surely, in construction and in arrangement, superior science must afford some advantages; and it ought to be remembered that it is only by the union of taste and knowledge, that economy can produce simplicity, or expense be well directed to attain magnificence.

It is no surprise that architects sought out each other's company to gain some comfort in numbers, nor that their very next impulse would be to ensure that they only admitted those who could show some evidence of standing as gentlemen, whether in the form of education, an ability to pay an entry fee or the recommendation of an existing member, or better still, all three, as the resolutions passed at the first meeting of The Architects' Club required (Mulvany 1846, cited in Jenkins 1961, pp113–14):

> That no man be proposed to be elected a member or an Honorary member of this Club, unless he be an Academician or associate of the Royal Academy in London, or has received the Academy's Gold Medal for Composition in Architecture, or be a Member of the Academies of Rome, Parma, Bologna, Florence or Paris. …
>   That every new Member do pay to the Treasurer five guineas as his sub-scription on admission.
>   That every candidate be proposed.

Two outcomes emerged from these concerns over status. The first would be the development for the first time in Britain of an architectural profession with a formal organizing body, the Institute of British Architects, established in 1834–5 (the body that later became the Royal Institute of British Architects (RIBA)), founded 'for facilitating the acquirement of architectural knowledge, for the promotion of the different branches of science connected with it, and for establishing an uniformity and respectability of practice in the profession' (Institute of British Architects of London 1835), with the last of these likely to have been the most significant; the founder members of the Institute cared deeply about prestige. The second outcome was the claim made by architects in general to a particular artistic sensibility. This sensibility placed them well above tradesmen and artificers on the social scale and, ideally, at least on a par with their cultured clients and other opinion formers in society. The claim would unleash the style wars of the 19th century, in what became a bid for particular status on the part of different groups of architects and their supporters. It was inevitable that in time these two views would clash as they developed and matured, informing, or obscuring, as you wish, much of the development of the profession from thereon. The establishment and development of the formal profession will be discussed further in Chapter 3.

While the architectural profession was gradually coalescing in the 17th and 18th centuries, working mainly in the service and for the needs of the state, the gentry and the church, other forces were bringing other professions into existence.

Surveyors

Unlike 'architect', which has always been reasonably well understood and has a relatively limited range of meanings, the term 'surveyor', right from its initial use by Fitzherbert, was capable of being attached to almost anyone who was involved in the management, sale and development of land, property and buildings.[6] Surveying had its roots in estate management and stewardship, legal services, design and construction, measuring and cartography. Although it continued to maintain a series of essentially separate disciplines within its broad scope, it ultimately came together in the UK, unlike elsewhere, under a single professional umbrella.

The origins of the role of the Surveyor lie in the offices of the King's Works, established after the Norman Conquest in the 11th century. The Surveyor was a minister in the King's service charged with the care of the royal estates of castles, palaces and land holdings as well as managing undertakings such as the nationwide collection of data for the Domesday Book (1086). The term spread during the following centuries to become the title of the official responsible for the estate management of a noble household, taking over part of the traditional duties of the Norman *seneschal* or steward. The estates surveyor's tasks included the regular updating of the manorial survey that was at the heart of mediaeval husbandry. This inventory was, in part, a detailed topographical description of all the features of the land (*demesne*) under a Lord's ownership or control, who farmed it, what crops it produced, the rents it generated and when. Some of these surveys would have included crude maps of the *demesne*. As these maps lacked the techniques to measure and survey land accurately, they had limited practical use. Words were often more effective. In the *Seneschaucie*, a 13th century handbook by an unknown author, the process of developing such a survey is described (see Lamond 1890, p85):

> The senescal ought, at his first coming to the manors, to cause all the demesne lands of each to be measured by true men, and he ought to know by the perch of the country how many acres there are in each field, and thereby he can know how much wheat, rye, barley, oats, peas, beans, and dredge one ought by right to sow in each acre.

The reference to 'true men' is telling in this passage, indicating an uncertain combination of competence, trustworthiness and respectability; all attributes that would come to play a major part in the development of professionalism.

Land surveying

The agricultural revolution and the enclosure movement of the 16th century required accurate estate plans. This demand coincided with the rediscovery of ancient triangulation techniques and the development of instruments capable of measuring both distances and angles with precision. For the first time, the means existed to provide accurate plans. Land surveyors were suddenly in high demand, even if the skills and knowledge to achieve an adequate standard were not always available. Those surveyors who mastered the new technology did very well, but so did the publishers of numerous teach yourself surveying guides such Leonard Dige's *Tectonicon* of 1556 and *Pantometria* of 1571, *A Preparative to Plotting of Lands and Tenementes for Surveighs* by Ralph Agas in 1596 and Aaron Rathbone's 1616 *The Surveyor in Foure Books*. So did any number of mountebanks and charlatans (Thompson 1968).

Age-old farming practices were dispensed with as land was consolidated into larger parcels, common land enclosed and new farming methods adopted. The enclosure movement in particular caused upheaval and violence as land-owners commandeered large areas for their exclusive use. The dispossession of rural families caused widespread misery and resulted in general condem-nation from both church and society. Surveyors were necessarily closely implicated in this process, while the burgeoning workload left them struggling to keep up with the technical and quasi-legal demands that it placed on them. The temptation to cut corners, to favour one side over another or to take bribes was immense and from the 1550s on a growing literature on surveying began to stress the need for accuracy, honesty and trustworthiness on the part of the surveyor. One early 17th century writer described 'the Land-lords that inclose their villages' where 'the Surveyor is his Quarter-master which goes like a Beare with a Chaine at his side' (Lupton 1632, p306) and the well-known surveyor Ralph Agas (1540–1621) in an undated printed advertisement for his services (cited in Darby 1933, pp531–2) declared:

No Man may arrogate to himself the name and title of a perfect and abso-lute Surveior of Castles, Manners, Lands and Tenements unless he be able in true forme, measure, quantitie, and proportion; to plat the same in their particulars *ad infinitum* and beat out all decaied, concealed and hidden parcels thereof.

During this boom time for their services, land surveyors railed against the dangers of the untrained and inexperienced newcomers undercutting them and damaging their reputation. Agas complained about his competition 'he was a plumber, and had learned from a painter' (1596) and Edward Worsop devoted a whole book, *Discoverie of Sundrie Errors and Faults Daily Committed by Landemeaters*, in 1582 to the subject of surveyors' mistakes, which he ended by proposing a system of 'training and examining every potential surveyor', which would provide surveyors with a licence to practise, 'as was done in the legal profession' (Thompson 1968, p12), anticipating the establishment of a surveyors' professional organization by almost three centuries.

Over 5200 Inclosure (Enclosure) Acts were legislated for between 1600 and 1914 and each act, according to a 1800 report to Parliament, *Enclosing the Land*, required:

a survey to be made ... and a map to be prepared from it; both which is generally done by a Surveyor, especially appointed for the purpose; who also frequently makes all the calculations for the Commissioners, and stakes out the several allotments; for all which the charge is one shilling and sixpence per acre, besides a guinea and a half per day for attending the Commissioners, and an allowance for making a reduced plan.

Surveyors were kept busy, especially as some counties required the use of two surveyors, to take a general and a detailed survey respectively, and 'in some instances another description of person is appointed by the Act, called Quality-Men, whose business it is to value the land' (Parliamentary papers 1800). Valuation and estate agency became natural extensions to the surveyor's business activities.

As a result of the enclosure movement several increasingly prosperous surveying firms were established in the 18th century carrying out estate management, land surveying and property agency. They included Cluttons, founded in 1765 by William Clutton (1740–1826) when he took over a Sussex-based firm from his father-in-law that had been trading from at least as early as 1744, and Player and Sturge, established in Bristol in 1772 by John Player (1769–?) and his nephew Jacob Sturge (1754–1811). Such firms rapidly expanded their areas of work across the country and found themselves very well placed, if under-resourced, to fulfil the new demands generated by the incipient Industrial Revolution. Surveyors were needed to assess, value and purchase land for major road building programmes, for the canal building boom that started in 1755 with the Sankey Brook Canal and finally for the railway building boom than began in 1835.

The bonanza of work generated by the Industrial Revolution again overwhelmed the availability of competent surveyors and resulted in many under-qualified and overenthusiastic workers being used to carry out surveys. *The Builder* reported (1845, p557), possibly in a fit of pique:

> Scores of the parties employed in surveying, levelling and mapping are utterly incompetent and yet are paid immense salaries for their services. We know youngsters hardly able to tell the right end of a theodolite from the wrong, who are receiving two, three, and four guineas a day, and their expenses.

At the same time landowners who wanted to stop the railway crossing their land were clashing with surveyors attempting to plot courses across their fields and sometimes through their houses, clashes that occasionally ended in bloodshed and with surveyors gaoled for trespass (Thompson 1968, pp112–13).

The surveying companies found they had overreached themselves and too frequently failed to deliver the quality of work required. The Ordnance Survey eventually took over the task of land surveying, while engineers rather than surveyors would carry out future surveys for major infrastructure works. The *coup de grâce* had been provided by Edwin Chadwick (1800–90), who in his historic *Report on the Sanitary Condition of the Labouring Population of Great Britain* (1842, p317) included a statement from 'a gentleman who is himself a surveyor of extensive practice':

> As regards the appointment of surveyors to the Commissioners of Sewers, I would observe that, in my opinion, very few of them are properly qualified by education or otherwise to perform the important duties entrusted to them in an effective and proper manner. A man to be a good surveyor of sewers should be a practical civil engineer.

It was a hard lesson to learn and land surveying has long since been overtaken by other surveying disciplines.

Buildings

A second strong aspect of the work undertaken by the Office of Works and the subsequent manorial surveyors was their work on the fabric of buildings. In this instance the term 'surveyor' came to mean someone who provided the necessary information to enable a building to be repaired and maintained or be newly built, a role that also entailed overseeing and valuing its construction. In a handbook of 1703, *The City and Countrey Purchaser & Builder's Dictionary*, its author, Richard Neve (using the alias T.N. Philomath), explained the process of building to 'Either Gentleman, or Workman' and attempted to persuade builders (p57):

> to make choice of such Surveyors, and Workmen, as understand what they are going about, before they begin the work, viz. Such as be Masters of what they pretend to, as a Surveyor that understands how give the Draught, or Model of a design; so as that when it is erected, it may answer to the end.

But elsewhere in his book (pp11–12) he defines an architect as:

> A Master-workman in a Building; 'tis also sometimes taken for the Surveyor of a Building, vis. He that designs the Model, or draws the Plot, or Draught of the whole Fabrick; whose business it is to consider of the whole Manner, and Method of the Building, and also the Charge and Expence.

Philomath's readership must have been left very confused as who was expected to do what, and under whose authority.

Such confusion led to the use of portmanteau terms such as architect-surveyor or surveyor-builder and thereon to growing reputations for double-dealing by architects who acted as contractors for their own designs and surveyors responsible for 'the charge and the expence' while also being the builder paid by the amount of work measured. The reputational damage caused by widespread fraud in the industry led architects to distance themselves from both contracting and surveying and surveyors to establish rules by which they could be seen to be acting fairly and transparently in their dealings with both builders and employers.

Nowhere was this more necessary than in the practice of measuring, a core area of the surveyor's work, involving establishing the exact quantity (hence 'quantity surveying') of materials used and work carried out by the builder in order to calculate the amount due for payment. Measuring had, as is clear from Wren's reference above, long been a concern and, as money was involved, one of the most contentious issues in any project. Craftsmen, artificers, builders, surveyors and architects all did their own measuring and then argued at length about who had carried out the calculation correctly. Chimneys in particular seem to have generated numerous disputes. Textbooks were published, including Venturus Mandey's *Marrow of Measuring* (1682) and William Hawney's *The Compleat Measurer* (1717), providing advice and instruction as to how to measure building works, yet agreed standards were missing, and were something that only a dedicated discipline could provide.

Thomas Skaife, a carpenter and joiner, bemoaned in his book of advice to working builders *A Key to Civil Architecture or the Universal British Builder* (1774, p51), the humble but vital status of measuring in the industry:

> It is greatly to be lamented that there is not a proper standard or certain pitch of perfection in this as well as many other learned professions, for a man to arrive at before he can be called either architect or surveyor, the one much inferior to the other, and those who had not merit to the former be deemed the latter, and those that had not pretentions sufficient for either be termed measurers. A gentleman would then know whom he had to apply to for masterly compositions and undertakings, there would be a visible difference in their professions though at present they are synonymous terms.

As measuring gradually matured into quantity surveying the status of the estimator and measurer gradually rose, but it would take a change in the method of contracting for buildings brought on by a crisis in public expenditure to cement its position and that of the new 'professional' role in the industry.

The outbreak of the French Revolutionary War in 1793 and the subsequent Napoleonic wars, lasting until 1815, created the need for a mass mobilization of British troops and the barracks and training facilities to house them. A new Barrack Office, essentially a revival of part of the Office of Works (by then the Board of Works), was established to provide them at speed. The work was handled by two architect-surveyors working part-time for the Office using external measurers to assess the quantities of work carried out, and, owing to the urgency of the work, a single builder was appointed to carry out each project as a whole, in violation of the standard method of dealing with parcelling out work to individual trades. The unorthodox methods applied and the high costs of providing so many barracks at such speed inevitably led to accusations of waste and extravagance and a Parliamentary Inquiry was held in 1806 focusing on the procurement of the Chichester Barracks. Instead the inquiry, having taken evidence from both the chief Office measurer, Thomas Bush and a recognized expert, independent London surveyor, James Pain, vindicated the Office's approach and effectively endorsed the practice of single-point contracting, or in the terminology of the time, 'contracting in gross' (Thompson 1968, pp82–4).

Such an approach remained controversial for over 20 years with individual architects, craftsmen and builders supporting opposing sides of the argument and the dispute, so far as public works were involved, was only settled in 1828 when another Parliamentary Inquiry (Parliamentary Papers 1828, p5) cautiously concluded:

> the method which appears the most prudent and economical for individuals to adopt could not prove disadvantageous to the Public, and they are therefore inclined to think that with precise specification and careful superintendence, and where all deviations from the original plan are avoided, the system of contracts in gross might be found to be the least expensive.

What the Inquiry's conclusions did not mention was that the system also required the employment of a high-quality quantity surveyor by the contractor.

The costs of such a surveyor could easily be afforded by the omission of so many individual measurers, and the introduction of a single, overall contract. This established the role of quantity surveying and when the Old Palace of Westminster burnt down six years later its replacement, designed by Charles Barry (1795–1860) came accompanied by a full Bill of Quantities and an accurate estimate prepared by Henry Arthur Hunt (1810–89). Hunt was only 26 at the time but already 'a surveyor employed very much by architects and builders in making estimates' (Parliamentary Papers 1836, pp170–5). The main part of the works was let as a contract in gross to the contractors Grissell and Peto (est. c1830).

Hunt had started in the office of Thurston and Sons and was then articled to John Wallen (The Surveyors' Institution 1891, p487), at the time the most prominent quantity surveyor in London and a magnet for aspiring trainee surveyors. His training shows how a formal system of education for quantity surveyors had developed by the early 19th century, a system recognized by the Office of Works, among many others. The Office was paying good salaries to their professional measurers by this time and were always on the lookout for 'young men who are well recommended, and who have been brought up as measurers; they are very difficult persons to get' (Select Committee on the Office of Works 1828, cited in Thompson 1968, p86). Hunt himself went on to superintend the construction of the 1853 Great Exhibition in Hyde Park and to become a consulting surveyor to the Office of Works where he was responsible for the Royal Courts of Justice, although his career was dominated by his work as a civil engineer for the railway companies and the early stages of the London underground system. He was a founding member of the Surveyor's Institution and also an Associate of the Institution of Civil Engineers and knighted for his services in 1876.

Organization

In this period several attempts were made establish organizations that would bring surveyors together. The earliest of these was made by a group of 16 city surveyors, who formed the Surveyors' Club in 1792 in direct response to being excluded from membership of The Architects' Club, established the year before. The professional distinction that the architects wanted to draw between themselves and everyone else involved in the commercial world of building, meant that surveyors had had little chance of meeting their entry requirements. It is hard to say whether this spurred on the surveyors to become more professional themselves or just that the architects had been premature in rejecting a group who were gradually organizing and would eventually become as respectable, but the separation has endured. The Surveyors' Club continued to exist, but only as a small dining club of no more than 25 members, well into the late 20th century.

An attempt at bring both sides together was made in 1834 in the form of the Society of Architects and Surveyors. After a series of meetings at the regular home of such machinations, the Freemasons' Tavern in Holborn, London, a group of architects and building surveyors proposed an Institution that:

> shall consist of such persons as have been educated for, and are practising, solely, the profession of an Architect and Surveyor, and have signified their

intention of becoming subscribers on or before the – day –, 1834, and shall be approved by a Committee on payment of their subscription: as subsequently of such other persons as shall be proposed as Candidates, and shall be approved by ballot as hereinafter mentioned.

This proposition (see Kaye 1960, p76) appears to have triggered a counter proposal from a breakaway group of architects involved in the initial meeting in the Freemasons' Tavern to form a society comprised solely of architects and entitled the Society of British Architects. They too produced a prospectus, this time suggesting an examination for new entrants covering design, construction and the business of architecture and a code of professional conduct (see Chapter 3).

Although attempts were made to bring the two parties together in a coalition at a joint meeting on 13th May 1834 it was clear that the Society of British Architects had drafted their prospectus specifically to exclude the surveyors and wanted the maximum distance to be placed between the two disciplines. A few more meetings of the Architects and Surveyors were held and aggrieved notes sent to their rivals, before it finally petered out with the publication of a final pamphlet of complaint and recrimination (cited in Kaye 1960, p78). The Society of British Architects met again on 4th June 1834 and resolved to form the Institute of British Architects. The surveyors were left out in the cold, empty-handed.

Twenty-three days later a different group of surveyors met at the Freemasons' Tavern. On 27th June a group of well-established London surveyors held the inaugural meeting of the Land Surveyors' Club. The initial membership, which rapidly expanded beyond London, was eager that the club should do more than just meet 'for eating and drinking, or backbiting their brethren' (Minute book of the Land Surveyors' Club 1834, cited by Thompson 1968, p94) and they were particularly keen that the new club should enhance their dignity and reputation for competence and integrity. They represented just one of the strands of the profession with its base in agriculture and estate management and it no longer reflected the greater part of the work or the interests of the majority of surveying firms. The club gradually lost momentum and had effectively faded away by the end of the 1840s. Francis Michael Longstreth Thompson in his major study on the growth of the surveying profession (1968, p95) suggests that the decline of the club was:

> not for any lack of recruits, not because of resistance to domination by London – for its exclusive London character was rapidly remedied – but because of lack of will and determination on the part of its members, lack of incentive to organize effectively on the part of the profession at large, and lack of help from the economic and intellectual environment.

Nonetheless the club was an important dress rehearsal for the eventual establishment of the Institution of Surveyors 30 years later with many of the same personnel involved.

When sufficient momentum culminated in a coherent and durable institution for surveyors in the 1860s it surprised everyone involved that it had taken so long and that the profession as a whole had not been recognized sooner for the achievements of its most significant and well-established members. These included surveyors such as John Clutton (1809–96), son of William, who by this time was highly respected and influential and ran a large office with separate departments for 'Drawing, Charges, Town, Country, Crown and Tithe Receiverships, Town and Country Management, and Town and Country Building' and represented many of the most prestigious clients in the country (Thompson 1968, p132). The editor of *Building News* probably overemphasized the general feeling when he wrote in 1860 (pp447–8):

> It will scarcely be deemed credible, when architects, builders, engineers, and even their subordinates, clerks of works, and foremen engineers, have their institutions, and when every trade down to the most mechanical handicraft has its mutual benefit association or provident club, that a profession which is in close connection with the practice of construction and engineering, standing midway between the two, should have no common centre for the reception and radiation of professional intelligence and assistance.

The greater likelihood was that in the boom years of social upheaval and the Industrial Revolution there had been no need for the surveyors to be organized. In a period of extreme growth they were easily busy enough neither to have time for such exigencies nor to care. Any temporary storms over incompetence or embezzlement could be weathered. It was only in the aftermath of Chadwick's attack in 1842 and the end of the bubble of railway work that the mood began to change. New scandals and accusations against surveyors arrived with alarming frequency and there was a strong feeling that disreputable practitioners needed to be driven out. An 1866 exchange of letters on 'Sham Surveyors' in *The Builder*, the most influential and widely read journal in the industry, called for an association to 'break down and cast aside the formidable barrier of quackery which threatens to overshadow us' (1866, p547) and attacked 'the mushroom growth of self-styled "surveyors", or "civil engineers and surveyors", whose intrusion into the profession tends to lower the status of the really educated and capable man' (p513). An institution had become necessary and inevitable, certainly to provide dignity and respectability to the profession, but above all to give a guarantee of competence, trustworthiness and integrity, to be enforced by a discipline system that was prepared to enforce harsh penalties as and when required.

Civil engineers

The third of the professions to emerge from the broad church of 16th century surveying was civil engineering. It was both the last of the three to identify as a separate discipline and the first to organize itself as a professional body.

Engineering, of course, has ancient antecedents and the mediaeval Latin word *ingeniator* has long been used to describe its practitioners. But while it is necessary to mention the triumphs of Roman building and infrastructure, as well as Brunelleschi's dome for Florence Cathedral, contemporary engineering's roots are more accurately located in military engineering and, particularly in the design and construction of siege engines and the means to resist them. It is where, over many centuries, the greatest resources were expended and where ingenuity was most highly prized and rewarded. Castles, moats, siege towers, naval harbours and naval architecture in general all exercised the mind of engineers throughout the mediaeval and early modern periods and it was only with the transition to relative peace and security in the age of empire that civil, non-military engineering began to dominate, at least in the home territories.

France, as the dominant western land-power in the 18th century, took the first steps in the transition from military to civilian engineering when the new regime acting for the boy king Louis XIV established the *Corps des Ponts et Chaussées*, a civilian force charged with inspecting and building bridges and roads,[7] in 1716 to focus the efforts of the state on building roads and canals to connect and open the country for trade as well as national governance. This was followed in 1747 by the opening of *L'Ecole Royale des Ponts et Chaussées*, effectively a school of civil engineering, to formalize the training of civil engineers in France and its colonies and to deliver the professional workforce required to carry out the State's ambitious programme of infrastructure building. The school produced not only a cadre of highly able engineers but also many authoritative textbooks on engineering practice, including works by Bernard Forest de Bélidor (1698–1761),[8] Émiland Gauthey (1723–1806)[9] and Jean-Rodolphe Perronet (1708–94)[10] (see Chapter 4 for more on French engineering).

In Britain the Royal Society published papers on engineering in its journal *Philosophical Transactions* from the 17th century on and recruited many engineer fellows from both home and abroad, but there was never an equivalent central organization to the French institutions. British developments in the discipline in the 18th century arose almost solely from the drive and determination of individual entrepreneurs that underpinned the Industrial Revolution, setting an alternative and independent non-state pattern for the industry, very different from that advanced by the Office of Works. From the start of canal building and through the building of the great factories, bridges and railway systems, project promoters found their engineers from wherever and whatever background they could. The canal and dock engineer, Thomas Steers (1672–1750) started his career as a soldier, John Smeaton (1724–92) as an instrument maker, Thomas Yeoman (1709–81) and James Brindley (1716–72) as millwrights and Robert Mylne (1733–1811) and Thomas Telford (1757–1834) as architects. Such engineers were largely self-taught or developed their knowledge by expansion and extrapolation. If they were seen as skilled and competent they were soon in great demand and many engineers established their own businesses as a result.

The nature of an engineer's work in the second half of the18th century, particularly if they were working on any of the many canal schemes crisscrossing the country, required them to engage with Parliamentary committees in London to ensure that the Bills for the new works were complete and submitted by legislative deadlines. This not only meant that they became well-known and feted public personalities, but it also brought them together, often on opposite sides of arguments in committee hearings. It wasn't too long before they started to organize together for 'conversation, argument and a social communication of ideas and knowledge' (Taylor 1837, p.xxi).

The 13 well-known engineers who formed the Society of Civil Engineers in 1771 had their first meeting at the King's Head Tavern in Holborn, London on 15th March and then held meetings each fortnight while Parliament was sitting. They included Smeaton, Mylne, Yeoman and Brindley (Watson 1988, p7). Although it was essentially a social and business club allowing its members to relax after a hard day of committee work, 1771 nonetheless marks the first stage in the formation of a recognizable civil engineering profession.

The driving force behind the Society was John Smeaton. Initially a mechanical engineer and physicist, he committed himself in his mid-thirties to the role of a civil engineer, a term that he invented to describe himself and that first appeared attached to his and Yeoman's names in a 1763 London directory. He designed and built numerous canals, harbours and bridges, with the third Eddystone Lighthouse, now rebuilt on Plymouth Hoe, for which he developed hydraulic lime (the forerunner of Portland cement), being his best-known creation. Smeaton was very aware of his professional status and responsibilities, remarking in one letter that 'I have never trusted my reputation in business out of my own hand, so my profession is as perfectly personal as that of any Physician or Counsellor-at-law' (Watson 1988, p6). After his death in 1792 the Society was reformed in his honour as the Smeatonian Society of Civil Engineers. It still runs as a dining society, but was effectively a false dawn for professional organization.

On Christmas Eve 1817 a very different group of engineers met. They were all young, with the oldest being 32 and the youngest 19 and all felt keenly the need for 'mutual instruction in that knowledge requisite for the profession' (Watson 1988, p11). Their leader, Henry Robinson Palmer (1795–1844), had originally been articled to Bryan Donkin (1768–1855) and by the time of the 1817 meeting had just started work in Thomas Telford's office. They were no longer a generation of autodidact pioneers and freebooting business entrepreneurs and needed a more formal means of acquiring the knowledge and skills for their chosen profession. Some of this they were clearly getting from being apprenticed in leading engineering firms, but beyond that they were seeking a training that gave them a wider knowledge and set of skills than was available in any single office.

By the end of their 1817 meeting they had adopted a set of 11 resolutions starting with 'First, that a society shall be formed, consisting of persons studying the profession of a civil engineer' and had agreed to meet two weeks later to elect the new society's officers. At their third meeting on 13th January 1818 they adopted the name of 'The Institution of Civil Engineers' with the stated purpose of 'facilitating the acquirement of knowledge requisite in their profession and for promoting Mechanical Philosophy' (Watson 1988, p12).

Institution

The Institution of Civil Engineers was to be the first new professional institution founded in Britain in the modern age (Corfield 1995, p181), predating the Institute of British Architects by 16 years and the Institution of Surveyors by 50. The founders of all three had similar motivations for banding together: the need to define their discipline and share their knowledge for mutual advantage and to establish a means to collectively defend and raise their standards as well as the advancement of the public interest. But while their initial rationale and the way they chose to express it built on each other's form of words they each added a new theme, an addition that may help explain their later development as organizations.

The civil engineers emblazoned the minutes of their first session in 1818 with the title 'Institution of Civil Engineers for facilitating the Acquirement of Knowledge Necessary in their Profession and for Promoting Mechanical Philosophy', clearly flagging up their intention to be a learned (and learning) society (Watson 1988, p10).

The architects in 1834 closely followed the rubric of the ICE with 'The Institute of British Architects has been founded for facilitating the acquirement of architectural knowledge, for the promotion of the different branches of science therewith', but then added a rider, 'and for establishing an uniformity and respectability of practice in the profession'. Respectability was of foremost importance to the men of the IBA and Joseph Gwilt (1784–1863), who delivered the address at the official opening of the Institute, went on to explain the actions, chiefly involvement with the building trade, that would lead to a member's expulsion: conduct that would be 'derogatory to his character as a gentleman in the practice of his profession'.

The surveyors added a further layer to both these, pithily expressed by John Horatio Lloyd in response to the inaugural address by the Institution's first president, John Clutton: 'The sponsors of this Institution aim, I conceive, at three things; first, intellectual advancement; second, social elevation; third, moral improvement' (Thompson 1968, pp151–2). The moral improvement he had in mind was professional and ethical integrity – leading to a reputation for competence and probity.

By the end of the 19th century these honest, intellectual gentlemen (and they were still all men – see Chapter 7) had succeeded in building three robust institutions that would define the way professionalism operated in Britain and beyond in the decades ahead. The nature of that definition is the subject of the next chapter.

Notes

1   Now in the British Library.
2   Guilds in many towns required 'testimonies of legitimate birth' for potential apprentices, not only proving that they were neither the sons of serfs nor members of ethnic or religious minorities, but also legitimate. Witnesses were required who would swear that they had seen the prospective apprentice's parents process to church on their wedding day at least nine months before his birth (Rowlands 2001, p57).
3   The Office of Works was initially established to look after royal projects including the defence of the realm. The relatively few churches built in the Tudor period continued to use the older master builder method. As the nation state developed in the 16th century the Office of Works gradually took on a wider range of projects including the building of the new City churches for the Stuarts after the Great Fire of London.
4   Part of the inscription on the Smythson Memorial in St Leonard's Church, Wollaton (1614)
5   Although Colvin also notes that after a short spell as architect and surveyor to the Octagon Chapel at Norwich, Brettingham in 1757 and 1760 re-advertised his services as 'a house and estate agent' (p138).
6   This section is indebted to F.M.L. Thompson's book, *Chartered Surveyors: The growth of a profession* (1968).
7   *Le Corps des Ponts et Chaussées* was formed in 1716 with a senior staff comprising: one *inspecteur général*, three *inspecteurs*, one *architecte premier ingénieur*, 21 *ingénieurs de généralités* (*Le Corps des Ponts: Histoire et Actualité*).

8   Bélidor was the author of *Nouveau cours de mathématiques* (1725), *La science des ingénieurs dans la conduite des travaux de fortification et d'architecture civile* (1729), *L'architecture hydraulique, ou l'art de conduire, d'élever et de ménager les eaux pour les différents besoins de la vie* (1737–53) and *Dictionnaire portatif de l'ingénieur* (1758). He was a Fellow of the Royal Society in London.

9   Gauthey was a mathematician, engineer and architect who built bridges, canals and buildings, mainly in Burgundy. His publications include the reference works *Mémoire sur l'application de principes de mécanique de la construction des voûtes et des Dômes* (1771) and *Mémoire sur les canaux de navigation* (1816).

10 Perronet, the first director of the École, was predominantly a bridge builder. He contributed articles to Diderot's *Encyclopédie*. He was also a Fellow of the Royal Society in London.

3
Foundations

When the engineers, architects, surveyors and builders, along with a range of other newly organized sectors of British society, began to organize into recognizably professional bodies in early 19th century Britain they were bidding to join a much older group, what Addison, writing in *The Spectator* in 1711, called 'the three great professions of divinity, law and physic'.[1] With hindsight a new tranche of professions looks like a natural development, but it was far from inevitable. The popular view of the learned professions at the time, if deferential, was not necessarily respectful, and mockery was rife. Professional status was neither well defined, nor a reliable way to become rich or achieve social status.

The satire boom of the 18th century saw the learned professions as fair game and they were frequently skewered as greedy, callous, know-nothing, verbose pedants. Contemporary prints showed priests either as fat and greedy or, worse, as lank and sanctimonious. Lawyers were portrayed as universally cunning and devious, prone to confounding even the simplest case in pursuit of additional fees. Doctors were shown as incompetent and pitiless, if constantly inventive in their diagnoses, as well as being invariably drunk (Corfield 1995, p43). Cartoonists such as Rowlandson, Cruikshank and Gillray, artists like Hogarth and authors including Sterne and Swift, didn't hold back when it came to eviscerating professional men, and even, albeit to a lesser extent, women. The characterization of professionals as self-serving continued to go down well long into the 19th century, as the works of Dickens demonstrate, and yet the professional model was the one that architects, surveyors, engineers and eventually builders, all deliberately chose to adopt. When Dickens targeted architects in the form of the 'hypocritical' Seth Pecksniff in *The Life and Adventures of Martin Chuzzlewit* (1844), as he had previously attacked lawyers, doctors and midwives, the profession was duly warned and could have guessed what was coming.

For all their perceived self-importance and arrogance – the underlying cause of much of the stinging criticism – and the long-standing nature of their position in society, neither the law nor medicine were in particularly good shape as occupations at the moment when the embryonic building professions were searching for suitable models for their own organizations.

Law

Barristers were relatively few in number in the late 18th century, approximately 280 in 1780 rising to 730 by 1805 (Corfield 1995, p91). All English barristers were

Figure 3.1
The March of Roguery,
Charles Jameson Grant
(C.J.G.), 1830

members of one of the four Inns of Court and were usually based within their central London precincts. Admission was overwhelmingly based on family ties and connections and although qualification ultimately depended on a lengthy period of pupillage, there was no formal programme of tuition with far greater importance being placed on attendance at regular Inn dinners. The clubbable and restrictive nature of the Bar sustained itself as an informal and closely knit arrangement until as late as 1851 when the Council of Legal Education was established and a course of study instituted, but it was only in 1872 that any formal test, in the shape of a written examination, was made obligatory. The General Council of the Bar (the Bar Council), the body that represents barristers in England and Wales and that upholds standards and matters of professional etiquette, was only established 22 years later again, in 1894.

In comparison to the relatively well organized Bar, the structure of the junior branch of the legal profession, solicitors and attorneys, was ramshackle and largely open to whoever chose to take it up and could convince clients to pay for the services provided. The historian, Penelope Corfield, in *Power and the Professions* (1995, p76) notes that:

theoretically, there were some safeguards against knavish lawyers. All those transacting legal business in England and Wales had to be enrolled in an appropriate law court, as specified since 1402 – attorneys in the common law courts and solicitors in Chancery. Hence a defaulter could be 'struck from the rolls'. But in practice the system was not operated strictly.

The two junior branches were unified by statute in 1750 and the system of articled clerkship leading to full legal qualification formalized with a requirement for official endorsement. This may have been used as a system for filtering out certain undesirable individuals and social classes, but it did not improve matters greatly. The memoirist William Hickey wrote about his experience of enrolment in 1775 (Spencer 1913–25, pp331–2), when he was cross-examined by a judge over his breakfast of French rolls and muffins:

> Breakfast being over, he asked me how I liked the law, how long I had been out of my clerkship, and two or three other questions equally unimportant, when a servant entered to announce the carriage being at the door, where-upon he desired his clerk to be called, upon his appearance he enquired whether Mr Hickey's Certificate was ready.

Hickey was sworn in and his name entered on the rolls later that morning.

The numbers of solicitors and attorneys grew rapidly during the 18th century, reaching 3666 across England and Wales by 1780 according to John Browne's *General Law List* (1780, pp35–138), as did their earnings and success in obtaining official positions. Nonetheless their reputation struggled – reflected in Samuel Johnson's comment in 1770 that 'he did not care to speak ill of any man behind his back, but he believed that the gentleman was an *attorney*' (Boswell 1827, p177) – and the profession became inordinately keen to improve its overall standing in society. As Corfield notes, 'Respectability was indeed the desideratum and that could not be achieved without some regulation within the profession' (1995, p80).

The first organization established to achieve this was the Society of Gentleman Practisers in the Courts of Law and Equity, founded in 1739 by a group of London attorneys and solicitors. At first a dining club like many other early professional organizations, it was nevertheless one with a definitive purpose, declaring in the minutes of its first recorded meeting that it was determined 'to detect and discountenance all male ['mal' = bad] and unfair practice' (Corfield 1995, p81). The Society, one of the forerunners to 'The Law Society', took its responsibilities for upholding the standing of lawyers gravely enough for it to take action in 1745 against one of its members, Landen Jones, who had been convicted of an offence sufficiently serious for him to be put in the pillory, and appealed to the courts to have his name struck off the rolls (Freshfield 1897, pp.iv–vi).[2] It also scrutinized new candidates for enrolment, advised on proposed changes to the law and made representations to Parliament on legal matters.

Although sporadic, these activities gave the Society several of the necessary attributes of a modern professional organization, but it was still a long way from organizing the growing numbers of solicitors and attorneys, who were easily attacked by one of their own, Joseph Day, for 'the numerous and serious evils with which the lower and unprincipled members of the profession torment society' (Robson 1959, p32), into anything like an ordered profession. That would only come when, in 1831, a newly established body, the Incorporated Law Institution of the United Kingdom (founded in 1825), was given a Royal Charter. The Law Institution rapidly absorbed the Society of Gentlemen Practisers and having changed its name to the cumbersome, but self-explanatory, The Society of Attorneys, Solicitors, Proctors, and others not being Barristers, practising in the Courts of Law and Equity of the United Kingdom, it took on the role of 'an independent, private body servicing the affairs of the profession like other professional, literary and scientific bodies'.[3] Colloquially it became known as 'The Law Society', a term first used in 1792 and only officially adopted in 1903.

Medicine

The medical professions started the 18th century with a multitude of belief systems, methodologies and branches. Theoretical knowledge had only recently started to accumulate with the breakthroughs of Vesalius (1514–64) in anatomy and William Harvey (1578–1657) on the circulation of the blood, but as the medical and social historian Roy Porter has commented (1997, p254):

> achievements proved more impressive on paper than in bedside practice; the war against death was stalled, and, to make matters worse, epidemics rained down on Europe in the decades around 1700 and mortality rates soared.

Progress was halting and gradual, but by the end of the 18th century medical practice had developed a scientific rigour based on the working methods of the leading hospitals and the exacting demands of a rapidly growing publishing industry, and by the 1790s the French political scientist, the Marquis de Condorcet (1743–94), even felt confident enough to declare that the 'The improvement of medical practice … will mean the end of infectious and hereditary diseases and illnesses brought on by climate, food, or working conditions' (Lukes and Urbinati 2012, p145). Despite inevitable setbacks, progress in medicine was very real in the active period of reform between the 1760s and 1840s, the point when anaesthetics began to be routinely used in surgery. Medicine had managed to consign the worst of the 'age of agony' to history.

The progress was achieved largely in the absence of any professional organization, for although there was a professional body, the College of Physicians (later the Royal College of Physicians), which had been granted a Royal Charter in 1518 to award licences to practice in London and its immediate vicinity, it offered little of relevance to the mass of medical practitioners across the country. The College, founded in the wake of other similar organizations in rival European capitals, including Dublin (1446) and Edinburgh (1505), rigorously limited its membership to a small metropolitan elite and deliberately chose to remain that way, typically with only 60 fellows and less than 100 licentiates until the mid-19th century. In the first 300 years of the College's existence it spent most of its efforts unsuccessfully fighting off incursions by surgeons and apothecaries into the territory it saw as the physicians' own, actions that finally exasperated the Lord Chief Justice, Lord Kenyon, who stated in a judgement in 1797: 'by what fatality it has happened that almost ever since this charter was granted this learned body has been in a state of litigation I know not', (cited in Clark 1965, p80) and only with the reforms brought in by the Medical Act of 1858 did it cease to be, in the words of one of its historians, George Clark, 'a solitary and half-secret body' (1965, p82).

Medical practice during the Georgian period was an open field with the work shared between the three rival occupational branches of apothecaries, surgeon-barbers and physicians, alongside any others who offered either care or cures, however ineffective. The title of doctor, originally designating someone with a higher university degree, was being used, in the mocking phrase of Tobias Smollett (1753, p249), by interlopers such as his anti-hero Ferdinand Fathom, who 'graduate themselves' above their immediate competition and who:

> without licence or authority, comes hither to take bread out of the mouths of gentlemen who have been trained to the business in a regular manner, and bestowed great pains and expense to qualify themselves for the profession.

A system of licensing did exist, controlled by local bishops, but it was even less effective at ensuring quality than that used for attorneys. There was however money to be made in medicine and the most senior and well-connected practitioners did well from it, but there was little appetite for professional organization. Dorothy and Roy Porter in their history of 18th century medical practice asked, rhetorically, 'why was medicine such a *hotch-potch* at precisely the time, when in other aspects of life, divides were solidifying between distinct cultures, high and low, patrician and plebeian, polite and popular?' and proposed, in response, that 'the answer, in part at least lies in the fact that its therapeutic efficacy remained hopelessly hit-and-miss' (1989, p27). As a result, the medically qualified general practitioner before the mid-19th century, although possibly well trained and with good hospital experience, was largely treated as a tradesperson rather than a gentlemen professional. As the fictional Lady Chettam remarks in George Eliot's *Middlemarch* (1871–2, but set in 1829–32), during a discussion about whether the town's new doctor was a 'gentleman', 'For my own part, I like a medical man more on a footing with the servants' (1991 edition, pp91–2).

If the European Enlightenment of the 18th century had a practical focus, it was in science and medicine and the period saw advances both in knowledge and its consistent dissemination. Traditional centres of excellence in France and Italy were being actively challenged by new rivals in Halle, Leiden, Edinburgh, Gottingen, Vienna and Philadelphia, as well as London, and information on the new techniques and treatments being developed were being constantly and methodically reported, discussed and shared in medical journals such as *The Lancet*, published weekly from 1823 on. Following Napoleon's reforms to France's public hospitals vast numbers of medical students (over 5000 at one point) went to train in Paris. Later, following Wilhelm von Humboldt's transformation of the Prussian education system and the development of specialist medical laboratories, similar numbers of students trained in Germany. By the mid-19th century physicians had access to high quality training and a remarkable and ever improving body of knowledge (Porter 1997, pp245–356).

A hugely improved foundation of knowledge and expertise allowed doctors to make the transition from a lowly status in the mid-18th century to the high social prestige represented by the ideal of the Victorian family doctor and the famous hospital surgeons of the late 19th century, but they did so 'with almost no professional organisation or identity and not by virtue of being doctors, but for their personal qualities' (Forty 1980, p39). The change almost certainly resulted from advances in skills and reliability and a much improved bedside manner, possibly born of greater professional confidence, but another important influence was the publication in 1794 (and in an expanded version in 1803) of Thomas Percival's (1740–1804) *Medical Ethics, or a Code of Institutes and Precepts, Adapted to the Professional Conduct of Physicians and Surgeons*. For the first time since Hippocrates and Galen, doctors had access to useful guidance on how they should behave in practice. Developments in knowledge and behaviour were far more significant for the successful growth of a professional ethos in medicine than anything that either the royal colleges or the many local medical societies promulgated or delivered. Yet well-trained and professional doctors still had no collective means of defending themselves from the vast numbers of unqualified competitors.

Professional medical bodies in the modern sense only began to appear in Britain in the early 19th century, alongside the institutions of the new professions like the ICE and RIBA, with the Provincial Medical and Surgical Association, founded in 1832 by the Worcester-based surgeon Charles Hastings (1794–1866) being far and away the most successful. By 1856 the Provincial had become a nationwide body and was renamed the British Medical Association (BMA). It started its own weekly publication, the *British Medical Journal*, the following year. Although intended as an association for the sharing of scientific and medical knowledge, the BMA rapidly became one of the strongest voices in a renewed campaign for the professionalization and regulation of medical practitioners.

As already discussed the standard British approach to the regulation of the professions was always to leave well alone and, if necessary, to allow them to regulate themselves. This approach led British governments in the 19th century to avoid and delay implementing any legislation on national licensing for surgeons and physicians despite a growing pressure for action from practitioners deeply concerned about 'quackery in a thousand shapes … the canker in the bud of our progress', as worded by Thomas Wakley, founder and editor of *The Lancet*, in May 1858 (cited in Roberts 2009, p41). Professional disagreement had proved a good enough reason for the government to avoid introducing a Bill in 1845 and another attempt in 1857 was prevented by the collapse of the government, but nonetheless the issue was becoming pressing. In the wake of the cholera epidemic of 1854 public health had become a national emergency at home, and the Crimean War (1853–6) and the subsequent campaigning of Florence Nightingale (1820–1910) had shown up the medical incompetence of the state abroad. The failure by the administrators of the 1834 Poor Law to deliver adequate minimal medical services had, by 1855, become a national scandal. As M.J.D. Roberts writes (2009, p43), in the hot summer of 1858, the year of the 'Great Stink', caused by the quantity of raw sewage in the Thames outside the windows of Parliament:

> the view that government might have a direct responsibility to intervene to guarantee standard minimum levels of protection for consumers, whether voluntary or involuntary, was becoming increasingly familiar to MPs.

The Parliamentary debate on the Medical Bill in June and July of 1858 rehearsed many of the arguments about professionalization that would take place elsewhere in the years to come as rival bodies lobbied for advantage and the operation of a free market in services was weighed against the 'public interest' in ensuring 'competence' and control over entry standards. For the government the aim, in the words of William Cowper MP (1811–88), President of the General Board of Health (Hansard 2nd June 1858), was:

1. To raise to a uniform and sufficient standard the education and acquirements of all persons who entered the medical profession.
2. To have an authoritative register, accessible to ordinary persons, clearly defining those who have attained the prescribed qualification.
3. The removal of all those local jurisdictions which restricted a competent man from practising in any other part of the country than that in which the licensing body which passed him had authority.

And he considered (ibid.) that any objections could only come from:

> those persons who were not members of the medical profession, but who wished to be supposed to belong to it – the class of uneducated ignorant quacks who practised on the credulity of the public, and who hoped, in the present disorganized, anomalous, and chaotic state of the profession to pass for what they were not.

But opposition was fierce and the free market view was expressed by John Chapman in the *Westminster Review* of April 1858 – an opinion cited in the debate on the Bill (Chapman 1858, p520):

> Only by suffering the penalties of employing fools or rogues as their physicians will the people be roused to acquire a knowledge of the simplest elements of physiology and of the laws of health, which would at once enable them to prevent a large amount of disease, and so far to understand what are the essential qualifications of trustworthy professors of the healing art as to be able to select those only who possess them.

Despite objections the Bill was enacted later in 1858 and it created the General Medical Council (GMC) as a statutory body 'to protect, promote and maintain the health and safety of the public'. In accordance with the government's desire to keep the professions at arm's length it was a self-regulating body, with representatives elected by its own membership. But having isolated the regulatory and standards function into a standalone body it left the BMA as only a quasi-professional organization, leaving it free to take on a trade union role on behalf of its members, a role that subsequent governments may have had reason to regret, and which few other professional organizations in Britain have attempted to replicate. In the event the Act satisfied relatively few, as either going too far or not far enough. By radically reducing the numbers of supposedly qualified medical practitioners in Britain it did succeed in providing an answer to the question, 'Why is our profession held in so little esteem by the higher classes?', posed by the *Medical Times and Gazette* in 1858. Respectability, status and esteem had been granted, at least for the time being.

Professional development

As the Napoleonic wars ended in the early 19th century, Britain could look back on decades of radical change which had transformed the country and its economic system, making it very different to almost anywhere else in Europe. The Industrial Revolution had been in progress for more than four decades and the country was increasingly wealthy. A new breed of industrial management had established methods for organizing large cohorts of workers in manufactories turning out cheap, mass-produced goods and an extensive canal system had been built across the country ensuring good connections for trading them. Together Britain's ruling class and industrial leaders had a supreme confidence in the mother country's imperial reach and the potential of its engineering capabilities to solve technical problems almost anywhere in the world.

In the late 18th century Britain began to develop new intellectual tools for running the economy. The Elizabethan theory of mercantilism that had been used to develop the Empire had run out of steam, unable to cope with economic growth. It was replaced by the 'classical economics' model advocated by Adam Smith (1723–90) who advocated a capitalist, small state model controlled by the 'invisible hand' of the market and David Ricardo (1772–1823) who, in 1817, published *On the Principles of Political Economy and Taxation*, which argued for the principles of free trade, taking Smith's arguments to their logical conclusion. As free trade policies developed in the post-Napoleonic period their shock troops, in Hobsbawm's (1977a, p290) phrase, were the 'British middle-class' and in particular the various, embryonic professional groupings.

The success of the new *laissez-faire* model and the official embrace of free trade, symbolized by the repeal of the Corn Laws in 1846, led to Britain's global trade dominance in the 19th century and despite the suffering of large numbers of its citizens and many more of its imperial subjects, it was not going to change course while there was money to be made, new parts of the world left to exploit and 'progress' to be served. But to make capitalism work both at home and across a worldwide empire with a relatively small civil service meant that an intermediate, semi-independent class of administrators, record keepers, decision-makers and advisers was required to run the self-regulating system. They needed to be trusted to follow and implement the rules with plausible impartiality – even if the presumption towards the state and powerful financial interests was naked and undisguised – and occasionally to moderate the demands of the market for the benefit of social stability. This was a very different model from the traditionalist French and Prussian ones for running the economy and industrial programme, with their large and hierarchical bureaucracies and stratified professional classes.

The British version of professionalism emerged and evolved to fulfil the need in the 19th century for an independent and self-directing workforce that could operate the complex mechanism of the free market. The professional system developed and prospered in relatively benign circumstances, before eventually establishing its own, separate and self-sustaining logic and becoming a substantial sector of the economy, capable of providing its practitioners with a good income even as economic needs changed; successfully generating economic, social and cultural value; establishing a new and immensely stabilizing sector of society; introducing a different way of working; and providing a successful means of rewarding skill and application. There was little downside to this new class, but nonetheless it did come with its own demands.

Above all the professional system, while supporting the workings of the free market and providing the necessary checks to its occasional exuberances, also challenged its dogmas. It demanded a degree of monopoly, regulation and licensing. It needed to be driven by other ends than pure financial gains and to operate outside, if alongside, the bounds of the market. The explicit requirement for exclusivity and autonomy was what exercised the free market parliamentarians so much during the debate on the 1858 Medical Act. At the heart of the issue was the normally unwritten and almost always unenforceable compact between society/the state and the developing professional classes that – in exchange for expertise, control over an area of work, social standing and a reasonable, if not necessarily high, income – professionals would provide socially and economically useful services employing skill, impartiality and judgement.

Model explanations

The inherently contradictory appearance of the new organizations in their ancient trappings and intentionally mediaevalist obscurantism, so central to the look and feel of the 19th century professional model, has led to a wide range of explanations for the system's origins and eventual dominance. For the social historian Harold Perkin, in his *The Rise of Professional Society* (1989, p22), the development of the idea of professionalism followed the logic of the Industrial Revolution with 'the familiar principle of the division of labour, which Adam Smith saw as the key to the wealth of nations in 1776'. Professionalism in this reading is part and parcel of the general segmentation of society into specialist functional units, for, as Perkin notes, 'specialisation leads directly to professionalism' (ibid.).

For Andrew Abbott in *The System of Professions* (1988), the development of professions represented the taking of control over vacant or weakly policed areas of work by organized groups of knowledge workers in order to achieve dominance or 'jurisdiction'. 'Professions compete within this system, and a profession's success reflects as much the situations of its competitors and the system structure as it does the profession's own efforts' (p33). In Abbott's interpretation the story is one of an expansion and demarcation of occupational territory in the early 19th century with claims and counter-claims being contested for professional control over areas and types of work.

Penelope Corfield (1995), writing exclusively about Britain, takes a similar position and sees the story as a social struggle for power: 'the social classes were not homogeneous blocs but drew upon a tessellation of different rival interest groups. It was within this pluralist context that the emergent professions jostled for power' (p8). Magali Sarfatti Larson (2013 [1977], p9), referencing Karl Polanyi's *The Great Transformation* (1944), sees that:

> in a society that was being reorganised around the centrality of the market, the professions could hardly escape the effects of this re-organisation. The modern model of profession emerges as a consequence of the necessary response of professional producers to new opportunities for earning an income.

She notes the paradoxical claim by the professions also to act as a necessary countermovement for the proper protection of society. It is this paradox, strongly suggesting a history of self-delusion on the part of the professions, that gets closest to both the success and contradictions of the professional position, one that tries to face both ways and serve multiple masters without apparent conflict or confusion.

F.M.L. Thompson (1968) in his history of surveying and the RICS expressed an optimistic version of the professional proposition when he wrote 'the transition to commercialism in building called for a transition to professionalism and specialism in the designing and supervising functions connected with it' (p80). But the more commonly held view was expressed by Adam Smith in *The Wealth of Nations* (1776), that 'people of the same trade seldom meet together, even for merriment and diversion, but the conversation ends in a conspiracy against the public, or in some contrivance to raise prices' (Book 1, Ch10, para 82), a view later re-iterated by the playwright and social reformer G.B. Shaw and others.

These various interpretations largely agree on the power shifts and economic struggles that lay behind the rise of professional organizations in the early 19th century and the establishment of bodies bringing together engineers in 1818, solicitors in 1825, doctors in 1832, architects in 1834, pharmacists in 1841 and veterinary surgeons in 1844 (Sarfatti Larson 2013 [1977], p246), but in practice the professional model evolved to serve the pragmatic needs of both the state and the market and was not deliberately planned or strategized in advance. It thrived by benefiting all sides in the *laissez-faire* economic model prevalent in 19th century Britain and North America and, if there was no single genius or even a significant theorist behind it, there were a great many who nudged it along because it suited their immediate interests.

Shaping the professional model

When the new associations started to emerge to support and represent the engineering, architectural and surveying disciplines in the 19th century they had a wide range of possible objectives to consider, including:

- Establishing a definition and the scope of the discipline
- Defining who could be, and under what circumstances, a member
- Cementing a hierarchy within the profession or the discipline, often age-based but sometimes skill- or experience-related
- Creating opportunities for conviviality and fellowship among peers in the same field
- Building society's trust in the profession and ensuring respect and social status for professionals in their discipline
- Defining standards of both performance and behaviour, usually to avoid bringing the new profession into disrepute
- Fostering training and educational programmes that extended beyond the bounds of apprenticeship or on-the-job training and with them the potential for granting/gaining qualifications and/or accreditation
- Building a repository for and a means of sharing and developing professional knowledge

- Eliminating 'unfair' (often price-based) competition and encroaching on the work of fellow professionals
- Restricting those who were believed to be unworthy or unfair competitors from the area of work considered to be under the aegis of the profession
- Providing a good livelihood for its members
- Representing its members' collective interests to outside groups.

When they wrote out their initial charters the new professional bodies each identified a selection of these aims rather than the full range, but nonetheless the new organizations gradually developed a common model for a professional institution, one also shared with other non-construction or property related disciplines. The standard model normally had several defining characteristics:

- The existence of an eligible membership group based on an occupational activity or particular expertise and often with a well-recognized title to describe it
- A means of deciding whether potential members were suitably accredited; later on this became based on an approved education system with a final qualifying examination
- A membership structure
- A set of corporate objectives, however loosely defined
- A decision-making body, usually a ruling council comprising active members
- A set of rules or by-laws governing the running of the organization including a means of disciplining or expelling members
- A code of conduct (ethics) governing members' behaviour
- The means of maintaining, developing and sharing (among members) a body of knowledge, commonly lecture series, a regular journal and a library.

To these, the three institutions voluntarily added a new goal, that of serving society and contributing to the greater public good, even if not always expressed in very direct terms. It is not clear where the demand for this additional goal came from and there is little or no discussion of such aims in the founding documents of the institutions. Possibly the impact of attacks from satirists and commentators on the traditional professions encouraged the new organizations to pre-empt similar treatment, or it may be that this was simply the result of well-meaning expressions of social purpose designed to smooth the way to general acceptance. Nonetheless it led to a significant point of difference between professional and other comparable bodies such as learned societies on one side and trade unions on the other, which may have encouraged its retention and development.

In the public interest

The long-standing example of the 'learned professions' may have provided a useful conceptual framework for the idea of professional status and provided important 'intellectual cover' for the new occupation-based associations, but they didn't offer much of a model for institutional organization as they had yet to develop their own professional structures. The forms and rules for professional organization needed to be developed anew, while maintaining the very clear impression that they had very ancient and strong roots that were well steeped in tradition. In Britain, as opposed to elsewhere in Europe, where professional formation was often state-directed (see Chapter 4), it was a process generally executed in the absence of any input or steering from government or, indeed, other public bodies. The UK government only intervened, as it sometimes eventually did (see the example of the GMC above, or other instances in Chapter 7), when its hand was forced.

The founding members of the ICE at one of their early planning sessions in Gilham's Coffee House in the Strand, London, prior to inviting Thomas Telford to become their first President, agreed that 'in order to give effect to the principle of the Institution' they needed 'to render its advantages more general, both to the members and the country at large'. Telford, having accepted their invitation, echoed this in his first address to the fledging institution in March 1820 stating that:

> it has in truth like other valuable establishments of our happy country, arisen from the wants of its society; and being the result of its present state, promises to be both useful and lasting.[4]

The RIBA declared in its Royal Charter (1837) that the Institute had a societal purpose at its core; 'the general advancement of Civil Architecture ... as tending greatly to promote the domestic convenience of citizens, and the public improvement and embellishment of towns and cities'. In 1868 the Institution of Surveyors also declared that it had been established 'to promote the general interests of the professions and to maintain and extend its usefulness for the public advantage' (1869, p.vi). Where did this enthusiasm for being socially useful and for acting for the public good come from?

None of Addison's three great professions: the doctors, lawyers and priesthood, saw themselves as responsible for attempting to improve the general public good. The contribution, if any, of their members to the commonweal was through individual acts of charity or, in the case of the clerics, through saving souls, one at a time. In any case only the church was adequately organized to attempt to make a difference at a large scale, and had chosen not to. Admittedly the College of Physicians had declared in 1688 that it was 'instituted for publick benefit', but did little to justify this beyond fulfilling its obligation to publish the *London Pharmacopeia*, of which it produced ten editions over 233 years, and running a 'Dispensary for the poor of London', in the face of strong opposition from the Company of Apothecaries, between 1696 and 1725 (Clark 1965, p80). In any event by the early 19th century such claims had long been downgraded and were no longer part of the narrative on the professions.

The grant of a Royal Charter had long been the preferred route for new bodies to gain official status and recognition from government and the crown. The charter document at the heart of the process of acceptance had a formulaic rubric that spelled out the ownership of the body, its membership structure and, critically, its objects – the purpose of the body and the goals that it intended to achieve. In the case of the early institutions these objects began to be used to commit the new organizations to ever higher aims as they sought royal favour and approval.

In the hundred years prior to 1828, the Privy Council, on behalf of the Crown, issued 75 Royal Charters. Of these, 18 were commercial ventures, including the Gaslight and Coke Company (1812) and the West India Company (1826), with relatively prosaic objects; 17 were learned societies, including the Linnean Society (1802 – 'for the cultivation of the science of natural history in all its branches') and the Geological Society of London (1825 – 'for investigating the Mineral structure of the Earth'); and just seven were for professional bodies, including four local societies for lawyers in Scotland and the Royal College of Surgeons of England in London (1800). But from 1820, as new charters were being drafted, their authors must have wanted to show more ambition than their predecessors and began to raise the ante. The only object in the surgeons' charter was 'that the Establishment of a College of Surgeons will be expedient for the due promotion and encouragement of the Study and practice of the said Art and Science', although there was also a side requirement, presumably a *quid pro quo* inserted by a government official, to provide premises for the dissection of executed murderers, 'for better preventing the horrid Crime of Murder'. It was clear from the early discussions amongst the founders of the engineers' and architects' institutions that they were looking something that produced a greater public impact.

As large numbers of soldiers and sailors returned home after the conclusion of the Napoleonic wars they needed to find their place back in society. They, like the demobilized forces at the end of more recent wars, would have been looking for recognition that the sacrifice they had made on behalf of their country had some purpose and reward. The result was a great clamour in Britain for social renewal – a demand for change that would eventually underpin the shift from Georgian to Victorian Britain. The demand for change would later be fed by the second phase of the Industrial Revolution (1840–70) encompassing the arrival of steam power and the emergence of modern capitalism, but in its initial stages it was expressed as violent insurrection amongst the working class, as occurred at St Peter's Field outside Manchester in 1819,[5] and repression under the draconian Six Acts passed later that year. In contrast to the violence taking place on the streets the upper classes expressed their desire for change through a sudden enthusiasm for Greek classicism. The Hellenic and then the Greek Revival movements swept across educated British society, catching large numbers of architects and builders in their embrace as they went.

By the end of the 18th century the Grand Tour had almost exhausted ancient Rome as a source of inspiration and it was Ancient Greece's turn once again to fuel the imagination of the well to do. A series of British travellers had managed to travel there in the mid-century, including the artists Stuart and Revett who had surveyed extant classical buildings on behalf of The Dilettanti Society of London. They were in Greece from 1753 to 1755, but their great publication, *The Antiquities of Athens*, only began to emerge in 1762 with the more significant buildings being illustrated in volumes that appeared in 1789, 1794, 1814 and 1830. The timing of the latter volumes couldn't have been better as Lord Byron, the greatest of all Greek enthusiasts, blazed across London society in the years 1811 to 1816. Prominent architects travelled to Athens including Robert Smirke (architect of the British Museum and University College, London) and William Wilkins (the National Gallery), who were both there in 1803–4, and C.R. Cockerell (Ashmolean Museum, Oxford) in 1810. Across Britain 'Greek' buildings started to appear in the form of country houses, churches, civic buildings and follies (Mordaunt Crook 1972).

With the hyper-charged revival of Greek forms came an interest in Greek political thought, in particular the importance of civic values and the idea that the good of the community, the *polis*, should take precedence over that of the individual. As Aristotle wrote in *The Nicomachean Ethics*, 'whilst it is desirable to secure what is good in the case of an individual, to do so in the case of a people or a state is something finer and more sublime' (349 bce, I, ii.1094b 6–10). This was a belief system radically different to the religious/monarchical political structure that had been the mainstay of European life for over a millennium, but critically it came with the enormous classical authority and respectability necessary to effect change without threatening to trigger the republican revolutions seen in recent memory in France and America.

It is not clear that the link between mediaeval city charters – the origin of the Royal Charters awarded by the Crown to trading companies and various corporate, public and professional bodies – and the idea of the self-governed Greek *polis* with its important civic values was ever made explicitly, but nonetheless the concept of the charter embodied the historic principle of the independently governed city with its ancient freedoms and citizenry rights. In many ways it was an ideal model for a membership organization with public service aspirations.

An unlikely player made the first link between ancient Greek philosophy and a Royal Charter. Thomas Burgess (1756–1837), then Bishop of St David's in Wales, was a man of polymathic interests, with nearly 100 publications to his name, including Christian tracts, language primers for Hebrew and Arabic, an essay inveighing against slavery and many translations of ancient Greek texts, including at least two volumes of Aristotle (Harford 1840, pp551–5). He was instrumental in founding numerous organizations including St David's College, the Odiham Agricultural Society, the Royal Veterinary College and the Royal Society of Literature (RSL), of which he was the first President. The RSL, in its early years, was devoted to making Greek classical works available to the British public and Burgess wrote a constitution for the Society (RSL 1823) that had as its principal object:

> the advancement of General Literature by the publication of inedited remains of ancient literature and of such works, as may be of great intrinsic value … by endeavours to fix the standard, as far as is practicable, and to preserve the purity, of our language.

The Society, originally founded in 1820, obtained royal patronage in 1823 and was granted its Royal Charter in 1825.

It was perhaps the spirit of public purposefulness, rather than the enthusiasm for 'inedited remains', that encouraged the civil engineers to use the RSL's constitution as a model when they began to draft their own charter in 1827 (Watson 1988, p19) and they chose to define their Institution as (ICE 1842, p321):

> A Society for the general advancement of Mechanical Science and more particularly for promoting the acquisition of that species of knowledge which constitutes the profession of a Civil Engineer, being the art of directing the Great Sources of Power in Nature for the use and convenience of man.

The ICE Charter then became the model for other professional institutions as each, in turn, sought to craft a suitably worded set of objects in their pursuit of royal benediction.

The Institute of British Architects in its 1837 Royal Charter assembled somewhat awkward phrasing, defining its purpose as:

> The general advancement of Civil Architecture ... as tending to promote the domestic convenience of citizens, and the public improvement of towns and cities,

and in 1868 The Institution of Surveyors stated that it had been established:

> To promote the general interests of the profession and to maintain and extend its usefulness for the public advantage.

The idea that British professional institutions should have a public interest purpose is now unremarkable. The key criterion today for any new chartered body laid down by the Privy Council (2016) is that it should operate in and for the 'public interest'. But it is not so elsewhere. In France the professional *Ordres* are private bodies whose status and organization are defined by a decree issued by the *Conseil d'Etat*. Similarly the German *kammern* are public legal entities. In America, in contrast, the main professional associations, the ASCE and AIA, are private not-for-profit corporations (see Chapter 4). The fact that chartered bodies have a public interest remit is a quirk of a particular moment when interest in classical Athenian political theory was at its peak and Thomas Burgess was on hand to make his modest intervention. It may well have suited the new institutions' quest for respectability that they were committed to a broader purpose and not only the self-advancement of their members, but the idea has nonetheless proved extraordinarily robust and has, in turn, served the idea of an independent and self-governing model of professionalism well.

A defined membership

While there was significant overlap between the various professions at the outset of professional organization, with some individuals such as the architect-engineer, Robert Mylne, and builder-engineer, Thomas Cubitt, participating in more than one of the emerging clubs and institutions, there was no confusion about which area of practice was represented by each grouping. The division of labour had already effectively occurred well before the start of the 19th century and the professional jurisdictions, whether engineering, architecture or contracting, had largely hardened up, at least for practitioners in the larger towns and cities. Surveying still remained fluid and a special case. It was one of the unspoken purposes of professional formation to confirm the sectoral divisions that had emerged and to cement the clear boundaries between the disciplines. The cause of the long period between the initial founding of the ICE and RIBA in 1818–34 and the Institution of Surveyors in 1868 was largely the lack of an adequate definition of the surveying profession, rather than an unwillingness to organize. The solution for the surveyors eventually came in the form of a collegiate approach embracing many different sub-disciplines, including land, building and cost surveying, valuing and auctioneering, within a single organization, and even today the membership of the RICS is a far more disparate group than those of the other, more centred, construction and property professions. The history of professional development in the construction sector largely corroborates Andrew Abbott's assertion that, 'of the various exclusive properties of professions, jurisdiction is the most important' (1988, p87).

There were many, then as now, who saw such divisions as unnecessary barriers to freedom of practice and the cause of much potentially damaging over-specialization. The Editor of *Building News* asked in 1860 (see Thompson 1968, p136):

> is there any valid reason why every architect who enters the Institute [the RIBA] should not qualify himself to receive a diploma, representing a certain amount of acquaintance with surveying; and any reason why surveyors, who could easily gain it, should not be admitted as such to all the privileges of the Institute?

Such concerns were swept away by the aggrandizing effect produced by the move to a system of closed membership, at least amongst those on the inside. The legacy created by a series of separate and mutually exclusive professional groupings would only become a serious problem far into the future. In the meantime it would prove difficult enough to hold each profession together under a single umbrella, as practitioners found they had less in common than they initially thought with fellow members of their chosen profession.

A more immediate problem was to define who could and, by default, who could not join a profession. Different approaches were possible, but selection was principally based on three qualities: the personal standing of a potential member, their job function and the nature of their work, and the set of skills and knowledge they possessed. Most institutions effectively required a mix of these qualification criteria before acceptance and accreditation, even if the defining elements, until fairly recently, were kept reasonably fluid and contingent. In the early days the degree of flexibility over who was admitted or excluded was strongly influenced by how confident institutions felt about recruiting an appropriate membership, by their need to recruit sufficient numbers of new members (along with their subscriptions) and by the degree of threat posed by 'sham' practitioners encroaching on their existing members' workload and clients.

In its Charter (1828, p321) the ICE used job function and project description to define the profession of a Civil Engineer at great length as:

> the means of production and of traffic in states both for external and internal trade, as applied in the construction of roads, bridges, aqueducts, canals, river navigation, and docks, for internal intercourse and exchange, and in the construction of ports, harbours, moles, breakwaters and lighthouses, and in the art of navigation by artificial power for the purposes of commerce, and in the construction and adaptation of machinery, and in the drainage of cities and towns.

Potential members were relatively straightforwardly defined as those who were 'engaged in the practice of a Civil Engineer' (whether full, London based members or corresponding ones with offices outside the area of the three-penny post), while Associates were 'those, whose pursuits constitute branches of Engineering, but who are not Engineers by profession' creating a circular logic barring them ever achieving full membership. The last category of membership, Honorary Members, initially limited to 40, comprised those 'persons who are not engaged in the practice of a Civil Engineer in this country, but who are men eminent for science, and have written on subjects connected with the profession' (ICE 1828, p300) – the nobility and a range of dubious overseas monarchs, including the Duke of Wellington, Prince Albert, Napoleon III, Pedro II of Brazil, Carlos I of Portugal and Leopold II, King of the Belgians (Watson 1988, pp118–23), were especially welcome in this category.

At the outset the definition of who was and wasn't an engineer may have seemed adequately clear to the founders, but they and their successors then spent much of the following decades debating and arguing the point and amending and re-amending their membership categories and criteria to suit. In 1838 the ICE Council eventually agreed the rules for allowing a candidate to be put forward for election as a full member of the institution (quoted in Watson 1988, p116):

1. He shall have been regularly educated as a civil engineer according to the usual routine of pupillage, and have had subsequent employment for at least five years in responsible situations, as resident or otherwise, in some of the branches defined by the Charter as constituting the profession of an engineer, or
2. He shall have practised on his own account in the profession of a civil engineer for five years and acquired considerable eminence therein.

According to Garth Watson, the official historian of the Institution, these criteria survived, albeit in the face of increasing criticism and with ever shifting membership categories, until reforms were introduced in the 1970s.

Like the engineers, The Institute of British Architects started with three classes of members. Fellows were 'Principals for at least seven successive years in the practice of Civil Architecture'; Associates were over 21 and had studied and practised architecture for less than seven years; and Honorary Members were to 'consist of noblemen, who shall contribute a sum of not less than twenty-five guineas, and of gentlemen unconnected with any branch of building as a trade or business, who shall contribute a like sum'.[6] There was no attempt to explain what 'Civil Architecture' was, with the Institute's Royal Charter, published three years later, only expanding on the wording slightly to explain that it was 'an art esteemed and encouraged in all enlightened nations, as tending greatly to promote the domestic convenience of citizens and embellishment of towns and cities' (Institute of British Architects 1837). The seven year rule will have ensured a reassuringly solid, respectable and competent set of members who undoubtedly had a very clear view of what a career as an architect meant.

At its foundation the Institution of Surveyors again made an attempt in its constitution (1869, p.vi) to describe the scope of professional activity that would be carried out by its members, defining:

that knowledge which constitutes the profession of a Surveyor, viz. – the art of determining the value of all descriptions of landed and house property, and of the various interests therein, the practice of managing and developing estates; and the science of admeasuring and delineating the physical features of the earth.

Unlike the engineers with their list of project types, this tried to define surveying through functional activity, although it scarcely captured the extraordinary breadth of surveyors' involvement in industries ranging from mining to transport to agriculture.

As with its fellow organizations the Institution had three classes of full membership: Members, Associates and Honorary Members, with the additional, and by then standard, category of student members. A Member was required to have 'acquired a practical knowledge of surveying in one or other of its branches', to be aged over 25 years and to have 'practised on his own account for more than 5 years; or be a member of a firm of surveyors established upwards of 10 years, or in partnership with a surveyor of 10 years' standing'. As with the architects, the main requirement was experience and a long-standing commitment to the business of surveying. An Associate could be younger, with a threshold age of 21 and 'not necessarily a surveyor by profession, but his pursuits must be such as to qualify him to concur with surveyors in the advancement of professional knowledge'. Honorary Members were not greatly different to the non-surveyor Associates except in eminence and experience. What and where students should be studying is not discussed in the absence of any available courses in surveying, the one certain criterion was that they were to be between 18 and 21 (Institution of Surveyors (1869, pp.vi–vii).

In the early years of the Surveyors' Institution, membership growth was precariously slow and relied heavily on close personal links with the founder members. Thompson records that 'Of the first 100 members to join after 15 June 1868, 53 were relatives of one or other of the original 49 members, or had been trained in one of their offices; no less than 22 had some professional connection with John Clutton'. In 1881 immediately prior to the grant of the Royal Charter there were only 357 members, 164 Associates and 14 students. This changed in the years immediately after Chartership in 1881, despite a tightening of membership criteria to include a new requirement for formal qualification for Members, now promoted to be 'Fellows', and the new class of Professional Associates. Within five years the membership of these two classes had doubled (Thompson 1968, pp158–9, 341). With a qualification hurdle, membership now meant something to surveyors, their clients and society. The Institution could put its struggles with recruitment behind it and it could begin to compete with the rest of the industry.

The memberships of the three institutions at the turn of the century were all on an upward trajectory. The ICE roughly tripled its membership every 20 years and had over 6000 members by 1896. The RIBA grew more slowly but reached a total of 1633 Fellows and Associates in 1900. The RICS after its slow pace pre-Chartership (521 professional members in 1881) accelerated markedly and had more than quadrupled membership to 2267 by 1896 (see Figure 3.2).

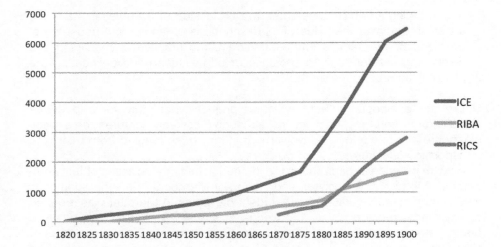

Figure 3.2
Membership profiles
(Full + Associate)
in the 19th century
for the ICE, RIBA
and RICS

Sources: membership
numbers sourced
from Thompson
(1968), Kaye (1960)
and Watson (1988).

The three institutions were not alone in the rapid increase in membership as the rolls of all professional bodies in Britain began to rise around 1880. The Industrial Revolution had largely run its course and society was evolving and reconstituting itself. Although inequality in a strongly class-based society was at its peak in Britain at this time, the middle classes were growing in numbers and strength, and professionalism was both their chosen vehicle for achieving social acceptability and the most effective tool for moderating the high Victorian capitalism that was the prime cause of the social division. It threatened a social revolution, as Harold Perkin has described in *The Rise of Professional Society* (1989, p123):

> From the 1880s by contrast, concomitantly with the accelerated growth of professional occupations of all kinds, it [the professional social ideal] began to take shape in a form that appeared to many landowners and business men to be an organised threat to the rights if not indeed to the security of private property and to the foundations of capitalist society.

The architecture, civil engineering and surveying professional organizations all ended the 19th century (just) with a similar structure of membership, arranged to recognize a core senior membership and with an aspirant 'Associate' class closely examined upon entry on their qualifications to join each profession. In addition, the creation of a student membership class encouraged the next generation to take an active interest in their future institution. Honorary membership was available for co-opting the great and the good to support the institution and facilitate its activities. So much was normal for the growing body of professional associations in all disciplines and set the model for several new bodies at the start of the new century, including the Institution of Structural Engineers (initially the Concrete Institute) in 1908 and the Royal Town Planning Institute in 1913. If there were variants to the basic model it was in the extent and scope of the entry examinations and whether institutions welcomed non-professionals as an additional and generally more diverse class of members.

Knowledge

The Royal Charters awarded to the professional institutions turned them into corporate bodies with both gravitas and influence, a status echoed in their purpose-built headquarters buildings lined with impressive portraits of past presidents. But each organization was also working to achieve more than just position. In addition to their public interest remit they all committed themselves to acquire and advance knowledge in their discipline, reflecting the 18th and 19th centuries' enthusiasm for learned societies.

The ICE in its 1828 charter declared the institution was being formed, *inter alia*, 'for the general advancement of knowledge which constitutes the profession of a Civil Engineer', while the Institute of British Architects' 1837 charter make a similar commitment 'for promoting and facilitating the acquisition of knowledge of the various arts and sciences connected therewith', and in 1868 the Institution of Surveyors stated it would 'secure the advancement and facilitate the acquisition of that knowledge which constitutes the profession of a Surveyor'.

It was never fully clear what was originally intended by 'advancing the discipline', but it was a usefully vague and all-encompassing description of a host of activities carried out by the institutions, including political lobbying and developing and maintaining standards. The idea of 'knowledge acquisition' was a far more straightforward proposition. In the early years of all the institutions learned papers were read at meetings and the first action by Thomas Telford, when he accepted the presidency of the ICE, was to donate a collection of books, including significant French volumes on engineering by Perronet, Gauthey and de Prony amongst others, in order to seed the institution's library. The institutions all invested in substantial libraries, archives and collections and these later became imposing spaces within their headquarters buildings often used to symbolize their commitment to learning and knowledge.

The production of learned papers for reading at fortnightly institution meetings led, in emulation of the Royal Society and its journal, *Philosophical Transactions*, to their publication in the journals published by the institutions. The *Transactions of the Institution of Civil Engineering* began publication in 1836 and the *Minutes of the Proceedings* the year after. The RIBA started publication of its *Journal of proceedings of the Royal Institute of British Architects* (later renamed the *Journal of the Royal Institute of British Architects* and then *The RIBA Journal*) from its foundation in 1834 and the *Royal Institute of British Architects Transactions* started life in 1835. The Institution of Surveyors started publishing its own *Transactions* in 1868 and complemented it with *Professional Notes* in 1886. These journals necessarily became the main point of contact between regional members and activities centred on London, carrying not only papers describing technical achievements or the results of research, but also reports from the institutions' council and committees. The journals demonstrated to the public and decision-makers the useful contribution made by the institutions to developing knowledge and expertise and helped substantiate their essential seriousness of purpose.

Although the quality of papers published by the journals was initially uneven and their appearance was intermittent, their contents gradually became more reliable and publication more regular. This enabled information and guidance to practitioners on both technical and professional matters to be consolidated and standardized and to become the professional approach to a range of issues. The activities of the institutions were also reported by a growing number of outside journals, periodicals and magazines including the *Architectural Magazine*, edited by J.C. Loudon (1834–8), the *Civil Engineer* and *Architect's Journal* (1837–68) and *The Builder* (1842 to date – renamed *Building*), which dealt with many of the same themes, but often took a less respectful line on matters in the industry and contained more commercial news, information and gossip.

Access to a nationwide discussion on current and professional matters enabled by the journals as well as the impact of the railways meant that the many new local study groups and professional clubs meeting across the country were able to participate in an often-lively professional debate. Architectural societies were established in Manchester (1837), Edinburgh (1840 – when the Institute of Architects of Scotland was founded), Liverpool (1848), Bristol (1858) and Nottingham (1862), developing into the Architectural Alliance in 1864 (Kaye 1960, pp90–1). Some, or even much, of the time may have been spent at convivial dinners, but the connected nature of the late 19th century was a powerful force not only for the developing education of professionals but also in the development of a considered collective view of what professional behaviour and activity could and should be.

The objective shared by all the institutions to advance knowledge also led them to undertake and commission research. They began to harness the breadth of experience and knowledge of their membership to inform society and government on matters of policy and strategy. Conferences and very popular annual *conversazione* events were held to disseminate research, research committees were inaugurated to steer each institution's efforts and they attempted to develop and maintain strong links with government ministers and departments to allow them to inform and influence policy (both in the public's and their own interest).

Governance

If the institutions were capable of being useful they could also be ceremonious, bureaucratic and prone to rivalry. A great deal of time and effort was put into their organization and governance and numerous internal battles were fought for their control. Each adopted a roughly similar corporate structure with a president, ideally a prominent senior representative of their discipline whom both the membership and public held in high regard. The president chaired a decision-making council, composed of senior members, which oversaw a range of committees and sub-committees. Together they formed the public face of the organization, while a small and much less visible executive team managed the day-to-day activities of the institution, including its internal organization and premises. Initially internal groups ran the selection of new presidents and even council members, albeit with some input from local associations. AGMs and SGMs were later used to consult and encourage more democratic participation from the wider membership and postal ballots were introduced at the end of the 19th century, with the ICE having its first full membership vote in 1896 (Watson 1988, p58). The basic institutional model has survived broadly intact to the present day, although the balance of influence within each body has gradually shifted away from the members and into the hands of the executive.

With organization also came a plethora of regulations, by-laws and procedures that governed the way institutions acted and could be persuaded to respond. These ultimately were devolved from the Charter and the requirements of the Privy Council but often evolved in an *ad hoc* manner. This is not the place to examine these in any detail, but there is one area where they directly relate to professionalism: the code of conduct, and with it the disciplinary process for those accused of breaking it.

Code of conduct

The notion of the public good at the heart of the institutions' missions may have harked back to an ancient version of civic virtue, but the idea of professional codes of ethics and standards of behaviour derived from a more contemporary stream of thought.

The turmoil in Europe that was brought to a conclusion at the Congress of Vienna in 1814–15 had disrupted almost all European thought and had a profound effect on future developments. There was a strong desire to change the world for the better, but great uncertainty as to what made for a good society or what was the best way to act. As the writer, Kenan Malik (2014, p201), has noted:

> Nowhere was the debate between individual psychology and social need more intense than in Britain. The debate about enlightened self-interest that Hobbes had unleashed with *Leviathan* was accentuated not just by the British obsession with empirical science but also by the rising power of the forces of commerce and the political needs of the nascent capitalism.

The immediate conflict in ethical thought was between the home-grown utilitarianism and continental rationalism. Philosophers David Hume (1711–76) and Jeremy Bentham (1748–1832) advocated the greatest happiness for the greatest number of people. Bentham stated that 'Nature has placed mankind under the governance of two sovereign masters, pain and pleasure. It is for them alone to point out what we ought to do, as well as determine what we should do' (1789, p4). In contrast Immanuel Kant (1724–1804) argued for a duty-based ethical principle, which requires that one should 'always act in such a way that you can also will that the maxim of your action should become a universal law' (Kant 1993 [1785], p.v).

The contingency and individuality of utilitarianism was never going to work well for organizations attempting to bring together large numbers of practitioners into a coherent body and persuade them to act in a consistent way, but Kant's newly minted ethical framework was ideal for them. It was a non-religious system that connected rationality with moral action based on what Kant termed 'categorical imperatives'. Such imperatives meant that there was a way to behave, whatever the circumstances, and rationality would allow you to determine what it should be. To reach this determination Kant proposed working through a sequence of three formulae (see Fewings 2009, p21):

- Act as if your action was a universal law of nature;
- Act in such a way for the action to be an end as well as a means; and
- Act as if you were a law-making member, i.e. for rational means that would integrate with others wanting to do the same.

Such an approach opened the way for generating open-ended rules of behaviour, which could be worked through to reach specific decisions. These were not precise commandments but a system that should yield a definitive outcome matched to particular circumstances. The closeness of the Kantian process to the development of professional codes of conduct, which largely left individual professionals with the ability to interpret them themselves, is marked, but wouldn't have been possible without the revolution in ethics that he instituted.

Several of the precursors to the professional organizations attempted to develop standards and codes of behaviour. At the founding of the Society of Civil Engineers in 1771 (later the Smeatonian Society), it was proposed that members should meet 'without jostling one another, with rudeness, too common in the unworthy part of the advocates of the law' (Taylor 1837, p.vi).[7] In 1796 Robert Mylne, also a founder member of the Society of Civil Engineers, presented a resolution to the Architects' Club (quoted in Kaye 1960, p59), that:

> if any Member shall solicit ... to be employed ... during the known employ-ment of any other Artist ... he shall be considered as acting directly contrary to the Established Practice & derogatory to the honour of the profession of an Architect.

Likewise the Land Surveyors' Club, founded in 1834, determined that members should 'enhance their respectability by acting in accordance with an agreed scale of charges, which would prevent unbecoming and undignified competi-tion' (Thompson 1968, p95).

These were relatively gentle moves leading towards a comprehensive professional code of conduct. Little thought was given to what rules, if any, might govern the behaviour of a professional. At the time the critical matter for professionals like *Middlemarch*'s Dr. Lydgate, was that they should be perceived as 'gentlemen' and act, and be allowed to act, accordingly. One of the few practitioners who were thinking about what this meant in practice, and particularly in relation to an architect's responsibilities to his clients, was John Soane. In the preface of his book *Plans Elevations and Sections of Buildings* (1788, p7), Soane wrote:

> The business of the architect is to make the design and estimates, to direct the work and to measure and value the different parts; he is the inter-mediate agent between the employer, whose honor and interest he is to study, and the mechanic, whose rights he is to defend. His situation implies great trust; he is responsible for the mistakes, negligence, and ignorance of those he employs; and above all, he is to take care that the workmen's

bills do not exceed his own estimates. If these are the duties of an architect, with what propriety can his situation and that of the builder, or the contractor be united?

Soane's last point concerns the great worry of architects at the start of professional formation that they could, in any way, be confused with builders and measurers, who were not to be considered gentlemen. At one of the final meetings of the Society of British Architects in 1834, just prior to the founding of the Institute of British Architects, one of its members, James Savage, drafted a sketch code of professional conduct including 'Disqualifications of Members and Causes for Removal'. These were, in their entirety:

- Measuring and valuing Works on the behalf of Builders, except those executed from the Member's own designs or directions.
- Receiving any pecuniary consideration or emolument from Tradesmen, whose works he may be engaged to superintend on the behalf of others.
- Having any interest or participation in any trade or Contract connected with Building.
- For any other unprofessional conduct.[8]

The link between this proposal and Soane's view on professionalism was explicitly drawn when Savage's proposed code was sent to Soane as part of an invitation to become the Society's first president. Soane declined, citing a conflict of interest with his position at the Royal Academy (Jenkins 1961, p117). At that time he was aged 81, very much at odds with the world and increasingly frail.

For many of the institutions Savage's proposed code would have been several clauses longer than they felt they needed. The final phrase 'unprofessional conduct' was generally enough and whether or not a member was guilty was a matter to be decided by his peers. The ICE's regulations in 1836 allowed for the expulsion of a member following a report to and a decision by the Institution's Council at a Special General Meeting (ICE 1842, p300). There is no discussion in the regulations for possible grounds for expulsion. The RICS took a similar position. Its 1864 by-laws allowed its Council to suspend a member and a Special General Meeting to expel them 'for reasons which they shall deem sufficient'. It was only in 1903 that a serious attempt was made to clarify what these reasons might be (Thompson 1968, p153) and it took until 1934 for the Institution's first formal code of conduct for members to be introduced. As Watson (1988, p59) quotes, the ICE revised their disciplinary procedures in 1896 to allow their Council:

by a two-thirds majority, to expel any member who may be convicted by a competent tribunal of felony, embezzlement, larceny or misdemeanour, or other offence which, in the opinion of the Council, renders him unfit to be a member, and in any case in which the expulsion of any member shall be judged expedient on any ground whatever.

They thus managed at one and the same time to be high level, specific and keep all options open.

The first formal institutional codes and their modern equivalents are included in Appendix B.

Conclusion

The creation of institutions in 19th century was not a carefully calculated man-oeuvre in response to a grand plan. They evolved from serendipitous gatherings of a few like-minded individuals who saw the benefit of working together to protect and promote their mutual interests. The organizations they founded were subsequently shaped by changing economic and social circumstances, but the institutions and the underpinning system of professionalism are still with us and have proved to be remarkably robust and highly adaptive to developments that have both supported and threatened their future. The system of professionalism in the UK is far from an inevitable response to modern commercial capitalism, as the next chapter on alternative professional models in France, Germany and the United States will show. But it has proved both successful and enduring to the extent that professional services are an ever growing sector of the UK economy and the number of professional bodies continues to expand.

This chapter has indicated that the professional model established in the 19th century had, at least, the following characteristics:

- A declared public interest purpose – if possibly expressed in mildly evasive terms
- A means of defining eligibility (and non-eligibility) for membership and for expelling undesirable members if necessary. In all cases this also meant different grades of membership to differentiate between experience and usefulness, among other attributes
- An idea of the knowledge and expertise that a joining member should have and would then be expected to maintain and develop. On entry this was usually tested by an interview process involving existing senior members, a process that could also be used to filter out any unwanted candidates.
- A system to collect, disseminate and share professional knowledge (probably involving libraries, journals of record, formal lectures and other means)
- A quasi-democratic structure that led to the selection of a president, who would act as part temporary leader of the institution and part figurehead, and a governing council by the membership. Quite how democratic this was in practice would vary from institution to institution and also over time
- A code of conduct, accompanied by more specific regulations, that would attempt to control the behaviour of members, enhance the reputation of the institution, its members and *marque* and ensure that members did not commercially undercut each other.

Underpinning the model would be the idea of the professional compact, that if a discipline organized itself into a self-administrating structure for delivering expert services its members would be rewarded with a range of social and commercial advantages. When this principle eventually shifted under political and commercial pressure the model would also change, as discussed in chapters 8 to 10, while still, and perhaps surprisingly, retaining the basic characteristics listed above.

Notes

1 Addison was not being entirely serious about the 'greatness' of the learned professions, as he went on to say 'how each of them [are] overburdened with practitioners, and filled with ingenious gentlemen that starve one another'.
2 E. Freshfield (Ed.) (1897), *The Records of the Society of Gentlemen Practisers*, Incorporated Law Society, pp.iv–vi, cited by Corfield 1995, p81.
3 The Society of Attorneys, Solicitors, Proctors, and others not being Barristers, practising in the Courts of Law and Equity of the United Kingdom (1845), The Charter of the Society.
4 ICE, Minutes of the proceedings of the Institution of Civil Engineers, 25th January 1820, cited by Watson (1988, p14).
5 Nicknamed Peterloo in mocking tribute to the battle of Waterloo.
6 J. Gwilt, Address of the Institute of British Architects, explanatory of their view and objects, and the Regulations adopted at a meeting, held July 2nd, 1834, Institute of British Architects, cited by Kaye 1960, p80.
7 The preface to the first volume of Smeaton's works, where this quotation comes from, was probably written by Robert Mylne or Thomas Yeoman.
8 J. Savage (c1834), Sketch of Original Constitution and Laws, Prospectus for the formation of a society to be called The Institution of British Architects, inc in miscellaneous papers, Folio, RIBA library, cited in Kaye 1960, p77.

4

France – Germany – United States

This book is largely a comparative study of just three British construction industry professions and doesn't even significantly stray as far as examining how professionalism is different and distinctive in Scotland, Wales or Northern Ireland, each with their own history and traditions.

In order to provide a broader perspective and understand alternative influences on the development of professionalism, however, it is important to look further afield and examine how professional systems arose in a range of other countries and jurisdictions. The roots of professionalism in Britain, France and Germany may be from a shared stock but they diverged many centuries ago. The divide between Britain and America occurred more recently, but again long enough ago for substantial differences to have developed. The divergences have, of course, also happened while cross-fertilization between systems and international conferences and discussions on professional subjects has continued, not least those addressing the subject of education (see chapters 5 and 7), leaving a hybrid system with many recognizable elements across different countries and fields while retaining other illuminating national characteristics and differences.

The field of education for construction professionals in particular illustrates the differences but also the strong relationships and interplay between them. Initially the French were dominant in the development of engineering, followed by the Americans, and both their approaches profoundly influenced advances in Britain. Architectural education was similarly dependent on first a French model (*Beaux-Arts*) then on German exemplars (Humboldtian and *Bauhaus*); but they were imported into the UK via the US after much modification and reinvention. Much has been learnt and exchanged.

France

Education

As discussed in Chapter 2 state sponsored engineering training began in the early 18th century in France and was consolidated with the establishment of the *L'Ecole Royale des Ponts et Chaussées* in 1747. This was only five years after 'the first recorded and substantially correct analysis of an existing structure in terms of its static equilibrium' had been carried out on the dome of St Peter's in Rome by the mathematician-physicist team of Thomas Le Seur (1703–70),

François Jacquier (1711–88) and Roger Boscovitch (1711–87) (Mainstone 1975, p283) and engineering as a science was still effectively in its infancy. 'Ponts', as the school was commonly known, was joined in 1783 by *L'Ecole des Mines* and together the two affiliated schools (after a hiatus caused by the French Revolution) became responsible in the Napoleonic period for training and qualifying the large numbers of *ingénieurs d'Etat* required to build and, more significantly to run, the hugely expanded physical infrastructure of France.

Ponts is the oldest of the prestigious *Grandes Ecoles*, schools run by the French state outside the university system whose graduates dominate, to this day, both the government and the private sector. By the end of the 19th century the graduates of the *Grandes Ecoles* had become a well-educated and privileged elite within the technocratic state and as Elliott Krause notes, became 'unavailable for run-of-the-mill engineering jobs'. This made it 'necessary to get most French engineers from another source – the *Ecole centrale*'. 'Addressed as "*Monsieur l'ingénieur*", this expert was clearly a member of the upper middle class (in terms of ideology and eventual job possibilities) and in symbolic alliance with management' (1996, p154).

The class structure of the system was clear and led to a tradition of engineers remaining outside the professional system of *ordres* established by the state for doctors and lawyers. No licence to practice was required. It was deemed sufficient that engineers had high social status and a prestige conferred on them by their *école* affiliation. They were supported by powerful co-alumni in other parts of the system, rather than any professional association, and as Krause concludes (p163):

> Engineers are not a *profession libérale* and are not likely to become one. The fate of each engineer can still be predicted according to the school from which he graduated and his standing within that school.

See Table 4.1 for details of engineering qualifications in France.

The formal education of architects in France was handled by a lower tier of school than the *Grandes Ecoles*, but nonetheless with the strong imprimatur of the state. The first school in the country, the *Académie Royale d'Architecture*, was founded in Paris in 1671 and began by holding public lectures two days a week delivered by Jacques-François Blondel (1617–86) on the theory of architecture as well as covering arithmetic, geometry, mechanics, military architecture, fortifications, perspective and stonecutting. In 1717 the lecture series was

Table 4.1 Qualification procedure for certified engineers in France

The title *Ingénieur Diplômé* (certified graduate engineer) is regulated and protected by the French State.
The *Diplôme* is only awarded to graduates of accredited *Grandes Ecoles*.
Courses are overseen and accredited by the *Commission des Titres d'Ingénieur* (CTI), a department of the French Ministry of Higher Education and Research *(Ministère de l'Enseignement supérieur, de la Recherche et de l'Innovation,* MESRI).
Misuse of the title *Ingénieur Diplômé* is a criminal offence. The title *Ingénieur* is unregulated and available to anyone.

Source: *Commission des Titres d'Ingénieur* (CTI).

converted into a two to three year course with students (*élèves*) enrolling in the *ateliers* of one or other of the school's professors. In 1793, after the Revolution, the *Académie* system was abolished by an order from the National Convention,[1] but with the support of the Minister of the Interior, who declared 'the school of architecture is of immense utility',[2] the school itself survived, albeit with very few students. The *Académie* merged in 1816, in a Napoleonic reorganization, with the *Ecole Royale des Beaux-Arts* and re-emerged, as a separate faculty, the *Section d'Architecture de l'Ecole des Beaux-Arts* in 1819 (Chaffee 1977).

The *Section d'Architecture*, with its distinctive system of individual *ateliers* ran under a set of regulations that would remain essentially unchanged until its closure in the wake of the *évenments* in Paris in May 1968. A student would first enrol in an *atelier* under its master, the *patron*. Bearing a letter of introduction from the *patron* and proof that he (and later she) was between 15 and 30 years old, the next step would be enrolment as an *aspirant* at the *Ecole des Beaux-Arts*. Preparation would then start for the annual entrance exams, testing *aspirants* on mathematics, geometry and architectural design (history and drawing were added to the exams in 1864 and 1867 respectively). Achieving a pass typically took two years although some *aspirants* carried on trying until, eventually over 30, they were barred from continuing. During that time they had access to the *Ecole*'s premises, library and lectures, but training proper only started with an exam pass and admittance into the 'second class' of the school, as an *éleve de l'Ecole des Beaux-Arts*.

The journey to the 'first class' commonly took two to four years, although it was possible to complete it in a year, and comprised entering a series of time-limited monthly competitions, *concours d'émulation*, focusing on three different areas: architectural composition, analysis of architectural elements and construction. This last area, which included detailed studies of stone, iron and wood construction along with mathematics, geometry and stereotomy, was considered the hardest to pass. For most French and foreign architectural students participation in this system was considered enough, as the title of *ancien éleve de l'Ecole des Beaux-Arts* was an honorific that far outweighed the lesser title of architect, but many hard-bitten souls carried on competing again and again until the cut-off age in the *Ecole*'s annual *grand concours* for which prizes and medals were awarded and, ultimately, although for French students only, the highest honour of all, the *Grand Prix de Rome*. The winner of the *Grand Prix* was sent to study in Rome for four or five years, but the real prize was the position at the end of the period of study as an *Architecte de Gouvernment* and the opportunity to design a major public building and fill a series of prestigious and life-long public roles including the strong possibility of election to the *Académie* itself.

The *Section d'Architecture*, as the leading architecture school in the world in the 19th century, attracted large numbers of overseas students eager to be trained under the *Beaux-Arts* system. British students included Lawrence Harvey, who studied first with Gottfried Semper (1803–79) in Zurich before going on to Paris in 1867 (Harvey 1870, p280; Mallgrave 1996, p365), and the Scottish architect, John James Burnet (1857–1938) who was part of Jean-Louis Pascal's (1837–1920) *atelier* in the 1870s. Of the many Americans who joined *ateliers* and enrolled at the *Ecole*, Henry Hobson Richardson, Charles McKim, Louis Sullivan and Bernard Maybeck (Draper 1977, p215 and see below) are perhaps the best known; but at least 11 others passed through Pascal's *atelier* alone towards the end of the 19th century. Ever the radical, Frank Lloyd Wright (1867–1959) turned down the sponsorship opportunity offered by Daniel Burnham (1846–1912) of Chicago to study at the *Ecole* for four years.

With the *Ecole* system, and by extension the French state, in control of the upper reaches of the architectural profession there was little concern over lower ranking practitioners. As Richard Chaffee notes, 'from the time of the Revolution until 1940, anyone who wished to be known as an architect needed only to buy a patent, and some so-called architects were but pretentious entrepreneurs' (1977, p85) and others were felt to be and were paid little better than craftspersons. The *Académie* and the *Ecole* controlled the distribution of commissions and the competitions for major projects and, at the end of the century, the *Section* was training approximately 25% of practitioners (Rodriguez Tomé 2006, p58). The power of the professional elite was consolidated further in 1840 with the establishment of the *Société centrale des architectes français* with a predominantly Parisian (72.5% in 1895) membership. Limited to 500 members and composed solely of professors of the *Ecole des Beaux-Arts* and senior administrators, it was scarcely representative of the wider profession and for several decades actively resisted the establishment of other organizations and in particular any accreditation process for the architects in France (Rodriguez Tomé 2006, pp64–72).

Professional societies and codes of conduct

The situation gradually changed in the latter part of the century as professional schools opened in provincial cities, including Valenciennes, Lille, Marseille, Toulouse, Lyon and Nice – all offering their own versions of academic training, and as local associations and societies of architects developed across France. The stresses on the wider profession struggling to make a living outside the charmed circle of the *Beaux-Arts* were beginning to be widely felt and were expressed both as antagonism amongst local rivals, as in Nice where architects were said to be 'popping up like mushrooms' (Pontremoli 1959, p59), and as a reason to organize and set up mutual associations to protect local interests against outsiders, including, as the *Bulletin de l'Association provinciale des architectes français* complained in 1890, moonlighting public officials 'who are given a fixed salary, are insured for a pension, do not pay patentees, are not responsible, often travel free of charge or reduced prices, and, through their official position, unfairly compete with the architects' (quoted in Rodriguez Tomé 2006, p64).

In the years from 1872 to the end of the century over 30 local groups representing architects emerged alongside at least six rival organizations and defence funds claiming to speak for them nationally. They all had different approaches but shared a desire to control access to the profession, with the French state acting as a guarantor of standards, and a common opponent in the *Société centrale*. The oldest of them, the *Société nationale des architectes de France* (est. 1872), lobbied hard for state recognition and a national diploma, arguing at their 1889 Congress that as soon as possible 'no one in France should be able to practice architecture if he does not have a diploma issued by the government stating that he possesses the minimum of knowledge necessary for the profession of an architect'.[3] Meanwhile the *Association provinciale des architectes français* petitioned the Minister of Public Works to prohibit surveyors working for government departments from undertaking additional architectural and construction work.

Such threats to their elite position were easily headed off by the *Société centrale*, which argued that the proposal for a diploma contravened the rights of both workers and employers to freely contract their services. But the position was changing rapidly. In 1884 the French Minister of the Interior, Pierre Waldeck-Rousseau (1846–1904), introduced new laws recognizing trade unions, and the unrepresented professions saw this as meaning that they too should be given official status. In 1889 Henri-Aimé Delmas-Azéma (1838–1911), architect-engineer of the city of Agen in south-west France, speaking at the Society of Architects of Aisne proposed a '*Société Générale des Architectes*'. He urged his colleagues to achieve a professional status comparable to that of engineers, while warning of the interlopers who were threatening their jobs:

> Engineers have shown us the way forward, let us follow it, all the more so because tomorrow's 'architect-eaters' are beginning to sharpen their beaks and nails to tear us apart and annihilate us.[4]

The bitter arguments between the new bodies and the establishment, represented by the *Société centrale*, were both geographic, with the architects of wider France ranged against those in Paris, but also connected closely with class, age and stylistic preference, with 'gothic' practitioners lining up against 'romans', in a reflection of the style wars also raging in Britain. The battles were fought in the specialist (and partisan) weekly journals such as *La Construction Moderne* and *L'Architecture* and especially over competitions for public projects. Denyse Rodriguez Tomé in her paper on the period notes one battle over a competition for a school in L'Orient, which the *Société centrale* urged its members to boycott while the *Union syndicale des architectes français*, one of whose prominent members was the chair of the jury, furiously replied, 'we must stand our ground to defend ourselves and preserve architecture'.[5]

In 1893 in the face of the bellicose chaos facing the profession and with a lack of any overall organization the *Société centrale* set up a Commission to define who was entitled to call themselves an architect and the scope of their professional duties. The *Société* chose one of their own for the task, Julien Guadet (1834–1908), an architect who had worked on the Paris Opera with Charles Garnier (1825–98), a winner of the *Grand Prix de Rome* in 1864, a professor of theory at the *Section d'Architecture* and a future president of the *Société*. What emerged two years later at the Congress of the *Société* held in Bordeaux was the first code of conduct for architects written in modern times, immediately referred to as the *Code Guadet* (1885, p288) and almost as immediately embraced not only by the *Société centrale*, but also by the great majority of the other warring parties. It didn't persuade them to join forces, but it allowed the profession to consolidate around an agreed, voluntary standard, one with no legal standing but with substantial moral authority.

The *Code Guadet*'s (see Appendix C for a full English translation) major concern is defining what an architect is not – 'a contractor, industrialist or the supplier of materials and articles used in construction … neither a businessman nor an agent … a contractor's clerk, quantity surveyor or measurer'; as well as ensuring that any rewards are transparent and above board – 'he shall be remunerated solely by payment of a fee'; and untainted by any suspicion of risk-taking and exploiting anything, 'which is commercial activity and runs counter to the liberal profession of the architect'.[6] This was, as Rodriguez Tomé has noted, more honoured in the breach and did not prevent many architects, such as Auguste Perret, maintaining family-run construction businesses in the background (2006, p75).

In between the injunctions intended to keep commerce and the corruption of money at bay the *Code* includes many other clauses aimed at raising the standards of the profession, including: preserving so far as possible a collegial approach that maintains respect for colleagues and seeks to learn from them; treating workmen well; giving clients the full benefit of any knowledge, skills and experience gained over a career and supplying them with proper and timely information; and not putting any responsibilities to clients before a clearly expressed duty towards the general public. In many respects, and notwithstanding the ban on commercial activities, the *Code Guadet* stands up well against today's standards.

Forty-five years after its publication the existence of a well-defined and generally well-behaved profession, bound to each other by their shared *Code*, allowed for the creation of the *Ordre des architectes*,[7] an arms' length body established in law by the state to regulate the profession. The establishment of the *Ordre* (see Table 4.2) brought respectability and placed French architects for the first time on a level footing with doctors and advocates in the eyes of government. It was a major achievement.

Table 4.2 Article 1, Title 1, Article 3 of the law instituting the *Ordre des architectes* and regulating the title of the architects' profession, 31st December 1940[8]

The profession of architect is incompatible with that of entrepreneur, industrialist or supplier of materials or articles used in construction.

Except where the architect enters, as an official in a public administration, he agrees with his client on the amount of his fees. He is prohibited from receiving, for the agreed work, any other remuneration, even indirect, from a third party in any capacity whatsoever.

The architect must observe the rules contained in the code of professional duties which will be established by a regulation of public administration.

This code will determine in particular the conditions in which the architect will have to take out insurance covering all the risks resulting from his professional responsibility.

Architects do not have the right to form unions governed by Book III of the Labor Code.

Table 4.3 Qualification procedure for architects in France

Qualification as an architect in France is governed by both French Law (Law n°77–2 01/03/1977) and the EU Professional Qualifications Directive (2005/36/EC).

The process conventionally consists of the following steps:

1) A three year course at one of the 20 écoles nationales supérieures d'architecture (ENSA) leading to an undergraduate degree in architecture – *diplôme d'études en architecture* (DEEA)

2) A two year course at one of the ENSA schools leading to a masters degree in architecture – *diplôme d'Etat d'architecte* (DEA)

3) A one year training course leading to the grant of a licence to practice – *habilitation à l'exercice de la maîtrise d'œuvre en son nom propre* (HMONP)

4) Registration with the Ordre des Architectes permitting use of the title of 'Architect'.

Germany

Education

> *When comparativists tried to apply the professionalization terminology, it soon became evident that categories derived from an Anglo-American model hardly fit the bureaucratic German pattern.*
> (*Jarausch*, The German Professions in History and Theory, *1990a, p.10*)

Unlike in Britain where the Hellenist movement opened up new opportunities for the professions in the early 19th century (see Chapter 3) a keen interest in ancient Greece in the German states led to very different results. The enthusiastic neo-humanism of men such as Wilhelm von Humboldt (1767–1835) and Friedrich Wolf (1759–1824) resulted in a rejection of the Enlightenment ideal and its interest in increasing technological specialization. Instead the focus was on a return to a classically focused education (*Alterumswissenschaft*) (Romba 2006, p2730). The neo-humanist movement defined the higher education of the German, and particularly the Prussian, middle classes for the rest of the century and meant that a general university degree far outdid any specialist discipline in prestige and popularity. As Anthony La Vopa has written (1990, p27):

> It was an emphatic Hellenism, not to say Graecomania, that initially inspired an intensely specialized research ethos and hence made the new discipline a precocious model for modern academic professionalism. Yet the same Hellenism was also the vehicle for a pronounced antimodernism, and at the heart of the antimodernism lay an indictment of specialization itself.

Despite the rejection by elite society of technical expertise in favour of the cult of *Bildung*, the pursuit of philosophical self-cultivation encouraged by von Humboldt, the rapidly industrializing economy still urgently needed highly trained and skilled individuals. The responsibility for producing architects and engineers was assigned to the state-run schools, including the Berlin Building Academy in 1799 (the *Bau-Akadamie*, with its famous Schinkel-designed brick home). This created an ambiguous position for professionals as they were no longer entitled to the title of master builder (*Baumeister*) but they also lacked the status of an academic qualification, of being an *akademischer Berufstand* – the standard equivalent of 'professional' in Germany at the time. The schools struggled to attract students and the few schools that survived only managed to continue once they had been granted self-government and a state-guaranteed income that allowed them to improve their facilities and employ salaried and tenured teaching staff. Otherwise, as the historian Joachim Whaley has noted (2012, p519):

> Few of the many attempts to establish commercial schools, industrial schools, drawing schools, mining or construction academies, or navigation schools, including some for girls, enjoyed any lasting success. Even where such institutions had government support, they rarely flourished, failing as often for lack of students as for lack of money.

The situation was exacerbated when architects and engineers, who had traditionally been educated together with similar curricula, found that their courses were being split apart, possibly in the face of greater technical demands on engineers, but certainly as the engineers' courses were proving more popular while architecture courses failed to attract students. The *Bau-Akadamie* itself was divided in 1823 when the engineering department was transferred to the jurisdiction of the Board of Trade, while the architecture department remained part of the Academy of Art. The future career prospects of both groups, although they were all principally destined for the civil service, were being clearly signalled. The architecture school collapsed and needed to be reintegrated with engineering, but the dissatisfaction remained.

The solution that eventually emerged was the importation of the French *école polytechnique* model, designed to provide high quality training in specialist subjects and also to confer prestige and status on graduates. An early example was the *Karlsruhe Polytecnische Schule* (1825), which taught mathematics, engineering and architecture as well as running preparatory and commercial departments, with other polytechnics opening in Württemberg in 1831 and Hanover in 1847. Eventually in 1849 the *Bau-Akadamie* in Berlin too, was reorganized along the same lines (Clark 1990, p145). But even this intervention was insufficient to solve the problem. In the face of the need to reinvigorate the economy, the schools were reorganized again in the 1860s and 70s into institutes of technology (*technische Hochschulen*) and given the status of universities, even if this meant that students had to have at least a basic knowledge of French and Latin before being able to apply. It was an arrangement that lasted to the 1970s and beyond, educating both engineers and architects in the same institutes, even if in separate, self-governing faculties. Only when the education system delivered the desired social status to graduates could the two disciplines thrive.

Professional societies and codes of conduct

From relatively early in the 19th century the professions gradually started forming their own associations with the founding of the Architects' Society of Berlin (*Architektenverein zu Berlin*) in 1824, the joint Architects' and Engineers' Society for the Kingdom of Hanover in 1851 and in 1856 the German Society of Engineers (the now all-powerful *Verein Deutscher Ingenieure* – VDI). In 1871, following national unification, many local organizations federated, in the architects' case to become the Union of German Architects' and Engineers' Societies (*Verband Deutscher Architekten- und Ingenieur-Vereine* – DAI).

The dominant engineering association, the VDI, started as a technical society and built up over the years an enviable network of specialist sub-societies and a substantial publication programme on a wide range of technical and engineering management topics (Krause 1996, p238). But at the same time a struggle over its control and direction resulted in a takeover by industrialists and managers to the detriment of the original professorial founders and, although a strict political neutrality was claimed on issues such as labour relations, the organization adopted a stance in favour of rapid industrial growth and in particular Wilhelmine Germany's push for nationalist glory and world power (Gispen 1989, pp48, 143):

The engineering society's manifest goal became national unification and national glory, and industrial and technological progress was merely the means to achieve it. As a consequence the engineering organizations, like so many other voluntary organizations of the emerging middle classes, initially exhibited few of the characteristics of a modern professional association.

In particular the dash for growth meant the encouragement and development of twin tiers of training and employment, with a new breed of technician training filling the gap left by the elevation of the institutes of technology to university equivalence. The new schools, mainly in the private sector and varying enormously in quality, pulled in large numbers of students and output 'diplomized engineers' after two years or less of training. Kees Gispen notes that 'between 1880 and 1940 they probably produced at least 50 per cent of the new species of engineer' (1996, p141). In the long run the twin track approach proved to be greatly beneficial to social mobility, but its original motivation was social control and the maintenance of the dominance of a small elite in the industry.

The role of the VDI in these developments led to two alternative organizations being formed in the early years of the 20th century, the *Verband Deutsche Diplom-Ingenieure* (VDDI) and the *Bund der technishchen-industriellen Beamten* (BtiB) with more egalitarian principles, especially in the case of the BtiB, which was actively anti-elitist. Both were violently opposed by the VDI and both were ultimately closed down after the National Socialists took power in 1933. The VDI survived the Nazi period and the Second World War despite the policy of *Gleichschaltung*, the process intended to establish full control and co-ordination over all parts of society, and the plans of senior party leaders. Hitler protected the VDI, deeming engineers essential for war work and ineligible for the draft (Krause 1996, pp238–40). It meant that the organization had to fully re-invent itself once the war was finished.

The main means of rehabilitation chosen by the VDI, after a period of intense self-examination from 1947 to 1950, was the abandonment of the policy of professional neutrality. Instead it adopted a concept of socio-political responsibility, published in 1950 as the 'Credo of the engineer' calling for the adoption of humanistic and cosmopolitan values as guiding principles over and above purely technical or economic considerations (Gispen 1996, p156). The credo called for, *inter alia* (Jarausch 1990b, p209):

- 'reverence for values beyond knowledge';
- 'humility before the all-powerful' creator;
- dedication to the 'service of mankind', in honour, justice and impartiality;
- 'respect for the dignity of human life', irrespective of origin, class or ideology;
- rejection of technical abuse through 'loyal use for human ethics and culture';
- collegial cooperation for the 'sensible development of technology; and
- placing 'professional honour above economic advantage'.

The VDI then began to rearrange itself into a more traditional professional organization 'concerned not just with the technical substance of engineering but with the material welfare and social standing of its members' and 'it also began to articulate political positions on matters concerning engineers, technology, the environment and the economy' (Gispen 1996, p156). It then worked

Table 4.4 Engineering qualification procedure in Germany

Engineering qualifications in Germany are awarded by:

- Technical Universities (*Technische Universitäten*), and
- Universities of Applied Science (*Fachhochschulen*) – currently being phased out.

They award the *Diplom-Ingenieur* – equivalent to a Master's degree.

Use of the titles *Ingenieur* and *Diplom-Ingenieur* is protected by the individual German *Länder* (States).

Full Membership of the *Verein Deutscher Ingenieure* (VDI) is available to those entitled to the title of *Ingenieur* upon providing proof of graduation.

to bring the engineering profession together into a single whole, coping even with the internal divisions caused by the educational system and the different levels of qualification and status (see Table 4.4).

For architects the DAI played a similar role to the VDI for engineers, indeed the two organizations worked together in many instances. But instead of being captured by private, capitalist interests, as the VDI was perceived to have been, the DAI had a reputation for representing and working exclusively in the interest of government-employed architects (*Baubeamter*), who comprised roughly half its membership (Yoskuhl 2014, p116). The building departments, which employed them, were responsible for city planning and infrastructure as well as all the civic buildings and other public projects. In the early part of the 19th century the *Baubeamter* constituted the vast majority of the profession, with private architects only forming 15% of those who had begun in practice prior to 1840. But by 1880 the burgeoning private sector attracted at least 42% of new entrants (Clark 1990, p145). These two streams, although they relied on the same training schools, were in practice completely divergent, and once an architect had worked in private practice it was effectively impossible to return to the civil administration (Gerber 2015, p50).

The split in the profession brought around 60 independent architects working in Berlin together in 1880 as a group called the *Vereinigung zur Vertretung baukünstlerischen Interessen aus Berlin* to lobby for commissions for major public projects to be open to all and to the founding in 1903 of the *Bund Deutscher Architekten* (BDA), originally intended to solely represent private architects and planners. The BDA's campaign met firm opposition from administrative groups including the influential *Vereinigung der technischen Oberbeamten Deutscher Städte* (the Association of Technical Chief-Public Servants of German Cities). One such, the building director (*Baudirektor*) of Hamburg, Fritz Schumacher, although a member of both organizations, exemplified the problem so far as the BDA was concerned, as he designed all the publics projects in the city himself for the duration of his 24 year long intendancy (see Figure 4.1).

The aftermath of the First World War presaged a period of radical ferment in architecture in Germany with the foundation of numerous architectural groups including the *Novembergruppe* (1918), *Arbeitsrat für Kunst* (1918–21), *Gläserne Kette* (1919–20), *Der Ring* (1926) and *Der Block* (1928) all arguing for a form of true architecture that would return Germany to some of its previous imagined glory and sense of community (Welter 2010, p67). The output of the participants in these groups and the school they founded, the *Bauhaus*, is now

Figure 4.1
The Hamburg paper
Die Hamburger Woche
published a cartoon
showing the private
architects of the city
complaining to Mother
Hamburg that the city
architect, portrayed as
a child in the nursery,
was not sharing his toys
with them

Source: Gerber (2015,
pp54–7).

what architecture in Weimar Germany is remembered for. Yet this expressionist and modernist group formed only a relatively minor part of the overall activity, if a well publicized one. The appetite for an architecture that served a social end was far greater than that generated by the *Bauhaus* and encompassed both the radical and traditional wings of the profession. The profession coalesced around an – often spiritually expressed – mission intended at one end of the spectrum to deliver 'people's housing as the means of bringing all the arts to the people'[9] to the desire expressed by Adolf Hitler in 1939 for an architecture that would 'awaken the national consciousness' and 'contribute more than ever to the political unification and strengthening of our people' (Broadbent 1979, p38) at the other. The rise of the Nazis to power in 1933 decisively won the argument for the latter view and they attempted to realize their vision through the *Kampfbund für Deutsche Kultur* organization, run by the 'frustrated architect' Alfred Rosenberg and Paul Schultze-Naumburg. The *Kampfbund* programme was delivered by Hitler's personal architect, Albert Speer (1905–81), with the support of a massive office of architects, engineers and technical staff. Other organizations were swept away.

Following the Second World War the architectural community appears to have felt little of the guilt and angst expressed by the engineers in their 'Credo'. They were simply in denial. The American sociologist Everett Hughes in a lecture in 1948 (1994, p183) recalled a discussion with an architect in Frankfurt am Main about the persecution of the Jews earlier in the year:

> The architect: 'I am ashamed for my people whenever I think of it. But we didn't know about it. We only learned about it later. You must remember the pressure we were under; we had to join the party. We had to keep our mouths

shut and do as we were told. It was a terrible pressure. Still I am ashamed. But you see, we had lost our colonies, and our national honour was hurt. And these Nazis exploited that feeling. And the Jews, they were a problem. They came from the east. You should see them in Poland; the lowest class of people, full of lice, dirty and poor, running about in their Ghettos in filthy caftans. They came here, and got rich by unbelievable methods after the first war. They occupied all the good places. Why, they were in the proportion of ten to one in medicine and law and government posts!'

If there was an equivalent statement from the architects of the period it was an appeal signed by 38 architects and others – including Fritz Schumacher (1869–1947), Max Taut (1884–1967) and Lilly Reich (1885–1947) – issued in the periodical *Baukunst und Werkform* in 1947, in the face of the destruction of the German countryside and cities:

A post-war appeal: fundamental demands

The collapse has destroyed the visible world that constituted our life and our work. When it took place we believed, with a sense of liberation, that now we should be able to return to work. Today, two years later, we realize how much the visible breakdown is merely the expression of a spiritual devastation and we are tempted to sink into despair. We have been reduced to fundamentals and the task must be tackled afresh from this point.

   All the peoples of the earth are faced with this task; for our people it is a case of to be or not to be. Upon the conscience of us, the creative, lies the obligation to build the visible world that makes up our life and our work. Conscious of this responsibility we demand:

1. When they are rebuilt, the big cities must be divided up into a new association of viable local sections, each of which is a self-contained unit; the old city must acquire new life as the cultural city core.
2. The heritage that has been destroyed must not be reconstructed historically; in order to fulfil new tasks it must be rebuilt in a new form.
3. In our country towns – the last visible symbols of German history – a living unity must be achieved between the old buildings and streets and modern residential and industrial edifices.
4. Complete reorganisation also demands planned reconstruction of the German village.
5. For dwelling houses and for our public buildings, for furniture and fittings, we call for the replacement of over-specialised or wretchedly utilitarian shapes by simple and valid designs.

For only the validly simple can be used on a multiple scale. Architecture can succeed only on the basis of a concentration of effort, of commercial endeavour in design office and workshop.

   In a spirit of self-sacrifice we call upon all men of good will.[10]

In practice the sacrifice was less than they imagined. As reported in *Der Speigel* (Leick *et al.* 2010):

It was a good time for architects, urban planners, entrepreneurs and communist building collectives. The task seemed endless and money was everywhere. The outcome, however, was less than impressive – mass produced buildings that compare poorly with the prewar buildings which they replaced.

And Speer's working group was back in action: 'Nazi-era architects helped each other land jobs and projects while keeping them out of the hands of former Nazi opponents' (ibid.).

The professional organizations re-formed themselves after the war and both DIA and BDA continue to thrive.

Since the 1970s the joint profession of architects, interior designers, landscape architects and planners has become one of the 'liberal' professions whose titles (only) are protected under the law. Practitioners who want to use the titles and have the authority that comes with them, including the ability to take responsibility for projects, are required to register with one of the 16 regional Chambers of Architects in the German *Länder*, regulated under each of the federal states' regulations. They are collectively represented by the Federal Chamber of German Architects (*Bundesarchitektenkammer*), which defines their role as follows:

> to play a decisive role in helping to shape the habitat and spatial environment for people as a whole and to provide for individuals the preconditions for an optimum quality of life, living and working conditions, and opportunities for development, while working to find the best possible solutions to avoid conflicts within society.[11]

United States

Engineering societies and codes of conduct

As with much else, architecture and engineering expertise was initially imported to the United States from Europe (surveying, excepting land surveying, has, until relatively recently, never existed as a separate discipline in the US). Professional men rapidly made the new country their own. This is exemplified by the military engineer Pierre L'Enfant (1754–1825), who trained in Paris and arrived in America in 1777 to fight in the Revolutionary War, and the British architect, Benjamin Henry Latrobe (1764–1820), who having done the Grand Tour across Europe, was apprenticed to the civil engineer, John Smeaton, then worked for the London architect Samuel Pepys Cockerell before emigrating in 1796. In particular, civil engineering developed as a significant industry during the 1840s, driven by an insatiable demand for canal, railway and bridge building and an urgent need to provide clean water and drainage infrastructure to the rapidly growing cities. The skilled personnel necessary to meet this demand were largely trained engineers from Germany (Griggs 2003, p112). One such was Albert Fink (1827–97), who had trained in architecture and engineering at the *Polytechnische Schule* in Darmstadt in the Grande Duchy of Hesse, before he arrived in the US in 1848 to start a career as a railway engineer and later entrepreneur. Such men inevitably brought with them the traditions of their home country including the training methodologies of their original schools and masters.

Latrobe, the architect of the United States Capitol as well as a prominent canal and waterworks engineer, portrayed himself, without modesty, as the 'first professional architect and engineer to practice in the United States' and championed a very 'English' idea of professionalism. He expected deference from those around him and was dismissive of both the gentlemen practitioners he encountered 'who from travelling or from books have acquired some knowledge of the theory of Art, [but] know nothing of its practice' and the building mechanics 'whose early life being spent in labor, and in the habits of a laborious life have had no opportunity to acquire the theory'. He saw the role of the architect to 'get in between' these two 'sets of men' and, as he explained in a letter to Thomas Jefferson (1743–1826) to achieve a 'simultaneous consideration of the purpose, the connection and the construction of his work'.[12] Notwithstanding his successes Latrobe clearly found the difficulty of being a professional in the US frustrating; in 1808 he complained to a friend that he had:

> not so far succeeded as to make it [architecture] an eligible profession for one who has the education and the feelings of a Gentleman and I regret exceedingly that my own son ... has determined to make it his own.[13]

While the development of the professions in the US started as a hybrid offshoot of various European traditions, with the 'learned' and other newly emergent professions taking their place in society in the familiar intermediate position between the state and commerce, a serious rupture to the process of professional formation arose in the period 1829–41, under the seventh President, Andrew Jackson (1767–1845, President 1829–37), and his successor, Martin Van Buren (1782–1862, President 1837–41). The two presidents, both keen proponents of the populist doctrine of *laissez-faire*, mounted a concerted attack on professional privileges that included the deliberate policy of lowering the standards of entry (Sarfatti Larson 2013 [1977], pp113–35) and abolition of 'monopoly licensure' (Krause 1996, p49). This had a scarring effect on the development of American professional organizations, as Charles Rosenberg has commented: 'No longer did maintaining the status of the learned professions play a part – a necessary part – in maintaining the stability of society itself' (1962, p160). Although the built environment professions were still in their infancy and no organizations yet existed in the sector, the actions of Jackson and Van Buren established a principle that the central 'federal' state would not be party to any regulation or special recognition of the professions. Any regulation that followed thereafter would be left to the individual states. This would deeply affect the development of professionalism in the United States, creating a wide variety of local, stand-alone licensing and accreditation systems but also ensuring that national professional bodies evolved and remained free of any governmental control and direction. A position, ultimately, not dissimilar to that in the UK, but brought about by very different societal and political pressures.

America brought huge opportunities and challenges for the relatively few architects and engineers in the country in the first half of the 19th century. The US census report for 1850 records only 512 engineers and 591 architects in the country at that time (Merritt 1969, p10). Although there was initially very little that separated the two disciplines, with men like Latrobe spanning both, they rapidly became differentiated with Latrobe's son, Benjamin Henry Latrobe II (1806–78), and almost all his contemporaries choosing to concentrate their efforts in one domain or the other. The relatively low numbers also meant that the first attempt in 1839 at forming the engineering discipline into a coherent organization made by the junior Latrobe was premature. Working from Baltimore, he brought together a committee of 17 engineers, with representatives from New York, Virginia, Maryland, Georgia, Connecticut and other states. The group failed to even agree a constitution, with the *American Railroad Journal* noting that several of the putative committee 'were hostile to any form of organization, others were indifferent and many were unknown to each other' (Merritt 1969, p99). As the historian, Raymond Merritt, has written, 'it seems obvious that the profession depended at that early stage upon a competitive economy; its members clung to an obsolescent agrarian individualism and identified with one section or another of the country' (ibid.). It was only when there was a sufficient density of practitioners in the growing urban centres that it was possible or practicable for professional organization to take place.

In 1848–9 both Boston (with 68 engineers resident in Massachusetts in the 1850 census) and New York (62 engineers in New York State, presumably largely based in New York City) started local associations adopting near identical constitutions that aimed at securing for their profession 'the same respect and reward which it obtains in every civilized country' (Merritt 1969, p100). The Boston Society of Civil Engineers with the railway engineer James Laurie (1811–75) as its first President, survived and thrived, and still runs to this day.[14] The New York Institution of Civil Engineers, although it boasted 90 founding members and managed the successful publication of a set of learned papers in its first year, didn't last long after a disastrous first annual meeting in January 1852, which its President, Charles Stuart, was too busy to attend.

Despite this initial set back, when the Bostonian, James Laurie, opened a new office for his firm in New York in 1852 he collaborated with a group of 11 other civil engineers in the city to form the American Society of Civil Engineers and Architects and was elected their first President. The new society was essentially a New York organization, but adopted nationwide ambitions, in emulation of the ICE in Britain, from the outset. When Laurie left New York, to look after projects in Nova Scotia and elsewhere, the Society went into hibernation and no meetings were held for 12 years until 1867 when he returned to the city (Wisely 1974, p24). In the meantime its relatively few architect members left to set up their own association in 1857 (see below). But starting from around 1872, when the renamed American Society of Civil Engineers (ASCE) held its first national convention in Louisville, it began grow rapidly beyond New York and ultimately developed to become the dominant civil and structural engineering professional body in the United States and now has over 150 000 members worldwide.

The objectives of the new Society, as agreed in 1852, were (ASCE Metropolitan Section 2005):

the professional improvement of the members, the encouragement of social intercourse among men of practical science, the advancement of engineering in its several branches and of architecture, and the establishment of a central point of reference and union for its members.

This outlines the four-fold mixture of professionalism, networking, promotion and knowledge development which describe the activities of the ASCE to this day. Nonetheless the objectives, and in particular the definition of professionalism, initially understood to be 'a strictly personal responsibility – a matter of honor', were only hazily formed in 1852. As the ASCE's official historian, William Wisely, noted, the Society would 'struggle for almost 80 years to define the context of its professional development policy and then devote the next 40 years to devising the mechanism for implementing that policy' (1974, p76).

This struggle can be described in the eventual appearance and development of the ASCE's Code of Ethics and then the step-by-step appearance of a public interest focus to the Society's interpretation of 'professionalism'.

Initially there was an expectation that the gentlemanly conscience of individual engineers would be an adequate safeguard for their behaviour. A newly elected Honorary Member, John Jervis (1795–1885), expanded on his faith in the rectitude of the profession at the ASCE's first convention in 1869 (Wisely 1974, p128):

> the engineer eminently depends on character. The interests of others, in various ways, are committed to him. On his capacity for his profession, and his integrity as a man, reliance must be placed. He will meet many difficulties of a physical, and not a few of moral nature.

Such attitudes presupposed that civil engineers were autonomous agents in society at liberty to follow their conscience, whereas in practice the great majority were corporate employees and had relatively little ability to exercise their principles. As the journal *Engineering News* put it in 1904, 'We see then that the businessman is the master; the engineer is his good slave' (quoted in Layton 1971, p110). Concerns over conflicts of interest that arose when employee and professional loyalties were at odds, were addressed in a proposal put forward in 1877 by the ASCE's secretary Gabriel Leverich (ASCE 2007b):

> Whereas: A Civil Engineer, in the practice of his profession, is sometimes restrained or overruled by his employers, in matters involving serious risk to property and life which he only, as the engineer, should determine; whence he must either discharge his duties in a manner contrary to his best judgment or resign his position;
>   Resolved: That in the opinion ... of the American Society of Civil Engineers ... it is unprofessional for a civil engineer to continue the discharge of his duties when so restrained or overruled.

Leverich's motion was rejected by the ASCE Board who formally resolved 'that it is inexpedient for this Society to instruct its members as to their duties in private professional matters'. It would be a long time before the duties of engineers as employees, rather than individual consultants in private practice, were addressed again. In the meantime as transport and technology pushed forward, a series of engineering failures to railway bridges and dams in the period 1870–95, resulting in heavy loss of life, revealed the incompetence of many individuals working as 'civil engineers' and the very limited safeguards against

Table 4.5 ASCE Code of Ethics (1914)

It shall be considered unprofessional and inconsistent with honorable and dignified bearing for any member of the American Society of Civil Engineers:

1. To act for his clients in professional matters otherwise than as a faithful agent or trustee, or to accept any remuneration other than his stated charges for services rendered his clients.

2. To attempt to injure falsely or maliciously, directly or indirectly, the professional reputation, prospects, or business, of another Engineer.

3. To attempt to supplant another Engineer after definite steps have been taken toward his employment.

4. To compete with another Engineer for employment on the basis of professional charges, by reducing his usual charges and in this manner attempting to underbid after being informed of the charges named by another.

5. To review the work of another Engineer for the same client, except with the knowledge or consent of such Engineer, or unless the connection of such Engineer with the work has been terminated.

6. To advertise in self-laudatory language, or in any other manner derogatory to the dignity of the Profession

Sources: ASCE 2007b; Wisely 1974, p134.

them offered by such organizations as the ASCE (Layton 1971, p115). Action to deal with the situation was widely seen as necessary and the adoption of a code widely anticipated.

When the ASCE next discussed proposals for a code in 1893 they were rejected, as they were again in 1902. The then president, Robert Moore, declared 'the Society and its members will continue to rely, as they have done in the past, upon these vital and moral forces, and not upon the enactment of codes or upon any form of legislation' (Vesilind and Gunn 1998, p56). As a result other parties began to agitate and legislate for action. Individual states, starting with Wyoming in 1907 (see below), enacted licensing laws for engineers matching similar legislation for architects that had first been enacted in Illinois in 1897. A protest organization of young engineers, the Technical League, comprising over 1000 members, was formed in 1909 to lobby for reform, with a slate of demands including: licensing the profession, a minimum of four years college education, a system of professional ethics, greater publicity and raising the dignity and *esprit de corps* of the profession. The League then submitted a licensing law to the New York Legislature in 1910 with the same purpose. This caused the ASCE to move against the proposal, deploying a range of prominent engineers, including their President, John Bensel (1863–1922), in a campaign of opposition to the Bill, which was ultimately defeated. The Technical League disbanded, but a series of radical reforming groups formed in its wake and over the ensuing years the ASCE gradually implemented its demands, starting with a Code of Ethics approved by the membership in June 1914 (Layton 1971, pp111–12) – see Table 4.5.

The adopted code was markedly different from Leverich's attempt to grapple with the ethical responsibilities of individual engineers and largely concerned itself with preventing internal competition and any affronts to the dignity and reputation of ASCE members. In Edwin Layton's phrase it 'treated engineers as if they were consultants' (1971, p114).

The Code developed incrementally over the years. In particular further anti-competitive clauses were added, for example a canon was inserted in 1949 that forbade participation 'in competitive bidding on a price basis to secure a professional engagement', a transgression that 14 members of the Society were expelled or suspended for in 1954 (ASCE 2007a). In the 1970s the Federal Government required the removal of both that canon and the original canon 3, as being in violation of the Sherman Antitrust Act, but otherwise, even after 60 years, it still had substantially the same set of clauses (Wisely 1974, p135). A canon confirming an obligation to care for the environment; 'Engineers should be committed to improving the environment to enhance the quality of life', was added in 1976 (ASCE 2008).

The change in the Society's 'prime objective' in its constitution took a similarly gradual journey over the same period. As noted above, the first objective laid out in 1852 was 'the professional improvement of the members', but very little was done to spell out what this meant, either in theory or practice. In 1875, the ASCE's Committee on Policy of the Society (see Wisely 1974, p168) noted that:

> the influence of the Society upon the public is unfortunately not very *rapidly* developed by the mere professional character of the members, although this is the ultimate basis of its influence and usefulness. There must be some connecting links between the Society and the public.

It was only in 1919 that the ASCE's Committee on Development, under pressure from progressive campaigners, eventually concluded that: 'the engineering profession owes a duty to the public which it is believed can be best discharged by every engineer in the civic work of his community' (Wisely 1974, p253).

In the wake of the Wall St Crash of 1929 and the Great Depression, while the US Federal Government responded with great engineering works like the Hoover Dam and the Tennessee Valley project to underpin the New Deal recovery programme, the engineering professions themselves appeared to be in denial. In Edwin Layton's analysis (1971, p225):

> as the 1930s wore on, engineers turned away from social responsibility as a means of advancing professionalism. While the public clamoured for leadership, the engineers were bankrupt of ideas, incapable of action, and obsessed with their own immediate selfish interest.

Even a petition from the members in 1937 calling for action to deal with 'professional practice and ethics' and greater 'usefulness of the profession', went unheeded by the leadership of the ASCE. If there was a response, it was exemplified by a 'Statement of Principles' drawn up by the ASCE President, Malcolm Pirnie (1889–1967), in 1944, which he hoped would lead to the 'reestablishment of free enterprise', a 'return to faith' and the 'codification of moral principles becoming the order of the day' (Layton 1971, p235).

It was only in the late 1950s when a classic definition of a profession as: 'a group of men pursuing a learned art as a common calling in the spirit of public service', propounded by the legal scholar, Nathan Roscoe Pound (1944, p203), began

Table 4.6 ASCE Code of Ethics (2014)

Fundamental Principles

Engineers uphold and advance the integrity, honor and dignity of the engineering profession by:

1. using their knowledge and skill for the enhancement of human welfare and the environment;

2. being honest and impartial and serving with fidelity the public, their employers and clients;

3. striving to increase the competence and prestige of the engineering profession; and supporting the professional and technical societies of their disciplines.

Fundamental Canons

1. Engineers shall hold paramount the safety, health and welfare of the public and shall strive to comply with the principles of sustainable development in the performance of their professional duties.

2. Engineers shall perform services only in areas of their competence.

3. Engineers shall issue public statements only in an objective and truthful manner.

4. Engineers shall act in professional matters for each employer or client as faithful agents or trustees, and shall avoid conflicts of interest.

5. Engineers shall build their professional reputation on the merit of their services and shall not compete unfairly with others.

6. Engineers shall act in such a manner as to uphold and enhance the honor, integrity, and dignity of the engineering profession and shall act with zero-tolerance for bribery, fraud, and corruption.

7. Engineers shall continue their professional development throughout their careers, and shall provide opportunities for the professional development of those engineers under their supervision.

8. Engineers shall, in all matters related to their profession, treat all persons fairly and encourage equitable participation without regard to gender or gender identity, race, national origin, ethnicity, religion, age, sexual orientation, disability, political affiliation, or family, marital, or economic status.

Note: ASCE's Code of Ethics, adopted in September 2014, incorporate the 'fundamental principles' developed by the Accreditation Board for Engineering and Technology (ABET).

to be quoted within the ASCE that the topic of professionalism re-emerged. A few years later in the preamble to the 1962 version of the Code of Ethics the idea of 'service in the public interest' was explicitly recognized and in 1974 the Society determined that its 'First Goal' was 'to provide a corps of civil engineers whose foremost dedication is that of unselfish dedication to the public' (Wisely 1974, pp76, 253–4). The ASCE's Constitution was amended in 2001 to read 'The objective of the Society is the advancement of the science and profession of engineering to enhance the welfare of humanity'. Today the principle of protecting the public interest is so central – to both the idea of professionalism and the founding of the ASCE – that even the Society's own guidance documents obscure the historical truth of its origins by declaring: 'The American Society of Civil Engineers, the oldest national professional engineering society, was founded in 1852 with an objective to enhance the welfare of humanity' (2016).

Engineering education and accreditation

The first American engineering school was intended to be for purely military purposes. In 1802 the 3rd US President, Thomas Jefferson, directed that a corps of engineers should be 'stationed at West Point in the state of New York, and shall constitute a Military Academy' (Military Peace Establishment Act

1802). In practice West Point trained its officers (Coalwell 2001, p2) in both civil[15] and military engineering and many West Point graduates, including the railroad engineer Herman Haupt (1817–1905), resigned their commissions on graduating to work in the civil sector or to become teachers and professors. The first civil schools awarding engineering degrees, the Rensselaer School (1832) and Union College, Schenectady (1845) – both, like West Point, in the Hudson Valley of New York State – were founded a few decades later. Significantly, like the French schools, these early colleges developed outside the university system and Rensselaer, in particular, drew on the *polytechnique* model and was established 'for the purpose of instructing persons, who may choose to apply themselves, in the application of science to the common purposes of life' (Rensselaer 2011). The American civil engineering education system would henceforth develop largely independent of both government and profession, if much closer to industry, having established its own traditions and freedoms from the outset.

From a slow start the demand for engineering education grew rapidly during the American Civil War (1861–5), a struggle in which engineers were heavily involved, facilitating a technologically and industrially driven conflict. Passed in 1862 as a wartime measure, the (first) Morrill Land-Grant Act provided the resources for a succession of industrial colleges to be established in each State using the proceeds of federal land sales. The Act was initially envisaged as a proposal for enabling agricultural colleges, but in wartime conditions their agreed aim was repurposed to focus on engineering and military tactics (Morrill Act 1862):

> each State which may take and claim the benefit of this act, to the endowment, support, and maintenance of at least one college where the leading object shall be, without excluding other scientific and classical studies, and including military tactics, to teach such branches of learning as are related to agriculture and the mechanic arts, in such manner as the legislatures of the States may respectively prescribe, in order to promote the liberal and practical education of the industrial classes in the several pursuits and professions in life.

The majority of the Land-Grant colleges, with engineering as an obligatory programme at their heart, became the State Universities of the US (plus two private institutions: Cornell University and the Massachusetts Institute of Technology) and covered the whole country, including, eventually, the former confederate states. The programme massively expanded the number of degree courses in engineering available and the output of trained specialists in all branches of the discipline. Prior to 1860 only 300 engineers had graduated in the US, trained at six colleges; but by the 1870s an additional 2249 engineers had been educated by 21 colleges and by 1911, 3000 engineers were graduating each year (Williams 2009, p19). With the development of these new colleges and universities, questions inevitably arose about how they should teach practical and vocational subjects. In particular, one of the leading voices in the development of the Morrill Act, Jonathan Turner (1805–1999) – a former Illinois professor and an advocate for 'industrial education' – was hunting for a system that could deliver a more focused, practical and scientific education than was available at the traditional 'Ivy League' universities, such as Yale, where he had studied classical literature.

The old universities had been established in the 17th and 18th centuries along English lines to prepare men for leadership roles and the learned professions. In the words of an 1828 Yale report, written at the time of President Jackson's reforms and intended to defend the traditional curriculum, the education they offered was only 'preparatory to a profession' (Yale College 1828):

> Our object is not to teach that which is peculiar to any one of the professions; but to lay the foundation which is common to them all. There are separate schools for medicine, law, and theology, connected with the college, as well as in various parts of the country; which are open for the reception of all who are prepared to enter upon the appropriate studies of their several professions. With these, the academical course is not intended to interfere.

Jonathan Turner rejected this hands-off approach and believed that he had found the vocational system he was searching for in the German system established under von Humboldt (see above), with its very different view of the knowledge to be gained while studying at university. In the words of James Morgan Hart, one of the first professors at the new Cornell University, in a memoir describing his time at Göttingen University in 1861 (1874, p46):

> When a young man attends the university, he is supposed to have some definite object in view; he wishes to fit himself for becoming a theologian, or a lawyer, or a physician or an historian, or a teacher in the public schools, or a chemist, or a mathematician. In other words he is to get his professional outfit.

The new Land-Grant colleges would henceforth follow a 'German' model, as understood and re-interpreted by Turner, who set out his aims in an 1853 pamphlet, *Industrial Universities for the People*:

> the objects of these new institutes should be to apply existing knowledge directly and efficiently to all practical pursuits and professions in life, and to extend the boundaries of our present knowledge in all possible practical directions.

Turner's was a radical, democratizing vision, which he, remarkably, realized through the growth of the new universities.

As a consequence of the Morrill Act and the development of the new colleges and universities in the second half of the 19th century the numbers of engineers increased rapidly, with the decennial US census reports recording 8261 in 1880 (a rise of nearly 1500% from 1850). For comparison, over the same period, the number of architects rose to 3375 (up 470%), while the ranks of the longer established professions of doctors and clergymen only doubled in size (Merritt 1969, p10). Many of these new professionals may have arrived from overseas, drawn by all the prospects for a new and prosperous life in America, but increasingly they were being educated and trained locally. The opening up of the professions to so many over a relatively short period was a great equalizer in American society with long-lasting consequences. The state universities with their Land-Grant endowments had at this time the freedom to expand and run their programmes as they saw fit, without reference to either government or industry. They fuelled the growth in the number of engineers available to industry while opening up a gap between the worlds of education, state administration and business that would soon require a new set of initiatives to close.

Even before the Civil War, the US was the prime beneficiary of the worldwide manufacturing boom linked to an extraordinary growth in the use of technology and the development of the modern industrial corporation. The boom started in approximately 1850 and, although it was temporarily halted by a financial crash in 1857, growth resumed in 1860, leaving the US as one of the great economic powers (Merritt 1969, p20; Hobsbawm 1977b, pp43–63). It was the period of great capitalist enterprise that came to be known as 'the gilded age' and engineers had no intention of being left behind. To be a civil engineer in the second half of the 19th century in America was to be a businessman, or the employee of one, rather than to be an independent professional. Hence the ethical challenge identified by Leverich (see above), when for the great majority of engineering professionals their chances of advancement were almost entirely related to their corporate status (Ansell 1985, p26). Merritt (1969, p65) cites the statistics for the 108 graduates of Rensselaer between 1850 and 1875, showing that by 1893, almost half of the graduates had become 'professional executives'.

The challenges this posed to the professional attitudes of engineers were manifold. While they were contributing to the transformation of the country, and as a result improving peoples' lives, they were also knowingly cutting corners, sacrificing quality for quantity and endangering the users of railways, bridges and buildings. This inevitably became both a public and a professional issue when bridges and other infrastructure projects failed. For some this was an inevitable consequence of capitalism. Edwin Layton (1971, p55) has noted that many:

> leading engineers in the 1880s and 1890s were deeply influenced by Herbert Spencer's social Darwinism … engineers gloried in competition and survival of the fittest, not because these principles in any way insured their eventual triumph, but because they were guarantees of their own success.

In a similar vein the electrical engineer C.O. Mailloux declared that the engineer's desire for more power and status was 'in a sense' selfish. But he maintained that 'fortunately, the real motive … is … quite altruistic, for the benefits which will result … for the engineering profession will be trifling in comparison with the benefits to the community, to the state, and to humanity in general' (ibid., p62). For others it meant tighter regulations and controls.

The ASCE appointed a committee to deal with 'the Means of Averting Bridge Accidents', which first reported in 1875 (ASCE 1875), but broke up in acrimony a year later. More significantly a number of individuals emerged including Albert Fink, Herman Haupt (both mentioned above) and Alexander Cassatt (1839–1906), who between them introduced checking procedures, engineering standards and better management. It was a process that reintroduced a professional ethos to the industry, but also improved and brought professional standards to the process of industrial management (Merritt 1969).

The reputation of engineering had been badly damaged by the scandals and from the beginning of the 20th century individual states started to introduce licensing and regulation for practising engineers. Wyoming started the process with a Bill requiring anyone who presented themselves to the public as an engineer or land surveyor to have been examined and licensed by a state board established for the purpose. By 1920 ten states had introduced licensure for engineers,[16] with very different requirements in each one, and all the American states had passed registration laws by 1947. For engineers, who had always travelled extensively in the course of their work, this meant that they either had to become accredited in each state they worked in or they had to restrict their activities to a much smaller area. It was an unsustainable system and in 1920 the Iowa State Board of Engineering Examiners wrote to the other licensing states calling a meeting to discuss the possibility of establishing 'a form of permanent organization to arrive at the best understanding and to facilitate the business of state and interstate registration' (Corley 2004, pp5–6).

The National Council of State Boards of Engineering Examiners (NCSBEE – later NCEES) emerged from a meeting between seven states held later that year, tasked with trying to resolve the issue. The ASCE contributed its own proposals for a remedy in 1925 and many of the different states had their own solutions, which they either wanted others to follow or saw as unique to themselves. By 1926 the focus had switched from individual practitioners to certifying engineering schools. Eligibility to be a school of 'recognized standing' was defined as (Corley 2004, p12):

> one which requires the equivalent of a high school or preparatory school diploma as an entrance requirement and demands the equivalent of a four year's course in engineering for graduation.

It was only when the professional societies became involved in 1932 that the impasse was broken with the formation of the Engineering Council for Professional Development (EPCD), a joint venture between various professional organizations and NCSBEE with an official purpose of 'upbuilding engineering as a profession', but charged with running the accreditation of the engineering schools nationwide (see Table 4.7).

Architecture

Architectural societies

In 1852 the Boston-born engineer and architect, Edward Gardiner (1825–59), was one of the 12 founding members who attended the meeting that established the American Society of Civil Engineers and Architects. Five years later, with the ASCEA apparently moribund, he was at another meeting, this time of 13 architects, which established the American Institute of Architects (AIA). Originally intended to be the New York Society of Architects, the name was changed to the more expansive title at the second meeting of the Society, following the suggestion of the prominent Philadelphia architect, Thomas Ustick Walter (1804–87), at the time the latest architect to work on the US Capitol Building and a veteran of an 1836 attempt to form an institution (see below). The original stated purpose of the AIA was to 'promote the scientific and practical perfection of its members' and 'elevate the standing of the profession'

Table 4.7 Qualification procedure for engineers in the United States

The individual states have maintained their independence with regard to licensing engineers. Although the rules in each state can vary in specifics, most use the model rules of the NCEES as their base code and with a qualification in one state it is generally possible to attain a license in another (if their rules are sufficiently similar).

The most common route to a state license involves:

1) Graduation from an engineering degree course accredited by ABET (previously the Accreditation Board for Engineering and Technology and before 1980 the Engineering Council for Professional Development (EPCD)).

The criteria encompass both technical and professional standards, including:

• an understanding of and a commitment to address professional and ethical responsibilities including a respect for diversity;

• a knowledge of the impact of engineering technology solutions in a societal and global context; and

• a commitment to quality, timeliness, and continuous improvement. (ABET 2017)

2) Examination

A pass in the Fundamentals of Engineering (FE) examination set by NCEES.

3) Experience

4 years of documented experience verified by licensed professional engineers.

4) Examination

A pass in the final NCEES Professional Exam (PE) – taken in two parts, one covering breadth of knowledge and the other depth in a specific, selected subject.

(Other routes, for example if an engineer has a different first degree but takes a master's degree in engineering, are possible.)

5) State registration

Each state runs its own professional licensing board, with requirements based on 1–4 above. Engineers wishing to practice in multiple states can apply for comity licensure.

6) Full membership of the ASCE

ASCE membership is available either when a License is granted or when a combination of a degree and/or sufficient years of engineering experience 'in responsible charge' has been attained and verified (ASCE 2014b).

Both ASCE and NCEES also require compliance with their separate Code of Conduct/Rules of Professional Conduct, including maintenance of professional competency and licensees and members are subject to their respective disciplinary procedures.

Rules of Professional Conduct (NCEES 2015a)

To safeguard the health, safety, and welfare of the public and to maintain integrity and high standards of skill and practice in the engineering and surveying professions, the rules of professional conduct … shall be binding upon every licensee and on all firms authorized to offer or perform engineering or surveying services in this jurisdiction.

Sources: ABET, ASCE, NCEES.

(AIA 2007), but the following year, as the formal version of the constitution was prepared, it was rethought and new objectives agreed (AIA 2013):

to promote the artistic, scientific, and practical profession of its members;
to facilitate their intercourse and good fellowship;
to elevate the standing of the profession; and
to combine the efforts of those engaged in the practice of Architecture, for the general advancement of the Art.

It contained very little directly relating to professionalism, as we would understand it today.

It wasn't the first effort to establish an organization for architects in New York. The Brethren of the Workshop of Vitruvius had been founded in 1809 and the similarly titled American Institution of Architects in 1836. They both flowered briefly before disappearing for lack of members or support. In 1850 the US Census recorded 181 architects resident in New York State and 394 in 1860. One hundred and eighty-one seems to have been just enough to support the new organization, which after a brief period in abeyance from 1862 to 1864 due to the Civil War, rapidly revived to hold its first convention in the city in 1867. As the century progressed regional chapters were added to the original New York core, which, in 1887, formed its own chapter, so that by 1900 there were a total of 23 individual chapters, spread out across the country (Biancavilla 2007, p3) with a headquarters based, from 1898, in Washington DC.

Architectural practice in the US at the outset provided its own apprenticeship-based education system and, as in Britain, the fees paid by trainees were an integral part of the income of many firms. But in the 1850s the model for education shifted from the English approach of in-work instruction to the *atelier* system of the *Ecole des Beaux-Arts*. In 1846 Richard Morris Hunt (1827–95) – later one of the 13 founders of the AIA – was the first of many Americans to go to Paris for his architectural training. On his return to New York in 1855 he set up both a successful practice and in a separate building a dedicated school based on his French experience. The school ran as a *Beaux-Arts atelier* and won a reputation as 'a center of progressive education'; training many prominent architects and educators of the next generation (Boyle 1977, p310). If Hunt kept his professional and educational activities largely separate, later returners from Paris, including Henry Hobson Richardson (1838–86), Charles McKim (1847–1909) and Stanford White (1853–1906) integrated their students and the collegiate and highly competitive spirit of the *atelier* into their offices, in particular fostering the *Beaux-Arts* relationship of intense loyalty between master and *éleve*. One alumni of McKim, Mead and White, Egerton Swartout (1870–1943), who worked there from 1892 to 1900, recalled that the 'old' office meant more to his generation of draftsmen than their own practices did later (Woods 1999, p144).

The East Coast architects who founded the early AIA and their equivalents in Chicago, who came together in 1884 to establish the Western Association of Architects (WAA), represented the architectural establishment in more ways than one. They dominated the cities they worked in, secured the majority of prestige jobs – from private houses for the wealthy to major corporate and civic buildings – and provided the main education system for subsequent generations of architects. The offices that they ran became large operations; in 1909 McKim, Mead and White in New York employed 89 architects and had a staff of over 100, and in Chicago, the office of Burnham and Root peaked with over 180 staff in 1912 (Boyle 1977, pp313–15). In the absence of public sector employment almost all architects were independent practitioners or their employees, and for all their strength in times of growth, they had no defence against the collapse of market.

The United States suffered economic depressions in 1837, 1857, 1893 and, most notably, in 1929. Each time the business of architecture suffered. Henry Van Brunt recalled the depression of 1857, the year of the founding of the AIA, as a period when architects were set against each other and regarded their books and drawings as trade secrets to be kept hidden from competitors

(Van Brunt 1895, cited in Woods 1999, p33). In the depression of 1893 McKim, Mead and White halved their staff roll from 110 to 55 and Adler and Sullivan in Chicago dismissed almost all of their assistants before ceasing trading entirely two years later (Woods 1999, pp146, 116). Each time they learnt and relearnt the need to be businesses first and foremost, dedicated to survival most of all. As Mary Woods has written in her history of American architectural practice in the 19th century (1999, pp82–3):

> Although entrepreneurial practice contradicted a fundamental tenet of traditional professionalism – aloofness from commerce – it did free architects from the constraints of either private patrons or government officials and became the touchstone of American professionalism during the entire nineteenth century. … While eastern and midwestern architects might disagree over the priority given to aesthetics or tectonics in design, they agreed that American architects were businessmen creating opportunities to ensure the survival of their practices. This conception of architect as artist, constructor *and businessman* was uniquely American.

Or as the New Jersey architect, J.F. Harder, stated in 1902 to a meeting of his colleagues (Boyle 1977, p317):

> The architectural opportunities fall to those who are preëminent for business rather than artistic ability, and thus it is they who build the architecture of the country, good bad or indifferent. The architect must be a businessman first and an artist afterwards.

There were different versions of this approach available. In New York it was also essential to be seen as a gentleman. John Carrère (1858–1911), who trained in the Mckim, Mead and White office before setting up his own New York-based partnership in 1885, wrote in a 1903 guide to *Making a choice of profession*: 'The architect, indeed the ideal man, is an artist, gentleman and man of business' (Woods 1999, p167); whereas Daniel Burnham expressed the brasher view from Chicago: 'My idea is to work up a big business, to handle big things, deal with big business, and to build up a big organization' (Sullivan, 1922, p285). As the contemporary critic Barr Ferree wrote in 1893, at the time of the Chicago World's Fair: 'American architecture is not a matter of art but of business, A building must pay or there will be no investor ready with his money to meet its cost. This is at once the curse and the glory of American architecture' (Burg 1976, p261).

Architectural education

While practices focused on the business of architecture they lost control of education. The advent of the Land-Grant colleges and universities encouraged the establishment of new architecture courses. They could be created relatively cheaply in or alongside the obligatory engineering schools and there was apparently an appetite for learning architecture away from the grinding conditions of many office *ateliers*. In 1867 the AIA discussed and supported the principle of a 'grand central school of architecture' that would combine the best of American, British, French and German practice, but the idea went no further when both the Massachusetts Institute of Technology (MIT) in 1865 and Cornell University in 1871 established their own courses. MIT hired William Ware (1832–1915), a prominent student of Hunt, who later set up a similar course at Columbia University. Cornell appointed Charles Babcock (1829–1913) one of the founders of the AIA and an apprentice, partner and later son-in-law to Richard Upjohn (1802–78) (another of the original AIA 13), to run their new school.

Ware and Babcock both saw the system of office training as inadequate to the task in hand and were sceptical of the 'paper and pencil' architecture of the *Beaux-Arts* system. They adapted Jonathan Turner's 'German' educational mindset and set goals, in Ware's words, aimed at training 'a body of generously educated architects, gentlemen and scholars', and in Babcock's of producing a 'man of science' and a 'master of building', who could 'at once get employment'. Even more consciously 'German' ideas were developed by Clifford Ricker at another Land-Grant institution, the University of Illinois, from 1873 in direct imitation of the Berlin *Bauakadamie*, the first to attempt to introduce an architectural engineering course. The fourth American school, at the Tuskegee Institute in Alabama, a historically black college created by the second Land-Grant Act of 1890, based its tuition on a Russian system of manual training also endorsed by Ricker. Tuskegee was run from 1892 by Robert Taylor (1868–1942), the first African American to gain an architectural degree (from MIT – also in 1892) and worked on the principle that students 'do everything we teach' (Woods 1999, pp67–75). The four, very different, schools all offered a marked alternative to the *atelier* system of learning through drawing, and all shared a practically driven attitude to education. It was an approach that struggled to survive as the universities moved *en-masse* in the 1890s to the formalist approach of the *Beaux-Arts*.

With the decline of some of the more eclectic branches of American architecture, such as the Romanesque of H.H. Richardson, following his early death at the age of 47, or the gothic revival, popular for both churches and houses – that Lewis Mumford (1955, p46) blamed on 'the collapse of judgement which marked the Gilded Age' – *Beaux-Arts* classicism was left as the default American architecture at the end of the 19th century. At the Chicago World's Fair,[17] 13 of the 14 main exhibition buildings known as 'The White City' were by prominent *Beaux-Arts* architects, with only the Transportation Building, designed by Sullivan and Adler, breaking the pattern. The power of the movement was exemplified by the influence of the Society of Beaux-Arts Architects, founded by Hunt and others in New York in 1894. Ostensibly an alumni association for American architects who had studied in Paris it actively promoted the idea of a centralized American school of architecture similar to the *Ecole* in Paris as well as the *Beaux-Arts* education system in general. It even set up its own educational programme, formalized in 1916 as the Beaux-Arts Institute of Design (BAID), in support of practices, like McKim, Mead and White and Skidmore-Owings, or a number of independent groups, who were running their own *ateliers* (Simon 1996). The Society was also responsible for the annual, much feted and influential, Beaux-Arts Ball in New York.

As the *atelier* system went into decline with the rise of university courses in architecture at the turn of the century (if only eventually dying out in the 1950s), the Society redirected its attention to the university courses themselves. The ready-made imprimatur and authority of the *Beaux-Arts* was invaluable to many of the new universities as they established their new courses, and architects trained at the *Ecole* were soon in great demand to direct them. Former *Ecole élève* Warren Powers Laird (1861–1948) was appointed Professor of Architecture at the University of Pennsylvania in 1891, where he taught alongside Paul Phillippe Cret (1876–1945); Alfred Hamlin (1855–1926) took over at Columbia after Ware's forced resignation in 1903; and Alexander Trowbridge (1868–1950) was appointed Dean at Cornell. All of them instituted *Beaux-Arts curricula*. One of the most influential of this new breed of educators was John Galen Howard (1864–1931), who had returned from Paris in 1893. He was on the Society of Beaux-Arts' education committee and in 1903 was appointed Professor of Architecture at the University of California, where another *Beaux-Arts* alumnus, Bernard Maybeck (1862–1957), was already on the staff. Together they reinvented the *Ecole* in Berkeley using Julien Guadet's *Eléments et théories de l'architecture* as their main reference text (Draper 1977).

The universities were triumphant, as Mary Woods (1999, p79) has noted: 'The *Ecole* transfixed Americans because it represented architecture as a fine art', and they had become its American voice, with all the kudos and respectability that that implied. Officialdom across America appeared confident that the *Beaux-Arts* method was the one true method and, as the AIA's Committee of Education happily reported in 1906, it was educating 'gentleman of general culture with special architectural ability'. By 1912 16 of the 20 recognized US schools of architecture had direct links to the Society of Beaux-Arts (Howell-Ardila 2010, p48) and in 1924 the Dean of the Columbia University School of Architecture and *Ecole* alumnus, William Boring (1859–1937), speaking to the International Congress on Architectural Education, declared (*The Washington Post* 5th October 1924, cited in Howell-Ardila 2010, p63):

> The time we have to instruct pupils is too valuable to devote to petty details of the business of an office. It seems better to inspire students to work for an ideal of beauty than to equip them to take positions as technical assistants.

In 1894 the *Beaux-Arts* approach was even exported back across the Atlantic to England, where it established itself as the paradigm for architectural tuition for many more decades (see Chapter 7).

Educating architects in universities away from the clubbish atmosphere of the *ateliers* meant that America opened up the profession to women and ethnic minorities long before anywhere else. As early as 1862 the President of Cornell announced that his institution was 'open to all – regardless of sex or color' and although it took several years to achieve, three women had graduated, 'with honors', from the architecture course by 1888. Sophia Hayden (1868–1953), one of the architects of the Chicago World's Fair buildings, graduated from MIT in 1890 and African American students like Taylor and Vertner Tandy (1885–1949) graduated from MIT and Tuskegee-Cornell respectively. The numbers were and remained small and despite progress elsewhere. Some schools, including Columbia and Pennsylvania maintained a bar against women until the 1940s (Woods 1999, pp75–7).

For all the satisfaction of the universities with the *Beaux-Arts* system, neither the hidebound approach to education nor the awkward traditions sat easily with the way other subjects were taught or with academic staples such as the pro-fessorial system, lectures, classes and grades. Innovation and experiment were both inevitable and antithetical to the academic classicism that had developed. Gradually a natural opposition built up against the system and by 1935 eight schools had started experimenting with various different versions of modernist thinking in their teaching (Howell-Ardila 2010, p55).[18] Ellis Lawrence, the President of one of the first of the schools to break ranks, the University of Oregon, described in 1934 how the school used design projects as (see Weatherhead 1941, p212):

> the vehicle for teaching sociology, politics, education, economics, and ethics, as well as the structure, hydraulics, illumination, and the laws of design. ... Architecture is a projection of the society it serves. To teach it well it cannot be separated from the ideals and standards of society.

It was an approach very different from that of the *Beaux-Arts*. By the time Walter Gropius (1883–1969) and Marcel Breuer (1902–81) arrived in 1937 at Harvard's Graduate School of Design to re-establish the *Bauhaus* course in America, a change in the system was already substantially under way. Architectural education could return to a German-inspired approach largely unimpeded.

By the late 1940s America saw itself on the path to renewal as the modern nation *par excellence* and, as announced by President Truman in his 21 Points speech of 1945, was determined that the country would overcome the social problems exposed by the Great Depression of the 1930s and the burdens of war; through a mixture of public works, post-war reconstruction and advanced social programmes. In parallel, Cold War competitiveness and the strong fear of being overtaken economically and militarily by the Soviet Union, were forcing technology and science to the front of US government funding priorities. America had decided to embrace technological progress and in that context the *Beaux-Arts* approach no longer had any place in the American mindset. The last *Beaux-Arts* course, at the University of California, Berkeley, would convert to modernism in 1948 after vehement protests from returning 'GI Bill'[19] students led to the resignation of Warren Perry (1884–1980), disciple of John Galen Howard and the long-standing Dean of the architecture programme (Littman 1995, p140).

The universities set a post-war course for a modernist, and very American, future. The schools evolved a hybrid pedagogy based on various English, German, French and American ideas that has stood them in good stead for many decades and did so while maintaining a high degree of independence from both practice and profession. It was an independence that also covered various architects and other practitioners who worked within academia and achieved a degree of fame or notoriety for their propositions. In the view of Mary Woods, Professor at Cornell (1999, p174):

> The star designers on university faculties are not advocates for the profession or for practice. Although they accept awards and speaking engagements from professional organizations, they rarely involve themselves with professional issues. Unlike nineteenth century leaders, their principal allegiance is to the university, not the profession. No wonder that many practitioners feel estranged from the schools, finding the theoretical discourse baffling and complaining that graduates lack the technical, production and managerial skills that account for 90 percent of a firm's time.

Codes of architectural practice

The profession of architecture in America at the end of 19th century lacked definition despite the arrival of formal courses in architecture. Until then anyone could, and did, call themselves an architect. Even the professional bodies, the

Table 4.8 Qualification procedure for architects in the United States

The requirements for licensing as an architect in the US vary from state to state, and have yet to be harmonized. The main common components required by the states are:

1) A professional degree from a course accredited by the National Architectural Accrediting Board (NAAB) – established 1940.

NAAB Conditions for Accreditation (2014) encourage and promote diversity of educational experience but access courses on the basis of:

    A. Collaboration and Leadership

    B. Design

    C. Professional Opportunity

    D. Stewardship of the Environment

    E. Community and Social Responsibility

2) Experience

The National Council of Architectural Registration Boards (NCARB) – established 1919 – runs the Architectural Experience Program (AXP). Trainee architects are expected to log and report a minimum of 3740 hours of work experience, broken down into:

- practice management
- project management
- programming and analysis
- project planning and design
- project development & documentation, and
- construction & evaluation.

At least half the time needs to be completed under the supervision of a qualified architect. The AXP is accepted by all states although some require additional experience.

3) Examination

NCARB also administers the Architect Registration Exam (ARE) – first set in 1965. The ARE is a multi-part, computer-based exam with the same six categories (divisions) as the AXP (NCARB 2016).

4) State Registration

Each state runs its own licensing board, with its particular requirements, mainly based on 1–3 above. Once licensed candidates can call themselves an architect, it then also becomes possible to apply for a reciprocal license for practising in other states, commonly through obtaining an NCARB Certificate.

The individual states also maintain their own Rules of Conduct for architects, which unlike the AIA's Code have the force of law behind them. Since 1977 NCARB has published a Model for such Rules of Conduct in 5 parts:

Rule 1: Competence

Rule 2: Conflict of Interest

Rule 3: Full Disclosure

Rule 4: Compliance with Laws

Rule 5: Professional Conduct.

They are intended to be 'hard-edged' and used for 'policing and disciplining members.

AIA and WAA, included in their membership many claimants to the title whose credentials were largely illusory. In 1885 at their second annual convention the WAA addressed the issue of professional standards in two ways. The Chicago architect Daniel Burnham called for the introduction of a code of conduct and Dankmar Adler (1844–1900) proposed a Bill for the state licensing of architects, which was adopted by the State of Illinois in 1897. The WAA had a definite programme to achieve professionalization for architects, an ambition the AIA notably lacked. When the two organizations merged in 1889 the programme began to develop across the country with demands for more states to license architects. The first AIA Code of Conduct was introduced in 1909.

Although the AIA was relatively weak at influencing the education system or the commerce of architecture and only had a hit and miss impact on government, it was very effective at developing common business procedures for the profession. These included the schedules of charges for architectural services first introduced in 1866, rules for architectural competitions in 1870 and standard building contracts, which were first published in 1888. All of these activities had an impact on architectural and construction practice that extended far beyond the membership of the organization itself (AIA 2007).

The pressure for a code had been building for many years and individual local societies, like the Boston chapter in 1895, had started to introduce their own, sometimes contradictory, versions. In 1909 the outgoing AIA President, Cass Gilbert complained: 'what is perfectly proper for a practitioner in one city under the Institute's rules becomes "unprofessional" for a practitioner in another city under the chapter's rules' (*The American Architect* 1909, p273). There was also a growing opinion that various powerful clients who expected free services from architects were exploiting their relative weakness and that this should be controlled (Fisher 2016). Despite these concerns the 1909 'advisory' code only attempted to formalize what was already the practice of the Institute by amalgamating individual resolutions into a single document.[20] It didn't advance standards of practice themselves.

The *Circular of Advice Relative to Principles of Professional Practice and the Canon of Ethics* published by the AIA in 1909 (pp273–4) was certainly cautious about demanding too much from its members:

> The American Institute of Architects, seeking to maintain a high standard of practice and conduct on the part of its members as a safeguard of the important financial, technical and esthetic interests entrusted to them, offers the following advice relative to professional practice.

But the preamble to the *Circular* gave the Institute the opportunity to formally make clear its view on the nature of professionalism, possibly for the first time:

> The profession of architecture calls for men[21] of the highest integrity, business capacity and artistic ability. The architect is entrusted with financial undertakings in which his honesty of purpose must be above suspicion; he acts as professional adviser to his client and his advice must be absolutely disinterested; he is charged with the exercise of judicial functions as between client and contractors and must act with entire impartiality; he has moral responsibilities to his professional associates and subordinates; finally he is engaged in a profession which carries with it grave responsibility to the public. These duties and responsibilities cannot be properly discharged unless his motives, conduct and ability are such as to command respect and confidence.

The principles laid out in the *Circular* (given, in abbreviated form in Table 4.9) were designed to govern the conduct of the profession.

Table 4.9 Principles of Professional Practice (abridged)

1) *On the Architect's Status*: An obligation to act with impartiality to both parties [client and contractor];

2) *On Preliminary Drawings and Estimates*: The importance of sufficient time for preparation of drawings and specifications, the duty to secure preliminary estimates and to meet such costs, but not guarantee any estimate or contract;

3) *On Superintendence and Expert Service*: All work to employ a superintendent or clerk of works to be selected by the architect;

4) *On the Architect's Charges*: The Schedule of Charges of the American Institute of Architects is recognized as a proper minimum of payment;

5) *On Payment for an Expert Service*: Under no circumstances should experts knowingly name prices in competition with each other;

6) *On Selection of Bidders or Contractors*: The award should be made only to reliable and competent contractors;

7) *On Duties to the Contractor*: Drawings and specifications are [to be] complete and accurate … and he should never … attempt to shirk responsibility by indefinite clauses;

8) *On Engaging in Building Trades*: The architect should not directly or indirectly engage in any of the building trades;

9) *On Accepting Commissions or Favors*: The architect should not receive any commission … other than his client;

10) *On Encouraging Good Workmanship*: The architect should make evident his appreciation of the dignity of the artisan's function;

11) *On Offering Services Gratuitously*: The offering of professional services on approval and without compensation … is to be condemned;

12) *On Advertising*: Advertising tends to lower the standard of the profession and is therefore condemned;

13) *On Signed Buildings and Use of Titles*: The display of an architect's name upon a building under construction is condemned, but the unobtrusive signature of buildings after completion has the approval of the Institute;

14) *On Competitions*: An architect should not take part in a competition … unless the competition is to be conducted according to best practice … as formulated by the Institute. He may not accept the commission … if he has acted in an advisory capacity;

15) *On Injuring Others*: An architect should not falsely or maliciously injure the professional reputation or business of a fellow architect;

16) *On Undertaking the Work of Others*: An architect should not undertake a commission while the just claim of a fellow architect, … nor should he attempt to supplant a fellow architect;

17) *On Duties to Students and draughtsmen*: The architect should advise and assist those who intend making architecture their career. … He should give encouragement to all worthy agencies and institutions for architectural education. While a thorough preparation is necessary for the practice of architecture, architects cannot too strongly insist that it should rest upon a broad foundation of general culture;

18) *On Duties to the Public and to Building Authorities*: An architect should be mindful of the public welfare and should participate in those movements for public betterment in which his special training qualify him to act. … He should carefully comply with all building laws and regulations, and if any such appear to him unwise or unfair, he should endeavour to have them altered.

19) *On Professional Qualifications*: The public has the right to expect that he who bears the title of architect has the knowledge and ability needed for the proper invention, illustration and supervision of all building operations which he may undertake. For that and other obvious reasons, such title should not be assumed without adequate qualifications.

Source: AIA (1909).

Table 4.10 AIA Canons of Professional Conduct

Canon I   General Obligations

Members should maintain and advance their knowledge of the art and science of architecture, respect the body of architectural accomplishment, contribute to its growth, thoughtfully consider the social and environmental impact of their professional activities, and exercise learned and uncompromised professional judgment.

Canon II   Obligations to the Public

Members should embrace the spirit and letter of the law governing their professional affairs and should promote and serve the public interest in their personal and professional activities.

Canon III   Obligations to the Client

Members should serve their clients competently and in a professional manner, and should exercise unprejudiced and unbiased judgment when performing all professional services.

Canon IV   Obligations to the Profession

Members should uphold the integrity and dignity of the profession.

Canon V   Obligations to Colleagues

Members should respect the rights and acknowledge the professional aspirations and contributions of their colleagues.

Canon VI   Obligations to the Environment

Members should promote sustainable design and development principles in their professional activities.

Source: AIA (2017).

The Canon of Ethics then repeats much of the wording of the Principles, but in balder terms,[22] and without any reference to the public interest.

The AIA's Principles and Canon were redrafted in 1947 as Document No. 330, *Standards of Professional Practice*, but while the language was updated it did not materially change the prohibitions. In 1972, as with the ASCE (see above), the US Department of Justice challenged the AIA on the anti-competition clause in the document and in 1978 the code was adjusted to allow architects to participate in design-build contracts, but it was only after the loss of a civil damages suit with an architect whose membership was suspended in 1977 for 'supplanting' that a major reconsideration of the code was undertaken (Gerou 2008, p5; AIA 2016, p23).

In 1980 the AIA suspended its code of ethics and appointed a task force to develop a new *Code of Ethics and Professional Conduct*, which was adopted at its National Convention in 1986. The new code, which, with minor revisions, remains in force, is based on five Canons. Canon VI, on sustainability, was added in 2007 (see Table 4.10).

Each of the AIA's Canons – outlining the broad principles – are supported by Ethical Standards – aspirational goals for professional performance and behaviour – and Rules of Conduct – mandatory standards, violation of which can lead to disciplinary action (AIA 2017).

The professional landscape in the United States has multiple players, most notably the individual states, and a long tradition of large and dominant private practices. The regulatory bodies favour a rules-based approach that defines and emphasizes the welfare of the public as its primary professional concern, whereas the professional organizations have, relatively recently, decided to strengthen the professional obligations of their members on social and environmental matters. The schools and professional firms continue to cherish their independence and attempt to operate as freely as possible from the constraints erected by the other professional parties.

Conclusions

The main purpose of this chapter has been to provide some points of comparability for the British system of professionalism in the construction sector but it is also clear that none of these systems have developed in isolation and they have all fed into one another and have deep, centuries-old, connections. These have ensured that many aspects of professionalism, whether the institutional arrangements, practice standards or educational procedures, are relatively interchangeable and allow for mutual recognition to have been organized. Yet differences remain, particularly in the degree of government involvement and, above all, the presence in Britain of the (construction) surveying profession, which is not formally recognized in the other countries examined.

This may be gradually changing; the RICS has members working in all three countries and in early 2018 listed 1392 members in France, 1729 in Germany and 1510 in the USA. In 2015 it appointed Martin Brühl, the managing director of a Hamburg-based company, Union Investment Real Estate, as its President. Undoubtedly the professional system is becoming more international and, as it does so, more homogeneous, but the national differences will still be current for some time to come.

State involvement remains a significant differentiator and while the general trend in western countries is for this to diminish as greater control is handed over to the private sector it is the states (and supranational bodies such as the European Union) that continue to underwrite the system through recognition, control and above all the maintenance of the rule of law. The importance of national legal systems to provide a stable basis for professional activity as well as the enthusiasm of individual governments for promoting a strong professional compact (see Chapters 1 and 11) remains considerable.

This chapter has looked at how the France, Germany and the United States have organized and controlled entry into their professions. The next chapter addresses the early development of recruitment, training and validation to the construction professions in Britain.

Notes

1 'All the academies and literary societies, licensed or endowed by the nation, are abolished …' / '*Toutes les academies et sociétés litéraires, patentés ou dotes par la nation sont supprimées …*' Collection Baudouin (1793), Vol. for August 68 1793, pp56–7. Cited in Chaffee 1977, p69.
2 Quoted by M. Bonnaire (1937), *Procés-verbaux de l'Académie de Ecole des Beaux-Arts*, Vol. 1, Paris, p273, cited in Chaffee 1977, p70.
3 '*nul en France ne pourra exercer l'architecture s'il n'est pourvu d'un diplôme délivré par le gouvernement et constatant qu'il possède le minimum des connaissances nécessaires à la profession d'architecte*'. See Avezar and Ferrand, *La Société nationale des architectes de France, Revue générale de l'Architecture et des Travaux publics* (1889), p.142, cited in Clark (1990, p145).

4 *'Les ingénieurs nous montrent l'exemple, suivons-le et cela avec d'autant plus de raison que ces architectophages de l'avenir commencent à se développer dans ce sens, aiguisant bec et ongles pour nous déchirer et nous anéantir'*, quoted in Rodriguez Tomé (2006, p60).

5 *'Ne fuyons donc pas le terrain sur lequel nous devons nous défendre et sauvegarder l'architecture'*, L'Architecture, 3rd March 1893, cited by Rodriguez Tomé 2006, p72.

6 *'entrepreneur, industriel, ou fournisseur de matières ou objets employés dans la construction ... gérants, hommes d'affaires ou mandataires quelconques de propriétaires ... commis d'entrepreneur, métreur, vérificateur', 'Il est rétribué uniquement par des honoraires ... qui est l'essence de l'entreprise, en contradiction avec l'exercice de la profession libérale de l'architecte'*, Guadet 1885, trans S. Foxell (2017).

7 The proposal for the *Ordre des architectes* was developed by Jean Zay when he was Minister for Public Instruction. He defected but was later arrested and then murdered by the Vichy Regime. The Regime nonetheless passed the proposal into law on 31st December 1940. It was updated in 1977 by a decree in the Council of State.

8 *'La profession d'architecte est incompatible avec celle d'entrepreneur, industriel ou fournisseur de matières ou objets employés dans la construction.*

   Sauf dans le cas où l'architecte entre., en qualité de fonctionnaire dans une administration publique, il convient avec son client du montant de ses honoraires. Il lui est interdit de recevoir, pour le travail convenu, aucune autre rémunération, même indirecte, d'un tiers à quelque titre que ce soit.

   L'architecte doit observer les règles contenues dans le code des devoirs professionnels qui sera établi par un règlement d'administration publique.

   Ce code déterminera notamment les conditions:dans lesquelles l'architecte devra contracter une assurance couvrant tous les risques résultant de sa responsabilité professionnelle.

   Les architectes n'ont pas le droit de se grouper en syndicats régis par le livre III du code du travail.'

   Article 1er, Titre 1er, Article 3 de la Loi instituant l'ordre des architectes et réglementant le titre et la profession d'architecte, 31st December 1940.

9 *Arbeitsrat für Kunst* (1919), 'Under the wing of a great architecture', circular, cited in Conrad (1970, pp44–5).

10 A. Leitl (1947), *A post-war appeal: fundamental demands*, Baukunst und Werkform, cited in Conrad (1970, pp148–9).

11 *'es, den Lebensraum, die räumliche Umwelt des Menschen, maßgeblich mitzuplanen und mitzugestalten. Dadurch sollen die Voraussetzungen für ein Optimum an Lebensqualität, Lebens- und Arbeitsbedingungen sowie Entfaltungsmöglichkeiten für den einzelnen geschaffen werden und gleichzeitig die dabei auftretenden, einander vielfach widersprechenden Nutzungsabsichten innerhalb der Gesellschaft zu einer bestmöglichen Lösung koordiniert werden.'*

12 From Latrobe's letter of 29th March 1804 to Jefferson (1984, Vol. 1, p472).

13 From Latrobe's letter of 20th November 1808 to Henry Ormond (1984, Vol. 2, p680).

14 The US Census reported that there were 68 civil engineers in Massachusetts in 1850, the greatest number of any state (New York had 62) (Merritt 1969, p11), but of the founders of the Boston Society of Civil Engineers only three of its members had any formal training with just one, Eben Horsford, having a degree (from Rensselaer) (Griggs 2003, p112).

15 See Dennis Hart Mahan's *An Elementary Course of Civil Engineering* first published in 1837. Mahan was both a cadet and Professor of Civil and Military Engineering at West Point from 1832 to 1871. See www.asce.org/templates/person-bio-detail.aspx?id=11192

16 Wyoming, Louisiana, Illinois, Florida, Oregon, Nevada, Michigan, Iowa, Idaho and Colorado.

17 Properly known as the World's Fair: Columbian Exposition, held in Chicago 1983 to celebrate the 400th anniversary of Columbus' arrival in the New World. The fair featured the 'White City': 14 buildings, laid out around a reflective pool, housing the main exhibition halls. They were designed by: Richard Morris Hunt, McKim, Mead and White, George Post, Solon Berman, Henry Van Brunt, Robert Peabody, Sophia Hayden, Adler & Sullivan, Henry Cobb, Jenney and Mundie and Charles Atwood.

18 University of Oregon (1914), Yale University (1919), University of Cincinnati (1922), Cornell University (1929), the University of Southern California (1930), University of Kansas (1932), Columbia University (1934) and Harvard University (1935).

19 The GI Bill – The Servicemen's Readjustment Act of 1944 – provided payments to returning Second World War veterans for tuition and living expenses to attend college amongst other benefits.

20 This definition in the 'advisory code' may have been helpful to architects like J.Q. Ingram of Elmira, New York who was reprimanded by the AIA in 1895 for using advertising that savoured 'more of the quack medicine and ready-made clothing' than 'a dignified card of a professional man'. Ingram later resigned from the AIA because he preferred, or needed, to be an entrepreneur more than the AIA's vision of 'a profession man' (Woods 1999, pp88–92).

21 The reference to 'men' in the Circular is used despite the presence of women members in the AIA. Louise Bethune (1856–1915) was elected as an Associate of the WAA in 1885, of the AIA in 1888 and as a Fellow in 1889. Julia Morgan (1872–1957), who gained her license to practice architecture in California in 1904, did not join the AIA until 1921.

22 The Canon of Ethics (AIA 1909), although only a 'general guide', states that:

> It is unprofessional for an architect
> 1. To engage directly or indirectly in any of the building trades
> 2. To guarantee any estimate or contract by bond or otherwise.
> 3. To accept any commission from any interested party other than the owner.
> 4. To advertise.
> 5. To take part in any competition the terms of which are not in harmony with the principles approved by the Institute.
> 6. To attempt in any way, except as a duly authorized competitor, to secure work for which a completion is in progress.
> 7. To attempt to influence, either directly or indirectly, the award of a competition in which he is a competitor.
> 8. To accept the commission to do the work for which a competition has been instituted if he has acted in an advisory capacity, either in drawing programme or making award.
> 9. To injure falsely or maliciously the professional reputation, prospects or business of a fellow architect.
> 10. To undertake a Commission while the just claim of another architect who has previously undertaken it, remains unsatisfied, or until such claim has been referred to arbitration or issue been joined at law.
> 11. To attempt to supplant a fellow architect after definite steps have been taken towards his employment.
> 12. To compete knowingly with a fellow architect on the basis of professional charges.

5
Training

At the time when the new professional institutions were being founded in the early 19th century architects and engineers were seeking new ways of gaining the skills necessary for their careers. Young architects in London started self-organized training courses (see below) and as discussed in Chapter 2 the young engineers who founded the ICE did so in pursuit of 'mutual instruction in that knowledge requisite for the profession' (Watson 1988, p11). But in practice, and despite the example of the clearly successful academic training established in France and Germany, their ambitions were confined to modest improvements to the existing system. Entry to the professions continued to depend upon gaining real-life experience through the office of a 'master', and as a result the disciplines floundered.

Scholarly and intellectual interest in built environment issues was alive in Britain and ran alongside the industry, but did not intersect with it unless it concerned historical styles. 'Practical engineering' was taught at a few universities from the late 18th century as a component of a gentleman's education. It was introduced at Cambridge University by the Reverend Isaac Milner (1750–1820), the inaugural Jacksonian Professor of Natural Philosophy, as a component of its Mathematical Tripos, an obligatory course for any student studying classics – approximately a third of the students at the university at that time. The course lasted until 1849 when the Mathematics Tripos was dropped as an obligatory part of the Cambridge degree, the very year when the Scottish philosopher, Sir William Hamilton, noted in the Edinburgh Review (Welbourn 2001):

> Cambridge stands alone in turning out her clergy accomplished as actuaries or engineers, it may be, but unaccomplished as divines.

No formal school of engineering was established at the University of Cambridge until 1875, but even then it was not seen as the start of a career as an engineer. As the Professor of Engineering, Charles Inglis (1875–1952), noted in his inaugural address to the freshmen in the department in 1914 (Welbourn 2001):

> Your fathers have sent you to Cambridge to be educated, not to become engineers; but they have sent you to read engineering because they believe that no better route to becoming educated exists.

In saying this Inglis was echoing the earlier views of John Stuart Mill (1806–73), who at his installation as the Rector of St Andrews University, pronounced (1867, pp4–5):

At least there is a tolerably general agreement about what an University is not. It is not a place of professional education. Universities are not intended to teach the knowledge required to fit men for some special mode of gaining their livelihood. Their object is not to make skilful lawyers, or physicians, or engineers, but capable and cultivated human beings ... What professional men should carry away with them from an University, is not professional knowledge, but that which should direct the use of their professional knowledge, and bring the light of general culture to illuminate the technicalities of a special pursuit. Men may be competent lawyers without general education, but it depends on general education to make them philosophic lawyers who demand, and are capable of apprehending, principles, instead of merely cramming their memory with details.

But Mill also noted (during the ellipsis, (pp 4–5)):

It is very right that there should be public facilities for the study of professions. It is well that there should be Schools of Law, and of Medicine, and it would be well if there were schools of engineering, and the industrial arts. The countries which have such institutions are greatly the better for them; and there is something to be said for having them in the same localities, and under the same general superintendence, as the establishments devoted to education properly so called.

Eventually the lack of schools of engineering, such as those in France or Germany, as Mill noted, held Britain back from building on its early promise. Engineering training remained resolutely grounded in experience rather than theory until the end of the century. As the engineering historian, L.T.C. Rolt, noted in sadness (1970, p170), describing the situation at the time of the Great Exhibition in 1853:

Theoretical training cannot alone make a brilliant engineer, but at a time when increasing knowledge was making the profession of engineering more of an applied science and less of an intuitive art, the lack of it could handicap a good one and have a disastrous effect on the performance of the profession as a whole. That other countries quickly recognised the importance of technical education at a time when Britain continued to rely on the myth of innate superiority is another reason why she so soon lost her pre-eminence as an industrial nation.

Another, more recent historian, Maria Malatesta, has written (2011, p90):

The modern profession of engineer thus arose in the United Kingdom in a context devoid of linkage with the university or with higher-educational institutes. The curriculum taught at Oxford and Cambridge was founded on abstract disciplines and produced a mentality suspicious of – when not outrightly hostile to – technology and science. The industrialists for their part were equally ill-disposed towards technicians with only theoretical educations. Finally the engineers were the first to acknowledge the importance of training on the job.

The industrialists and engineers were not alone in their suspicions of tertiary education. In a paper, 'The Education of the Surveyor', delivered to the fledgling Surveyors' Institution in December 1868, William Sturge (1820–1905), later to be the RICS's 6th President, contemplated the dangers of an Oxford or Cambridge education for a young surveyor (Sturge 1868, p51):

> If he acquire the manners and tastes of gentlemen, he may also acquire the desultory and expensive habits, if not the vices, of too many of his associates; and instead of reading, he may waste his time in frivolity and dissipation. Even if he avoids these evils, the tastes and habits he will form will probably render the drudgery of a surveyor's office peculiarly distasteful to him.

If the influence of British universities on engineering was extremely limited during the 19th century there would be more a far more intrusive academic interest in architecture, if only one concerning questions of style and religion. In 1839 three Cambridge undergraduates, John Mason Neale (1818–66), Alexander Hope (1820–87) and Benjamin Webb (1819–85), set up a society to promote their interest in 'the study of Gothic Architecture, and of Ecclesiastical Antiques'. The Cambridge Camden Society they founded and the journal it published, *The Ecclesiologist*, would establish a powerful and baleful hold over both architecture and architects for several decades. The Society advocated a return to a mediaeval purity, intended to mask the perceived ugliness of the 19th century, and operated an early advice service for architects and church builders. Such was their reach that A.J.B. Beresford Hope, as Hope was later known, although never an architect, became President of the RIBA in 1865.

The Ecclesiological Society, following its own renaming in 1845, was not interested in training architects, but it had a great deal to say about their belief system (*The Ecclesiologist*, cited in Clark 1928, p143). It considered that 'good men build good buildings' and:

> If architecture is anything more than a mere trade; if it is indeed a liberal, intellectual art, a true branch of poesy; let us prize its reality and meaning and truthfulness, and at least not expose ourselves by giving to two contraries one and the same material expression.

The influence of the Society was all-pervasive. The art historian, Kenneth Clark, likened them to the inquisition, as well, for good measure, as Calvin and Robespierre, for 'they allowed no human accidents to weaken their zeal for truth' and 'were ruthless and infallible tyrants' (1928, pp146–7). According to *The Ecclesiologist* (Clark 1928, p143):

> the deeply religious habits of the builders of old … resulted in their matchless works: while the worldliness, vanity, dissipation and patronage of our own architects issue in unvarying and hopeless failure.

The ethical fallacy at the heart of this conflation of belief and building design has had a long and scarring effect on architectural culture, one that passed directly into the credo of modernism in the 20th century. It continues to present a difficulty in discussions of value in architecture and professionalism even today.

Ecclesiologist zeal contributed to what the architectural historian, Nikolaus Pevsner (1902–83), described as 'a stodgy and complacent optimism … prevailing in England about 1850' where 'charity, churchgoing, and demonstrative morality might serve to settle your accounts with heaven and your conscience – on the whole you were lucky to live in this most progressive and practical age' (1960, p.40). The spirit of complacency meant that while there was an open debate in 19th century Britain about whether the construction professions should receive a cultural or technical education (or both) and whether that should be at a university or specialist school, in practice very little action was taken. In an edition of *The Surveyor, Engineer and Architect* published in 1841, and in the course of a long article about rot in timber, the author complained that training is different in France where 'the principle of engineering … formed part of their general scientific education many years before there was a single pupil in Britain' (Mudie 1841, p82).

Routes into the professions

Away from the universities many construction professionals in the early 19th century entered from the building trades. The engineer and 5th President of the ICE, William Cubitt (1785–1861) did his apprenticeship as a cabinet-maker (Stephen 1888, pp268–9) and the (unrelated) member of the Carpenters' Company, Lewis Cubitt (1799–1883), the youngest of the Cubitt brothers, set up in practice as an architect in the 1830s (Colvin 1978 pp242–3). F.M.L. Thompson records the not uncommon case of the undertaker George Head who in the mid-19th century capitalized on his access to newly empty properties to move from conducting funerals to building and plumbing work before establishing himself, and later his son, as an estate agent and urban surveyor (1968, p204). Even though the professions were keen during the period to establish a distinction between themselves and builders it was reasonably straightforward to make the transition between the two until late in the century.

Another, if less travelled, route into the professions was, just as it had been in the 18th century, as a well-to-do amateur, converting enthusiasm into a permanent career, if not a full-time job. One remarkable untrained architect was Sarah Losh (1785–1853) from Cumbria, who, having returned from a tour of Europe, designed a series of buildings including houses, schools and a church in the villages around Wreay (Uglow 2012). Another untutored architect, the Earl of Lovelace (1805–93), invented a new method of steam bending roof trusses and, as a result, joined the ICE as an associate member (Tudsbery-Turner 2003). But if such well-endowed and connected amateurs taking up professional work became increasingly rare and less influential during the course of the century the same could not be said for those who made their way into the leadership of the professional institutions. Neither the first president of the RIBA, Earl de Grey, discussed in Chapter 1, nor Beresford Hope, mentioned above, was an architect. Similarly, the ICE selected the lawyer and self-taught hydraulics and weapons manufacturer, William Armstrong (1810–1900) – later Baron Armstrong of Cragside, as President in 1881. The institutions continued to be ambivalent about the true value of training and qualifications.

But despite any alternative routes in, the predominant and approved means for achieving professional status was to serve articles of either three or five years as a pupil with an established firm or under a respected master. The system was commonplace across numerous different crafts and dated back to mediaeval practice. In the case of the construction professions the pattern had been set by the Office of Works in the 17th century. But by the 19th century the system of articles, which had successfully taken pupils from a wide range of backgrounds, especially those with building skills, into the professions, had become affordable only for the better off. Most of the larger firms took in pupils in a formal arrangement in exchange for a premium – a training fee, which varied according to the prestige of the office. In John Soane's office, which kept meticulous records, pupils such as his first, John Sanders (1768–1826), paid a £50 premium in November 1784 for five years pupillage; two years later it had risen to £100 and by 1795, £150. In one instance his pupil, Thomas Sword (pupil 1799–1804), paid a premium of £175 for five years training (Bolton 1924 pp11–19).

A helpful publication from 1842, *The Complete Book of Trades or the Parents' Guide and Youth's Instructor*, provided an idea of how the training system for architects worked (Whittock *et al.* 1842, pp11–12):

> The youth desirous of becoming an Architect, should be liberally educated, and in addition to the Latin language, he should be master of French and Italian; have some knowledge of mathematics, geometry, and drawing. The premium required with a pupil by a respectable master, is from two to five hundred pounds: the youth will also require a considerable sum for the purchase of books, instruments, and drawing materials. He must, during his apprenticeship, learn to make architectural drawings from admeasurement, also to sketch picturesque buildings, columns, etc., he must be careful in observing the proceedings of workmen in every branch of business connected with buildings. When he is out of his pupilage, if he can afford it he should spend a few months in Italy, to study the remains of the ancient masters, and the works of masters of a more recent date. On his return, if he have no private connexion, he will wait for an opportunity of competing with other Architects for the execution of a public building. If his design be selected, and he complete the edifice, satisfactorily, his reputation becomes established, and he seldom lacks highly lucrative employment. But it is almost impossible for a man in the middle walk of life to afford the money to enable a youth to work his way in this arduous pursuit. If he have not the advantage of capital to live on till he succeeds in business, the pupil after he is out of his time, obtains employment as a drawing clerk in a Architect's office: and, during his leisure hours, makes plans and drawings for small builders, or is employed to measure and value their work. Some by this means get into extensive business.

John Soane's apprentices were expected to work 12-hour days and life as a pupil in Augustus Pugin's (the elder Pugin, 1762–1832) household and office was, according to Benjamin Ferrey (1810–80), no better (Ferrey 1861, pp27–8):

> First came the loud ringing of the bell to rouse the maids, then in quick succession, the bell to summon the pupils from their beds, and the final peal requiring their presence in the office by six o'clock. A pitiable site indeed it

was to see the shivering youths reluctantly creeping down in the midst of winter to waste their time by a sleepy attempt to work before breakfast. ... the pupils continued to work incessantly at the desk till eight o'clock. The only leisure afforded them was from that hour till ten, when they retired to rest. Nothing could exceed the stern manner in which this routine was carried out; and excellent as was the course of studies pursued in the office, the cold cheerless and unvarying round of duty, though enlivened by the cheerful manner and kind attention of the elder Pugin, was wretched and discouraging.

Such a regime would not even have given pupils the opportunity to attend the few evening lectures available, such as the six annual lectures at the Royal Academy or the school proposed in a letter from *An Architectural Student* in *The Builder* in 1846 'at which the student might attend after office hours, in which architecture might be exclusively taught' (1846, 26th September, pp464–5).

It was a system widely disparaged as 'pupil-farming' (White 1885), that habitually kept apprentices working long hours for the chance to enter the profession and provided substantial income and cheap labour to those offering 'training' and apprenticeships. Charles Dickens acerbically exposed the system in his novel *Martin Chuzzlewit* (1844, p12):

The brazen plate upon the door ... bore this inscription, 'PECKSNIFF ARCHITECT', to which Mr. Pecksniff, on his cards of business added 'AND LAND SURVEYOR'. In one sense, and only one, he may be said to have been a Land Surveyor on a pretty large scale, as an extensive prospect lay stretched out before the windows of his house. Of his architectural doings, nothing was clearly known, except that he had never designed or built anything; but it was generally understood that his knowledge of the science was awful in its profundity.

Mr. Pecksniff's professional engagements, indeed, were almost, if not entirely, confined to the reception of pupils; for the collection of rents with which pursuit he occasionally varied and relieved his graver toils, can be hardly be said to be strictly architectural employment. His genius lay in ensnaring parents and guardians, and pocketing premiums.

The situation in surveying was similar, with, in Thompson's words, pupillage remaining a 'chancy, haphazard affair, a little better than pot-luck for individual students, since reputations of good and bad principals presumably had some currency, but not geared up to any organized or systematic training' (1968, p209).

Engineering training

Engineering also used articles as its main means of training and, as discussed in Chapter 3, in 1838 the ICE made a successful pupillage followed by five years in 'responsible situations' its principle qualification for membership. Even then other voices considered that working oneself up through the construction trades was a better preparation. Thomas Telford in a letter of 1830, advising a Mrs Malcolm about her son's career, wrote (quoted in Watson 1988, p153):

the way in which both Mr Rennie and myself proceeded, was to serve a regular apprenticeship to some practical employment, he to a millwright and I to a general house builder … In this way, we secured the means, by hard labour, of earning a subsistence, and in time, by good conduct, obtained the confidence of our employers and the public, eventually rising into the rank of what is called civil engineering … This course, although forbidding to many a young person who believes it possible to find a short and rapid path to distinction, is proved to be otherwise by the two examples I have cited.

Outside the ranks of the Institution this informal route into engineering would have been the most common and it was necessarily the alternative way to achieve ICE membership, through 'five years spent on his own account in the profession', a route that later accrued a requirement for 'considerable eminence', but otherwise remained intact for much of the rest of the century. At the other end of the spectrum some engineers, including Isambard Kingdom Brunel in 1822, attempted to enrol at one of the French *Grand Ecoles*, but were disqualified by their foreign birth. They had to make do with studying the series of technical books that emerged from the continental academies.

For those going the approved route of an apprenticeship, the same 1842 guide for parents and youths quoted above (Whittock 1842, p211), at the end of a long section on engineering that mainly discussed engineering techniques instead of career opportunities, advised:

The young gentleman intended for this business, is generally articled for a certain number of years, and a stipulated sum is paid according to the respectability of the Engineer whom he is to be articled to; such amount may be from one hundred and fifty to five hundred pounds.

Sir John Rennie (1794–1874), President of the ICE 1845–8, showing the scepticism suggested by his inclusion in Telford's homily above, described the system at greater length and was more forthcoming (1875, pp433–4):

A youth leaves school about the age of seventeen or eighteen, without any previous training, and his parents, thinking he has got a mechanical turn, as it is termed, decide at once to make him a civil engineer, whether he likes it or is fit for it or not. They then send him, with a considerable premium, to an engineer of some standing and practice, who, unless special conditions are made (and very few engineers will make them), will not undertake to teach him the profession. The pupil is sent into the office, and placed under the direction of the principal assistant, whom directs him to do whatever is required, if he can do it, whether drawings, writing, or calculating, or anything else; and if he wishes to learn anything, he must find it out himself: neither the principal nor assistant explains the principles or reasons of anything that is done. If he prove to be steady, intelligent and useful, keeps the regular office hours, and evinces a determination to understand thoroughly the why and wherefore of every kind of work that is brought before him, and by this means acquires some practical knowledge, he will soon attract the notice of his employer, and will be gradually transferred from one department to another, until the expiration of his pupilage, which varies from three to four years.

Given the reluctance of civil engineers to accept the idea of a school of engineering along the lines of a *Grand Ecole* or the Berlin *Bauakademie*, it is surprising that a number of engineering schools did exist in 19th century Britain, although they only served other jurisdictions and purposes. The first to be established was the Royal Indian Engineering College, founded in 1809 and based in South London to train the East India Company's own recruits as engineers (Smith 2001, p276). It was joined in 1812 by the Royal Engineers Establishment in Woolwich (later to become the Royal School of Military Engineering) and authorized, by Royal Warrant, to teach 'Sapping, Mining, and other Military Fieldworks'. A school of civil engineering was even started in 1838 at the Polytechnic Institution in Regent's Street, London, although it was part of an effort aimed at educating the public in 'practical knowledge of the various arts and branches of science connected with manufacturers, mining operations and rural economy' (Cayley 1837) rather than a school for budding engineers.

The first civil engineering school proper was the private College for Civil Engineers based from 1840 in Putney, southwest London, an initiative that won the scorn of *The Civil Engineering and Architects' Journal* which announced that 'we have done our duty both to it and our readers, by unsparingly denouncing what we consider an erroneous and inefficient system of education, and a certain delusion to those who have the misfortune to be its victims' (Laxton 1840, p59). The College closed in 1850, although, as a minor postscript, when its alumni tried to establish a 'Society of Engineers' in 1857 and applied to the Privy Council for a Royal Charter, their efforts were determinedly squashed by the intervention of the ICE (Watson 1988, p46). Several universities followed hard on its heels.

Chairs in civil engineering were established at Durham University in 1838 and the University of Glasgow in 1840. The latter year also saw the start of a course at King's College London on the 'Arts of Construction in connection with Civil Engineering and Architecture' (Jenkins 1961, p167) and, in 1841, University College London inaugurated the Department of Civil Engineering and Architecture, appointing T.L. Donaldson (1795–1885) – the secretary of the Institute of British Architects – to be Professor of Architecture (Kaye 1960, p93). Such courses, even though they were not intended to compete with the institutions, meant that there were suddenly alternative ways of achieving vocational qualifications and recognition.

The ICE itself was sceptical about the new university courses. James Walker in his Presidential Address to the ICE in 1841 expressed the concern that the output of engineering graduates would exceed the potential of employers to hire. Committees were appointed in 1849 and again in 1865 to tackle the issue but they made slow progress, culminating in 1867 when a student membership category was eventually introduced. In practice they had no need to worry about the number emerging from the various engineering courses: by the 1920s the total number of engineering degrees awarded to individuals in England and Wales was only just over 200 (Albu 1980, p69).

By the 1860s the impact of British complacency concerning engineering training was beginning to affect the ICE. The 1865 inquiry was tasked to investigate the methods being used to train engineers in other countries. Sir John Fowler (1817–98), the Chairman of the Inquiry, referred to its findings when he became Institution President the following year (Mackay 1900, p385):

It is true that nearly all continental nations have an advantage over this country in the power which the nature of their government gives them of concentrating, in one recognised official school for the preparation of civil engineers, all the best available talent of their country.

This plan does not exist in our country, and on the whole we rejoice that it does not; neither does the inducement of government employment form the chief stimulus to our exertions, for which we are also thankful: but at the same time no good reason can exist why the opportunities of acquiring theoretical preparation in this country should be inferior to those of the Continent; and I have the confident hope, that even in the theoretical branches we shall shortly have to acknowledge no inferiority to any other nation. In the practical branches we are admittedly superior.

Unfortunately Fowler advanced no plan to remedy the situation he so ably analysed and it would be more than two decades before any determined action was taken. Although it had become clear that the existing system of pupillage was a poor way to prepare students for a discipline that was rapidly becoming more specialist and reliant on high-level expertise, it required a disaster to shake the system out of its complacency. In this instance it was the collapse of the Tay Bridge in Scotland, designed by the engineer Thomas Bouch (1822–80), and, at its completion in 1878, by far the longest bridge in the world. Just over a year after the bridge's opening, and Bouch's simultaneous knighting, it failed in a winter storm, taking a full trainload of passengers with it. The subsequent public inquiry found (Rothery 1880, cl.120):

that the bridge was badly designed, badly constructed and badly maintained and that its downfall was due to inherent defects in the structure which must sooner or later have brought it down. For these defects both in design, construction and maintenance Sir Thomas Bouch is in our opinion mainly to blame.

Within seven years, in 1887, the Council of the Institution introduced an educational qualification for membership, a studentship examination – similar to the RIBA's preparatory examination, intended to ensure a basic competency in English grammar and composition, elementary mathematics and elementary mechanics of solids and fluids. It was a low threshold and could be by-passed by producing evidence of General School Certificates or passing university matriculation. It can have done little to ensure basic engineering competence. Finally, in 1897, an examination was introduced for Associate Membership principally covering the scientific aspects of engineering. Eight university engineering courses immediately won exemption from the examination for their graduates: Cambridge, St Andrews, Glasgow, Edinburgh, Victoria (Manchester), Dublin, the Royal University of Ireland and the University of Wales (with mathematics passed in the final examination). McGill University in Montreal was added to the list (with certain reservations) in 1904. It had taken a considerable time but the civil engineers in Britain had eventually accepted the importance of skills taught in the lecture hall and seminar room and had at last put them on a par with the practical knowledge gained in the workplace (Watson 1988, pp158–9).

Architecture training

The six annual lectures on architecture organized by the Royal Academy (see Chapter 2) provided some very limited opportunities for architectural education in the early 19th century, but they were inadequate to the growing need for training in London, let alone elsewhere. Several self-organized study associations founded around the same time were a little more useful, including the London Architectural Society (founded in 1806) and the Architectural Society (1831–42), and there were moves by Royal Academy students to create a separate Royal Academy of Architecture in 1810 and an Architectural Students' Society in 1817 (Kaye 1960, p63), neither of which came to fruition. The Architectural Society in its *Laws and Regulations* (1835, p11) laid out its aims as follows:

> The primary objects of this Society are the advancement and diffusion of Architectural Knowledge, by promoting the intercourse of those engaged in its study; the ultimate desire being to form a British School of Architecture, with the advantages of a Library, Museum, Professorships, and periodical Exhibitions.

The part-time university courses at King's and University College represented a decisive move forward in the UK's effort to overtake France and Germany, but the student numbers remained small and the most successful breakthrough came from an independent group of students led by Charles Gray (1828–?) and Robert Kerr (1823–1904). They founded the Architectural Association (AA) in 1847, absorbing the earlier Association of Architectural Draughtsmen (1842) as they did so. The draughtsmen organized a meeting once a fortnight and the AA took over this tradition, alternating between a design class where members brought along drawings for discussion and criticism and lectures from invited outsiders, frequently of the highest calibre (Jenkins 1961, p169). The numbers of AA members increased rapidly to 166 by 1851 and in 1852 Kerr was considering establishing a fully fledged academy. Financial troubles may have prevented that but in October 1855 the AA President, Alfred Bailey, was proposing 'a diploma to be taken by the students at the end of their period of instruction' based on the French *Diplôme d'architecte* (Kaye 1960, pp97–8).

At an RIBA Council Meeting on 19th November 1855 a reformist member, J.W. Papworth (1820–70), read a paper describing the current position of architectural training in France and Prussia and then presented a memorial from the students at the AA that spelled out their proposals for reform (RIBA 1890, p148):

> Your memorialists, representing the younger members of the architectural profession, beg to lay before the Royal Institute of British Architects their desire for the establishment of an Examination, which may eventually serve as the basis for the issue of such a diploma as shall certify that the holder thereof is fully qualified to practice as an architect. ...
>
> The want of proper knowledge on the part of the architect, combined as it is with a want of information on the part of the public, leads to many of the anomalies which are now so frequently observable in the practice of the profession, and to the presence in its ranks of many who have not the power, and in some cases of those who have not the will, to uphold its credit. ...

members of the Institute must, from their position, lie fully cognisant of the evil results of the present system; and, therefore, do not doubt that the Council will take an early opportunity of organising an Examination such as shall be found best calculated to aid and direct the student, and to bring the real qualifications of the architect before the Public.

Goaded into action by the AA's memorial, the RIBA re-opened discussions on an entry examination. The subject had been simmering dangerously for many years and, following consultation with various regional architectural societies, a Special General Meeting of the Institute was convened in January 1861 to address it. The meeting agreed to a proposal for a 'voluntary professional examination' on the basis of the condition put forward by the RIBA that it should be 'confined to "practical" subjects (construction, surveying, mechanics and the like), and did not attempt to trespass into the mysteries of design' (Jenkins 1961, p170) and that the Institute would 'take upon itself the labour of constituting an Examination tending to promote a systematic professional education' (*California Architect and Building News* 1895). The move towards professional qualification and accreditation had begun, even if it took many more years for it to truly take root.

The RIBA like the other institutions found that the idea of setting examinations went down badly with many of their members, who believed that the existing system worked well for them and provided an important income stream from premiums as well as free student labour. But for architects the issue was further complicated by the on-going battle of the styles being encouraged by the Ecclesiologists and their primary cheerleader, the art critic, John Ruskin (1819–1900), who in an address to the RIBA in 1865 on the *Study of Architecture* (1865, p85), hyperbolically declared:

You may teach imitation, because the meanest man can imitate; but you can neither teach idealism nor composition, because only a great man can choose, conceive, or compose; and he does all these necessarily, and because of his nature … take a man fed on the dusty picturesque of rags and guilt; talk to him of principles of beauty make him draw what you will, how you will, he will leave the stain of himself on whatever he touches. You had better go lecture to a snail, and tell it to leave no slime behind it.

For the combatants in this long-running struggle it was essential that the examinations only tested technical competence and not artistic taste or 'the mysteries of design'. The agreement by the proponents of examinations to stay strictly neutral was warmly welcomed by one architect (*RIBA Transactions*, 1855 VI, 39 – quoted in Kaye 1960, p98):

It was very gratifying that the last meeting received with satisfaction the idea that taste cannot even be a question for the examiners; but that readiness with the chalk, the brush, and the pencil, are indispensable to the man who lays claim to the title of architect.

The RIBA's new voluntary examinations, eventually introduced in 1863, comprised two levels: 'proficiency' and 'distinction', and involved the initial submission of a number of specific drawings to the examiners. If acceptable these would be followed by written tests covering seven subjects: Drawing and Design, Mathematics, Physics, Professional Practice, Materials, Construction, and History and Literature. Significantly, before sitting the examination candidates were required to sign a personal assertion regarding their future professionalism (RIBA 1863, p6):

> I declare that I intend to practice as an Architect, and that I will do nothing in my practice to bring discredit on the profession. In case of my failing at any time to fulfil to the satisfaction of the Royal Institute of British Architects as advised by its Council, any part of the above declaration, I permit the erasure of my name from the list of passed candidates.

In the event the idea of a voluntary examination was less than successful. In the 18 years following its introduction only 43 students passed the Proficiency level and just three the Distinction (Kaye 1960, p100). Meanwhile the argument between those who believed that the workplace was the best place to learn the trade and the supporters of a diploma course continued. In evidence to a Royal Commission investigating the Royal Academy in 1863, the then President of the RIBA and MP, Sir William Tite (1798–1873), previously an enthusiast for the AA's diploma proposals, claimed that:

> to learn practical architecture effectively it must be learnt in the office where the work is currently going on; where the working drawings are actually made by which the workmen execute everything. That can only be learnt in the office of an architect.

Meanwhile, another witness stated the academic view that the reason why English architecture was in such a poor state when compared with that of France, was that it was not taught as a fine art (Kaye 1960, p103) – implying that he believed a *Beaux-Arts* training system should be introduced (see Chapter 4).

The failure of the voluntary exam led to a new campaign for change, this time led by the Architectural Alliance, an umbrella organization representing many of the non-metropolitan architectural societies. The Alliance carried out a survey in 1869 that discovered that only 5% of architecture pupils in London were attending a school of art for additional lectures during their training, although matters were conducted much better elsewhere: in Glasgow the percentage was 80% and in Liverpool about 50%. They concluded 'the absurdity of a profession like that of Architecture being without systematic examination for entrance, and the course to be pursued to render such examination legally necessary before an Architect can practise' (*The Architect* 1869, pp271–2). In 1877 the RIBA eventually made the lasting decision that (*The Builder* 1890, p60):

> All gentlemen engaged in the study or practice of civil architecture, before presenting themselves for election as Associates shall, after May 1882, be required to pass an examination before their election, according to a standard to be fixed from time to time by the Council.

The initial examination introduced in 1882 covered 'Professional Study and Practice only' and was a formidable affair, taking five days including an interview with existing fellows of the Institute, but it was very popular and in its first eight years 350 candidates passed the test to become Associates of the RIBA (Kaye 1960, p130). Its success was such that it led the RIBA to call a General Conference of Architects in 1887 to discuss and agree (Kaye 1960, pp130–1):

1. That it is desirable that the guidance and direction of the education of those entering the architectural profession should be undertaken by the Royal Institute of British Architects
2. That to realise this end the Royal Institute of British Architects should prepare a scheme of a complete system of education
3. That such system should comprise:

   1st, Preliminary: for pupils entering the profession, as a test of general knowledge – those passing this to be 'Probationers, RIBA'.
   2nd, Intermediate: for pupils in their third year or earlier, for the general principles of Art and Construction – those passing this to be 'Students, RIBA'.
   3rd, Final: pass examination to qualify for ARIBA, at age 23 or earlier.

None of the existing Associates or Fellows were expected to take any examinations to confirm their level of competence.

On the back of the RIBA's 1887 resolution, schools and universities across Britain began establishing courses in architecture.

In 1891:

- The Architectural Association set up a new course specifically to prepare students for the RIBA examination
- The Sheffield Society of Architects and Surveyors introduced classes for architectural pupils, as did the Municipal School of Art in Nottingham
- The Manchester Society of Architects started two courses, one on history and development of architecture, the other on materials and construction
- A reorganized three year course under the architect, surveyor, historian and MP, Banister Fletcher (1833–99), to suit the examination, began at King's College, London.

In 1892:

- The Birmingham Architectural Association and the Birmingham School of Art collaborated on a four year course.

In 1894:

- The University of Liverpool started a 'School of Architecture and the Allied Arts' with a two year course leading to a Certificate in Architecture.

By 1895 the floodgates were well and truly open and the RIBA President, F.C. Penrose (1817–1903), could happily claim that 'it will be seen that every allied centre now possesses local education facilities of some kind, more or less complete, for architectural pupils' (Kaye 1960 p142).

Table 5.1 Content breakdown of the RIBA Examinations, 1887

Preliminary Examination – Exemption granted against Certificates of Matriculation from English universities and other examinations – Eight Subjects:

- Writing from Dictation;
- A short English Composition;
- Arithmetic, Algebra and Elements of Pure Geometry;
- Geography of Europe and History of the United Kingdom
- [One of] French, German, Italian or Latin
- Geometrical Drawing or Elements of Perspective;
- Elementary Mechanics and Physics;
- Freehand drawing from the Round.

Exemption granted for certificate of matriculation from an English university.

Held twice a year in March and November, Fee – One guinea

Intermediate – Exemptions granted to certain Universities

Testimony of Study, comprising: Eleven sheets of drawings of classic details, mouldings, freehand sketches and joinery details and;

Examinations (time and maximum marks)

Art- Orders of Greek and Roman Architecture (1½ hours, 50)
- Varieties of Classic Ornaments (1½ hours, 75)
- English Architecture from 1066–1500, and thereafter (1½ hours, 50)
- Characteristic Mouldings and Ornaments of Each period (1½ hours, 75)

Science - Materials (1 hour, 40)
- Calculation of Strengths (1 hour, 40);
- Elementary Principles of Construction (1 hour, 40)
- Elementary Physics (1 hour, 40)
- Mensuration, Land-Surveying and Levelling (1 hour, 40)
- Applied Plane Geometry (1 hour, 40)

50% of the total marks (490) were required to pass. Fee – Two guineas

Final Examination (for those aged 21 years and over)

Further more detailed testimonies of study and;

Examinations (time and maximum marks)

Art - Design of a Set Building (12 hours, 275)
- History (3½ hours, 100)
- Architectural Features, Mouldings and Ornaments of Each Period (3½ hours, 150)

Science - Principles of Hygiene Applied to Architecture, Ventilation, etc. (3½ hours, 100)
- Nature of Materials (1½ hour, 75)
- Strength of Materials (2 hours, 75)
- Construction (3½ hours, 100)
- Specification and estimates (2 hours, 75)
- Professional practice (1½ hours, 50)

50% of the total marks (1000) were required to pass. Fee – Three guineas

Source: Kaye (1960, p131).

The RIBA's decision that passing a rigorous examination to prove competence was a necessary qualification for all future architects marked a turning point in construction professionalism in the UK and the other institutions would shortly follow suit. But it wouldn't take long before there was one last memorable backlash against the very idea of mixing examinations and design. In 1891, incensed by a failed submission to Parliament intended to introduce statutory registration of architects, a senior group of architects including Reginald Blomfield (1856–1942) and Richard Norman Shaw (1831–1912), who according to his later collaborator, Thomas Jackson (1835–1924), 'took it up hot' (Jackson 1950 pp223–7), wrote a letter to the RIBA stating (Prior *et al.* 1899, pp220–1):

> We maintain that no one is entitled to be declared an architect merely because he has answered the questions of an examiner in such subjects as admit of examination, inasmuch as, without imagination, power of design and refinement of taste and judgement, he can have no true claim to the title; and these are qualities that cannot be brought to the test of examination. Consequently we deprecate any attempt to make examination and diploma conditions of admission to the pursuit of architecture, because a legislative sanction would thereby be given to a false and delusive idea of the true qualifications for the art.

Shaw, together with Jackson, a well respected, establishment architect, followed up the attack on the RIBA by producing a volume of essays entitled *Architecture: a Profession or an Art: Thirteen short essays on the qualifications and training of architects*, which praised the artistic conscience of the true architect, who 'must go his own way; his art must be absolutely free, unfettered save by the canons of truth and nature, the limitations of human sense, and the possibilities of his instruments and his materials' and simultaneously railed against the Institute, accusing it of fostering 'a dead level of mediocrity' (Shaw and Jackson 1892, p3).

Shaw's campaign had all the characteristics of a last ditch action by a small group, opposed at all costs to change, and, although the controversy rumbled on, later causing Shaw to turn down the proposed award of the RIBA's Gold Medal, it had very little effect on the future of the profession beyond establishing, as an enduring trope, the question that formed the title of the 1892 book. Having set art and professionalism against each other, the question has refused to go away. To this day it is still discussed as an issue,[1] with architects occasionally feeling the need to take sides (see for example, MacCormac 2005, amongst many others who have made declarations in support of one side or the other).

Surveyors

The 1860s saw not only the RIBA and ICE worrying about the need to test students before they were allowed to become qualified: the new Institution of Surveyors was also concerned. In the paper delivered by William Sturge in the Institution's first year, on 'The Education of the Surveyor' (Sturge 1868), already quoted above, he argued that:

> the policy of the Institution should therefore be to require a high standard of education, both general and special, as a qualification for membership in

order that the fact of being a member may be an earnest of superior competence and skill, as well as of good character …

The aim of the Institution should be to maintain a standard of its members sufficiently high to ensure an average superiority of its members over other practitioners. In short, membership should be considered a distinction, not only by the profession, but by the public at large.

Unfortunately for the cause of checking competence and skill, Sturge ended his address with the caveat:

The test of fitness for membership must however be practical and not theoretical. I do not see it possible to institute an examination in professional knowledge, however desirable it may be thought by some. The only men in the profession, capable of conducting an examination are too much occupied to undertake so arduous a duty; and unless it were well done, it would be better not to make the attempt.

The initial reasoning may have been compelling, but the pressure to maintain the long-standing system of pupillage must have been overwhelming. Although there is little precise evidence, pupillage seems to have emerged during the early days of estate stewardship. Documents in the Essex Records Office suggest that the cartographer Israel Amyce (1548–1607) trained the surveyor John Walker (c1550–1626) of West Hanningfield in Essex before he, in turn, passed his knowledge onto his son, also John Walker (1577–1618). Geoffrey Holmes, following his work with the Stamp Office Records now in the National Archives, reveals how professional groups in the period 1680–1730, including surveyors, usually demanded a lengthy apprenticeship of their recruits, including one Norfolk surveyor who required a seven year period of indentured training (Holmes 1986, p335).

The system wasn't a requirement but after surveyors suffered extensive reputational damage during the canal and railway building booms of the 19th century (discussed in Chapter 2), a time when numerous doubtful advertisements were being placed to train surveyors, there needed to be a corrective. Advertisements included one from a surveyor at Shepperton Cottages, Islington, offering recruits:

for a fee of three guineas each to perfect them in the business in about fourteen days, and afterwards secure them constant employment at a salary of two guineas per week, upon payment of a further premium of five guineas.

As the editor of *The Builder* noted 'It may be laid down as an axiom, that the man who professes to make a surveyor in fourteen days, is not to be trusted any farther than he can be seen' (1847, p449).

But nonetheless, being able to point to a period of training in a reputable firm was important for a surveyor's career chances and details of traineeships were recorded in directories like Who's Who, published from 1849. The Institution also took an interest and F.M.L. Thompson notes that of the 36 building and quantity surveyors among its 200 founding members 28 provided details of their formal training process and all the other eight were practising architects, whom he pointedly suggests: 'presumably felt that the adequacy of their training could be taken for granted' (1968, p205).

This is not to ignore the large numbers of surveyors who lacked formal training and who entered the profession from construction, valuing and agriculture, but their numbers decreased steadily during the 19th century as the Institution became increasingly unwilling to accept them as members. An examination was the clear way of enforcing a base standard of entry and there was a clear if only implicit recognition that this would be introduced as a membership requirement when the Institution's Royal Charter was being negotiated in 1878 with William Sturge as President. The problem lay with the substantial number of senior members who were in disagreement with the idea. These included Sturge's successor as President, Edward Ryde (1822–92), who considered that the business of surveying was 'not reducible to a set of fundamental principles and a basic body of knowledge' that could be sensibly tested (Thompson 1968 p183).

Not for the first time it was the example of France that forced the issue, although in the surveyors' case in a singular way. In the run up to his presidency, Edward Ryde participated in and acted as spokesman for the Institution of Surveyors at the first international meeting of surveyors, the *Congrès International des Géomètres-Experts*, held in Paris as part of the 1878 *Exposition Universelle Internationale*. The agenda for the meeting consisted of 14 questions – the first of which asked 'is it appropriate to urge the government of each country to require surveyors to hold diplomas of competency, with the profession providing unrestricted access for surveyors so qualified?'.[2] Unwilling to admit the relative impotence of a British institution in the midst of the official surveyors' organizations from the other countries, Ryde found himself committing the Institution of Surveyors to the position that 'it is desirable in order to raise and maintain the status of the profession, that all surveyors should undertake some examination and be required to hold a diploma of competency'. There was no going back, even if it took the Institution another 13 years to make the Fellows' exam fully obligatory (Thompson 1968 pp183–5).

The structure of the examinations, based on those used for the British Civil Service and first held in 1881, was similar to the RIBA's. Initially it consisted of only two levels: Preliminary – designed to ensure a basic educational attainment and avoidable on the presentation of school or matriculation certificates; and Proficiency – covering technical and professional issues. In 1913 the Proficiency level was split into two parts; Intermediate and Final. Initially the Proficiency exam was divided into three different disciplines: land agency, valuation and building, with further disciplines appearing later in the 20th century.

If the Surveyors adopted qualifying examinations in 1878–81, this, initially, at least, did not make them any more receptive to universities. Agricultural colleges were preferred for providing training for land agents and the Institution ignored the few classes in urban topics at the Birkbeck Institute and City of London College. In 1902 proposals from Reading College and the Regent Street Polytechnic for course recognition were firmly rebuffed. It was intended that the pupillage system would be maintained and new recruits were encouraged to enter surveyors' offices 'in which alone they can obtain an adequate and practical knowledge of their profession' (*RICS Council minutes* 1901). Inevitably, in place of college-based training facilities, a range of other organizations and individuals offering private tuition for the examination emerged. The firm of surveyors E.J. Simmons, soon added instruction for the other professions to this new, profitable sideline and in 1891 Richard Parry established a coaching firm, Parry, Blake and Parry, to train pupils for the RICS examinations. By 1914 Parry was reputed to be preparing two-thirds of all entrants (Thompson 1968, pp213–15).

Qualifying associations

The move by the Institutions to become organizations prepared to guarantee a level of competence in their membership was the second significant stage in the process of professionalization in the UK. It wasn't a step that they took willingly, but eventually all did so despite the tremendous pressure to maintain the *status quo*. This resistance came particularly from the many practitioners who gained financially from the pupillage system but also from factions that believed, despite all evidence to the contrary from France and Germany, that the special character of their art or discipline would be irretrievably tarnished by the influence of the universities and technocrats. Nonetheless the pressure for change from prospective employers, from local societies and from society as a whole meant that it was both necessary and significantly overdue.

The construction professions were not alone. Over the course of the 19th century most professional bodies started setting examinations to test the proficiency of their new recruits. As a result professional organization after professional organization took on the responsibility of becoming a qualifying association (Millerson 1964), a role that if it had existed previously at all, had been the responsibility of practitioners. The Royal College of Surgeons started the process in 1800, examining all future practitioners in England and Wales (1900, p10). The Society of Apothecaries followed in 1815, The Society of Attorneys, Solicitors, Proctors and others not being Barristers, practising in the Courts of Law and Equity of the United Kingdom (later the Law Society) in 1836, the Pharmaceutical Society in 1841, the Royal College of Veterinary Surgeons in 1844 and the Institute of Accountants in 1878. By 1880 there were 27 qualifying associations including the four separate Inns of Court for barristers. Between 1880 and 1900 another 21 associations were added to the list, including the RIBA in 1882, the Institution of Electrical Engineers (1888), the Chartered Insurance Institute (1897), the Institution of Surveyors (1891) and the ICE (1897) (Perkin 1989, p20).

The implications of the new examinations-driven system of professional qualification and the subsequent rise of university education for the professions will be explored in Chapter 7, while the next chapter addresses the affect of changes in society on professional practice as the public sector rose in significance.

Notes

1   See Crinson and Lubbock 1994 or Joseph Rykwert's 2014 RIBA Gold Medal Lecture – available on YouTube.
2   *'Y a-t-il lieu d'insister auprès des pouvoirs des diverses nations pour que des diplômes de capacité soient imposés aux géomètres, la profession restant d'ailleurs de libre exercice pour les géomètres pourvus de diplômes?'* Congrès International des Géomètres-Experts (1878), Programme.

6

Private sector – public sector

Public and private

In October 1942 during the height of the Second World War the architectural historian, John Summerson (1904–92), published a memorable article in the literary and arts magazine *Horizon* (1942, p233). In the context of *Horizon* his topic dealing with the changing employment patterns in architectural practice and, by extension, other construction professions, was unusual enough – it shared space with poetry by Laurence Binyon and Anne Ridler, a portrait study of a louche Victorian baronet and discussions on Stéphane Mallarme and Paul Klee – but its subject matter was in urgent need of attention, wherever it was published.

Summerson's essay looked at professional issues when almost all public discussion about architecture was dominated by the on-going style wars between modernism and traditionalism and the hopes and fears for *a properly planned and managed future following the projected end of the war.* The short article – print capacity was strictly rationed in 1942 – was entitled 'Bread & butter and architecture'. It examined in vivid language the transition that had occurred between 'a student leaving an architecture school in, say, 1925 with one leaving in 1938' (p233):

> In 1925 the prospect looked something like this. You passed your final examination and entered the office of a F.R.I.B.A. at £3 or £4 a week. In two or three years it might become £7, in ten years £12; then, perhaps a partnership. But, with luck you would not have to wait for that. An uncle or a friend (pigmy shade of the eighteenth-century patron) would step in with a commission – perhaps a £1,000 house. From this you 'worked up a practice'. You put up a brass plate. You won a competition or two. You made good.

Summerson recognized that this was just 'the idea and the ideal' of the profession. He was aware that although 'the proportion of the building work which, before the war, was handled by [municipal] architectural departments was very big indeed', nonetheless 'for 107 years' the RIBA had 'upheld the ideal of the independent artist-constructor-business-man acting in a fiduciary relationship to his client'. The essay worried about the strength and extraordinary persistence of that one vision of professionalism, brass plate and all. It was the world of the fictional James Spinlove ARIBA, the central protagonist *of The Honeywood*

File (Cresswell 1929), a humorous and cautionary history of a house-building project told through contractual letters running from 1924 to 1926, but it wasn't the future.

In contrast to the sketched-out career of his 1925 graduate, Summerson's imagined 1938 student (pp233–4):

> took a different view. Behind him was the depression of 1929–31, an event which caused the vista of private practice to shrink horribly and any glamour it had retained to become very dim. … He had to be more of a realist. … But there was an alternative – permanent salaried employment.

This would have been in one of the 'architectural departments' such as the public sector offices run by Lancelot Keay (1883–1974) in Liverpool, Richard Livett (1898–1959) in Leeds and John Henry Forshaw (1895–1973) at the architectural department of the Miners' Welfare Commission. By even describing a career in one of these offices Summerson was addressing a topic that was clearly in plain sight, but was little discussed as a professional issue. In doing so he started a decades-long debate on how to adapt the system to include and provide for those professionals who were working for a large, non-professional organization, lacked personal professional autonomy and, most worryingly, exchanged their labour and skill for a salary. He saw the attraction, noting that public employment gave young ARIBAs 'those three essential things for any born architect – bread, butter and the opportunity to build', and pointed out the challenge to organizations like the RIBA that 'has never interested itself much in the status of the departmental principal, still less that of the "salaried" man in a humbler position' (p235).

Despite the supposed newness of the challenge, building professionals have taken positions in the public sector at least as far back as the Office of Works. British city administrations began to appoint city architects, surveyors and engineers in the late 18th century with Thomas Warr Atwood (c1733–75) informally holding the offices from the 1760s onward of Bath City Architect and Surveyor (albeit controversially as he was a plumber and glazier by trade). Attwood's positions were formalized after his accidental death during a building demolition, when his assistant, Thomas Baldwin (c1750–1820), was appointed to replace him. The posts, despite their titles, appear principally to have been civic positions with little distinction between the two roles. Both Atwood and Baldwin took on other responsibilities within the Corporation of Bath – Mayor and Deputy Chamberlain respectively. Nonetheless the appointments provided a series of prestigious and lucrative architectural commissions in the city for them and their successors. Perhaps more notably, Baldwin also established a separate tradition of public sector corruption and financial irregularities for which he was dismissed from all his positions in 1791–2. His biographer, Jane Root (1994), has commented that 'he had a history not merely of imprudence, but of deliberate dishonesty' (pp80–103). Although this put an end to the diversity of the roles he assumed (inspector of baths, new town planner, etc.) Baldwin continued practising as a successful private architect in the city until his death at the age of 70 (Colvin 1978).

Following Bath's example, William Paty (1758–1800) was appointed City Surveyor in Bristol under a 1788 Act of Parliament; and William Sibbald (–1809) became Superintendent of Public Works in Edinburgh in 1790, jointly responsible for laying out the New Town (Colvin 1978). As with Bath, such appointments were often combined with political appointments as a burgess or councillor and were part and parcel of the normal business activities of the city fathers. Only as the municipal arrangements of Britain's cities settled down in the 19th century and as public authorities took on the challenges of servicing and controlling the rapidly growing metropolises did recognizably professional public sector departments arise.

Metropolitan Board of Works

Foremost amongst these was the Metropolitan Board of Works (MBW), established in 1855 to principally deal with the sewage crisis in the capital. The arrival of imported guano (bird droppings) from South America for use as fertilizer had undercut the price of human and animal waste collected daily from London's two hundred thousand cesspools, resulting in the suddenly 'worthless' waste being dumped in the street and the rivers. The urgency, danger and unpleasantness of the situation was compounded by the increasing use of flushing water-closets, first introduced in the late 18th century. The WCs both diluted and greatly amplified the problem, leading to the pollution of water sources and regular outbreaks of cholera. Following one outbreak in London in 1832 and with another known to be approaching across Asia, the social reformer Edwin Chadwick (1800–90) proposed (1838, pp103–5) that the 'Causes of Fever in the Metropolis', then thought to be caused by 'poisons disseminated in the air' arising from 'putrefactive decomposition', 'might be removed by proper Sanatory [sic] Measures'. He calculated that the problem could be dealt with by:

> a perfect system of sufficiently sloping drains or sewers, by which from every house and street all fluid refuse shall quickly depart [and] a plentiful supply of water to dilute and carry all such refuse.

Such a proposal, affecting every house and street in the capital, required significant civic governance and organizational experience that simply did not exist at the time. London had no central authority and was administered instead by 300 different local bodies run by about 10 550 commissioners, many determined to thwart any activity that involved centralization (Owen 1982, p33). Action was required involving considerable professional expertise, but it would happen only when the environmental crisis made it inescapable.

A total of six Royal Commissions were established in the years 1848–55 to tackle the problem – each one moving slowly forward one step at a time. The first appointed the Ordnance Survey to prepare an accurate survey of the existing drains, but caused London basements to be flooded by overloaded drains when it abolished the use of cesspools. The second commission appointed three army engineers to find a solution, but the alternative plans they developed only led to arguments, a lack of agreement and then an open call to engineers to send in their own plans. There were 137 responses and the submissions only added further to the confusion. The third commission was dominated by railway engineers including Robert Stephenson (1803–59), who arranged the appointment of Frank Forster (1800–52), who had worked with him on the Britannia railway bridge across the Menai Strait, as Engineer to the Commission. When the 137 submissions were all eventually deemed unsatisfactory Forster was instructed to design his own scheme. Forster's death[1] two years later led to the project being inherited by his deputy, Joseph Bazalgette (1819–91). By the time the sixth commission was wound up little had been achieved, but a scheme combining the drainage of both sewage and surface water had emerged. More importantly, it was accompanied by the realization that it would need a semi-democratic body with substantial decision-making and executive powers to deliver it (Halliday 1999, pp35–57).

Over the course of the six commissions the ICE had become increasingly involved with both the deliberations and the proposals. The Institution organized papers and debates on drainage, for example Robert Rawlinson's (1810–98) paper, delivered to the Institution on the 23rd November 1852, *On the Drainage of Towns* (ICE 1853, pp25–109). The reading was followed by a technical discussion that ran over four evenings. The ICE was also thought be organizing its own slate of engineers to take command of the project, drawn from the many offices clustered around its headquarters building in Westminster. This group, known as 'the Great George St. clique' worked closely with a growing number of contractors based in the area and formed a formidable and powerful group – referred to by at least one engineer outside the circle as a 'monopolistic network' (Porter and Clifton 1988, p333). Bazalgette, who had his office in Parliament Square, next to Great George Street, was considered part of this network and was resented for it.

When the Metropolis Local Management Act, effectively the charter of the MBW, was finally passed by Parliament in 1855 there was no doubt that it would be a professionally staffed organization, if one overseen by the Board, comprising 44 locally elected and endlessly disputatious board members. The Board, also known as the 'Senate of the Sewers' and the 'Parliament of the Parishes', relied on a small, 'expert' staff of about 50 working under the clerk, architect and engineer as well as a larger workforce of draftsmen and site surveyors (Owen 1982, p44).

Interest from the press was intense over all matters concerning the MBW and on the 2nd February 1852 *The Builder* published a somewhat enigmatic announcement in its *Miscellanea* column:

> **Metropolitan Board of Work**s – At a meeting of this board, held on the 25th ult. The election of engineer was proceeded with, when Mr. Bazalgette, Mr. Rawlinson and Mr. Henry were called upon to urge their respective claims to the office, after which Mr. Bazalgette was elected. We have received several letters on the subject, mostly personal. Afterwards the board proceeded to the choice of an accountant, when Mr Hatton was elected by a large majority. As to the Superintending Architect, our advertising columns will show the required qualifications and duties. The salary is not large enough.

Bazalgette, as the historian of the MWB, David Owen (1982, pp43–4) has noted:

> was the obvious candidate … [although] he had been identified with a particular plan of sewerage and with 'the Great George Street clique of

engineers,' who were thought to be using his candidacy as a means of getting themselves in the saddle. But whatever his former loyalties, the fact is that in Bazalgette the Board was getting a great engineer, better than it realized.

The choice of an architect, needed to oversee the design of a range of projects, the creation of streets, bridges and public parks as well as managing the district surveyor system of building control, was less successful, perhaps because of the reportedly small salary. Shortly after Bazalgette's position was confirmed, the board appointed the architect and RIBA member, Frederick Marrable (1819–72) to the post. It was at a meeting dourly described by *The Elector* journal as having 'the dull business of that dull man, Mr Marrable on the carpet' (cited in Owen 1982, p43). His appointment would turn out to be a significant element in the MBW's ultimate undoing.

The MWB, having been installed as a quasi-government for London, even if it didn't have the resources, organization or powers to adequately fulfil the role, was never popular, or at least not in its lifetime. A centralizing force at the time of an officially endorsed creed of individualism and entrepreneurialism, it was cursed by the fact that its work would cause apparently endless disruption to the life of Londoners long before they would see the benefits of the improvements. The installation of the new drains, demolition of buildings and the enforcement of regulations and standards were heavily resented, as was the increased taxation that was needed to pay for it all. The newspapers picked up on and then inflamed the resentment (Owen 1982, p170):

> Occasionally, especially when the Board had executed a feat of particular distinction – completing the main drainage of the Embankment for example – the press would burst forth in an anthem of praise. Generally, however, any failing, however slight brought a slaying from the newspapers, and any achievement brought leading articles emphasizing that 'in spite of the Board's second-rate character, it did a reasonably good job here.' ... From the 1860s onwards, press and public could always picture Spring Gardens [the offices of the MWB] as at best the center of discrete jobbery, or, often, as a cave of outright corruption.

Inevitably the engineers, surveyors and architects who worked for the MWB were caught up in the mudslinging. In 1864 Bazalgette was accused of favouring a particular contractor relating to a deal they had both been involved with in Odessa, but was found not guilty by the select committee that investigated it. The press continued pushing and probing, looking for any peccadillos associated with the work of the Board and in 1886, over 20 years later, The *Financial News* 'sensationally' uncovered the story that two architects and RIBA members, F.H. Goddard (the Chief Valuer) and J.E. Saunders (a Board member), and Thomas Robertson (an Assistant Surveyor) were corruptly disposing of surplus land. Further investigation revealed that Goddard was taking commissions on compensation deals over compulsory purchased land and this in turn implicated the long retired Marrable and another architect member of the board, Francis Fowler. All five were excoriated in the press with demands that they should be prosecuted. The scandal galvanized the government, which instigated an inquiry into the dealings of the Board and its employees. The findings were published in 1889, but the MBW had already gone, unlamented by a hostile media, to be replaced by the London County Council.

Inevitably the story spread beyond the MBW and in 1887 the *Financial News* widened its attack to include the RIBA (quoted in Owen 1982, p187):

> [the] facts are no secret to Mr. Saunder's brother architects. They have, we are assured, frequently formed a discussion in professional circles, both public and private. They have been the occasion of comment by officials of the Royal Institute of British Architects. Doubtless the Metropolitan Board of Works takes a less squeamish view of them.

This forced both the Institution of Surveyors and the RIBA to take action. The Institution suspended Robertson and accepted the resignations of two other fellows while the President of the RIBA, John Whichcord (1823–85), addressed the issue in his presidential address of 1879–80 (quoted in Clifton 1992, p155):

> a Fellow or Associate of this Institute, if he be elected a member of the Metropolitan Board of Works, ought not from that moment to have any professional connection whatever with the purchase of land offered for sale or lease by the board; nor should he be professionally engaged in the superintendence of buildings to be erected upon land which is the property of ratepayers, whose agent and representative he is.

Successive presidents returned to the subject in speeches in 1881 and again in 1883, but it was clear that this was new and uncertain territory for the professional institutions and they didn't yet have the professional apparatus for dealing with work done within and by public offices. It was the first time they had been challenged to examine the responsibilities of those of their members who worked in the public sector and the institutions had yet to come to terms with how they differed from self-employed private practitioners. A new dimension of professionalism had been opened up involving all three of the institutions. It would be many decades before they managed to sufficiently explore the implications and make sense of them.

Institutional frictions

If the experience of combining the industry's early version of professional ethics with the power of the public sector hadn't been a happy one, the work of relatively unsung teams of building professionals working for the new London County Council (LCC), central government departments, school boards across the country and the many municipal authorities largely redeemed the reputation of public sector construction professionals. The LCC, with new powers over potential housing land, built the first council housing, the Boundary Estate, on the site of the infamous Old Nichol rookery in Bethnal Green under its new chief architect Owen Fleming (1867–1955) and followed it with a series of popular suburban style estates at Totterdown in Tooting, Roehampton, Bellingham and Becontree. By 1938 over 76 000 housing units had been built and the architects' department had become the world's largest architectural office with a staff of over 1240 (Harwood 2014b).

Although the leaders of the architectural profession had struggled to adapt to the idea of architects working as employees and designing buildings for their employers rather than independently for commissioning clients, in practice the main concern of the RIBA was with the projects themselves (e.g. hospitals, schools and police stations), work on which was being denied to the private firms that comprised their main and most vocal membership. In 1904 the Institute sent out an officious circular to local authorities on the subject reminding them that the municipal engineers and surveyors they were using for design work were 'fettered by a lack of expert knowledge possessed by architects',[2] a message that appears to have been comprehensively ignored. The Institute was provoked to return to the topic in 1912 when the LCC announced that it would take on additional staff to tackle its upcoming schools programme and an RIBA Official Architecture Committee was tasked to investigate the issue. The report, when it was produced, made the cautious recommendation that the work should be shared between public and private practices. There seems to have been little realization that there were other professional issues at stake beyond the apportionment of projects.

Despite the forthright statements made by professional leaders in the 1880s responding to the financial scandal involving architects and surveyors working for the Metropolitan Board of Works, the institutions gave very little attention to the different professional requirements of the growing numbers of their salaried members. This was despite the RIBA coming close to schism as breakaway groups (including the Society of Architects founded in 1884 and the Architects and Surveyors' Assistants' Professional Union (1919), later the Association of Architects, Surveyors and Technical Assistants – AASTA – representing junior members (Associates) and assistants respectively) were successfully established as alternative bodies to represent and support these otherwise neglected groups of professionals. The Society of Architects was eventually amalgamated back into the RIBA in 1925 but the AASTA remained a vigorous alternative and oppositional voice to the Institute until 1942, when it took on the explicit role of a trade union for architects and technicians as the Association of Building Technicians.

One outcome arising from the general neglect of and even antipathy to the public sector by the established institutions was the formation of a separate body representing the newly defined discipline of town planning, the Town Planning Institute (later the Royal Town Planning Institute), in 1914. *The Housing, Town Planning, etc. Act* was introduced by the Asquith government in 1909 to control standards and layouts of houses and housing development, and this work brought together a group of surveyors, architects, civil engineers and lawyers engaged with local authority planning. From the outset these professionals felt that they needed an independent body that could cut across their various disciplines and allow them to include interested outsiders and amateurs in their deliberations. They began meeting in 1910 and by December 1913 they had elected a Council, were ready to launch the new institute a month later and elected their first President, Thomas Adams (1871–1940), in March 1914.

The RIBA, having played an influential role in lobbying for the introduction of planning regulations, made a series of attempts to claim jurisdiction over the new discipline and retain control. In the wake of the introduction of the first Planning Bill in 1907 the RIBA created a special committee, the Development of Town and Suburbs Committee, under the chairmanship of its prominent Past-President Aston Webb (1849–1930). But instead of making common cause with local authority officers, the RIBA decided to develop its own proposals for expanding the suburbs around large towns, believing that it could do so in a more rational manner than the Local Government Board (Mace 1986, p.xxi). Although the RIBA remained influential over planning for several more decades and many of the key figures in the Town Planning Institute, including its second President, Raymond Unwin (1863–1940, TPI President 1915), were also active in the RIBA, the older institute lost what would be the first of many jurisdictional struggles as new disciplines defined themselves within the industry.

The response of the Institution of Civil Engineers to the issue was a conference organized in 1947 involving representatives from the RIBA, RICS and the Institution of Municipal Engineers. It concluded, possibly predictably, that work in the field of town planning 'should properly be entrusted to engineers, architects and surveyors with the necessary technical qualifications and experience. The [ICE] Council saw no reason for the creation of a separate profession covering town planning' (Watson 1988, pp84–5). There was little that they could do about it though.

The attitude of the RICS to the emergence of planning as a distinct discipline was one of even less interest, although only a short time before surveyors would have considered anything to do with land development as being fully within their fief and, as with the architects, several prominent members of the Town Planning Institute, including Thomas Adams, were RICS members. The RICS Council, according to their official historian, F.M.L. Thompson (1968, pp300–1), 'took the view that the Institution was primarily concerned with questions affecting the administration and costs of town planning provisions, particularly with matters of compensation to private owners'. Again, this took a typically private sector view. Thompson describes the approach as a 'self-denying ordinance by which the Institution disclaimed any wish to determine the objects of planning' and notes that this approach was only officially revised with the publication of the *Schuster Report on the Qualifications of Planners* in 1950, by which time, with the passing of the *Town and Country Planning Act* in July 1947, the new discipline was firmly embedded within the overall municipal system.

The RICS took a firmly private sector perspective even while their membership was growing fast in the public sector, notably in 'the Inland Revenue, Board of Trade Transport and the Service Ministries' of central government (Thompson 1968, p236). But their approach did not cause any immediate waves and it wasn't until 1959 that the first member rebellion occurred. In contrast, the acrimony between the public and private sectors within the RIBA in the 1920s and 30s was intense and frequently broke out into the public domain. An article in *The Architects' Journal* in August 1927 (quoted in Walford 2009, p16) laid the chief blame:

> at the door of private practising architects. They are the natural leaders of the profession; their members hold every position of importance and influence; … and although actually a minority in the profession they dominate and guide its policy.

> They regard themselves as the only representatives of true architecture; official and salaried architects and assistants they anathematize as strange beings who have wrought infinite harm to the profession, i.e. to themselves. … where prudence would have urged assimilation, fear has prompted successfully their exclusion.

The RIBA appeared to have addressed the growing acrimony across the private–public sector divide when it established a Salaried Members' Committee in 1928. The Institute invited AASTA representatives onto its ruling council and committees and in 1931 made Raymond Unwin its first public sector president. Unwin, a former member of William Morris's Socialist League and the Fabian Society, had left the successful private partnership he had built up with his brother-in-law, Barry Parker, to join the Local Government Board in 1914 in order to actively engage with the 'large scale questions of town planning' (Hawkes 1978, p329). His career then took him to the Ministry of Munitions during the First World War; an influential position on the Tudor Walters Committee on Working Class Housing in 1918; and an appointment to be Chief Architect at the Ministry of Health in 1919, a role that later became the UK government's Chief Technical Officer for Housing and Town Planning. He should have been the ideal man to establish a dialogue and find a degree of common purpose between the various factions.

A fragile peace held for a few years, but then disintegrated in 1934 when a series of AASTA proposals, including a salary scale for assistants in private offices, was rejected. In response AASTA cut its formal links with the RIBA and hostilities resumed (Walford 2009, p17). The RIBA's response, a new Committee on Official Architecture chaired by Unwin, hardly calmed the situation when it reported in 1935 (*Journal of the RIBA* 1935, p862):

> that local authorities were likely to stifle good architecture by favouring administrative above design abilities in their staff, and that to secure innovative work they should rely on outside architects rather than those 'cumbered about with much serving.

These comments were then further exacerbated by the election to the RIBA Presidency of H.S. Goodhart-Rendel (1887–1959, RIBA President 1937–9), who used his presidential address to condemn the standard of architecture produced by salaried officials, describing it as being like machine-dispensed chocolate, 'repetitive and slightly stale' (Powers 1979, p51), generating, according to John Summerson (1942, p236):

> a volcanic row. County architects, city architects, the chief architects to this and that, were wounded and indignant. … What the President evidently could not see was that 'official' architecture *need not be* either as second-hand or as tepid as his simile implied; nor, for that matter, did he see that the scales had already turned which would bring the brains and enthusiasm of the young down on the side of the departments.

William Hamlyn FRIBA, chief architect to the LMS Railway railed against 'the enormity of Mr Goodhart-Rendel's offence against professional decency', questioning in the newly launched journal *Official Architecture*:

Is it not the fact that for many years the policy of the Institute has been directed against official architecture, and that by every means, direct statements both verbal and written, by innuendo and suggestion, attempts have been consistent to undermine the confidence of employers in their official or staff architects?[3]

The rift, at that moment, must have seemed irrevocable.

Growth of the public sector

The numbers of building professionals employed by both public authorities and private companies grew steadily and stealthily throughout the early years of the 20th century. Both towns and cities employed Borough Engineers at senior levels within the municipal hierarchy to look after their increasingly complex infrastructure needs. The Borough Engineers then employed, in large and influential departments, specialist engineers, architects and surveyors. Larger cities also had architects' departments, which took responsibility for projects including civic buildings, housing, schools and hospitals, sometimes working directly for powerful committees within city councils. The Architect's Department of the London Passenger Transport Board, working under the leadership of Frank Pick, was only the best-known example of such a body at this time. Government ministries ran departments producing large numbers of building projects including defence buildings, post offices and the newly necessary telephone exchanges. Similarly many large private companies in the 1930s, including major banks and chain stores, also ran professional works departments, as did several not-for-profit organizations, including the Girls' Public Day School Trust (Franklin 2009, p14). In practice, as the RIBA calculated in a 1949 study, the architects working in the public sector in 1938 employed 31% of their total membership of approximately 10000 (Landau 1968, p39). This was far too large a proportion to be ignored.

In some respects the public sector and other large organizations were capable of acting like professional bodies. The development of a body of knowledge was important to them and they were good at commissioning and undertaking research and developing technical standards, knowledge and expertise, even if this was often kept in-department. The Survey of London represents a high water mark in the careful acquisition and sharing of construction knowledge by public bodies. Started in 1894 by the LCC, it has survived as an on-going project for recording and learning from existing buildings in depth. Importantly, this major publication programme, operating from 1900 on, became a significant resource available and open to all. The LCC also ran its own Research and Development Department with a remit to examine and develop how society could benefit from new and existing technologies. Official offices also trained their staff and maintained standards of behaviour. They were less good at letting go of control.

The divisions between the employers and the employed were one fault line in the sector. In the private sector there was always the expectation that employed professionals would progress in their careers to the point when they put up their own 'brass plate', as Summerson had put it. Working in the public sector was very different; it was a far more deliberate career choice. Some chose this path because there were few other options (especially in the febrile economic conditions of the late 1920s), but many had explicit reasons of conscience, solidarity and ideology. The fault line running through the professions in the inter-war period has been described in many ways: politically as a left–right divide; stylistically as a conflict between modernism and tradition; and technologically as innovation pulling away from the ingrained and habitual. It was all of these, but there was also a profound difference over the idea of professionalism itself.

In the 100 or so years of the institutions' existence they had developed to bring together many small private businesses to act roughly in concert: enabling them to command increased respect, develop technical and behavioural standards, advance expertise and knowledge, run a system for educating new recruits, control the number of entrants to the profession, provide mutual support, avoid internal competition wherever possible and to generate, as a result, an enhanced level of fees. It was a system configured to provide a degree of strength and moral force though loosely co-ordinated behaviour and gave some protection to professionals involved in the commercial transaction of procuring buildings against being driven solely by short-term concerns. The point of greatest vulnerability for the private sector consultant was at the selection of the professional team and the terms under which they were appointed. This led the institutions to focus on who should be given the opportunity to work on projects. Such issues were of far less concern in the public sector.

Although there were many shared issues with their private sector colleagues, public sector building professionals' particular concerns needed an equal degree of support from their institutions in the position of independent intermediaries. In the public sector professional esteem and respect were still essential; but, once gained, an individual's rank within an organization probably counted for far more than the qualification itself. Promotion, in particular, was unlikely to be linked to institutional membership. The multi-disciplinary nature of many municipal and government departments also meant that the jobs of many professionals in the public sector were more fluid than those working for even quite large single discipline firms. The RIBA's 1904 circular to local authorities (see above) was clear in its disapproval of architectural work being done by engineers and surveyors, but as the century progressed the Institute offered little support when architects themselves diversified into other built environment tasks and responsibilities.

Public sector employees may well have found themselves working on planning, engineering, architecture and surveying without being able to see any joins between the disciplines. In evidence given to the RIBA's Future of Architecture and the Architectural Profession Special Committee in 1918 the public sector architect, John Murray (1864–1940), Crown Architect and Surveyor to the Commissioners of Crown Lands, but earlier in his career an architect in private practice, argued that 'there was no clear line of demarcation between the design and construction of building work as practised by the architect, surveyor and engineer and that a combination of these professions was desirable' (Mace 1986, p.xxii).

The point of greatest vulnerability in the public sector was not the need to secure projects, but the nature of the relationship between the individual professional and the organization he or she both worked for and represented.

As employees, individuals could be threatened with losing their professional autonomy and the ability to make independent judgements and decisions in the face of pressure from above. These issues eventually caused friction in the London County Council where (Clifton 1989, p21):

> by the last decade of the LCC there was some feeling among the staff of the larger departments, such as the Architect's, that the bureaucracy was growing too rigid, and it was becoming increasingly difficult to change established procedures.

But equally, given sometimes extraordinary levels of power and jurisdiction, they were also faced with the temptation of taking personal advantage, just as Marrable had been at the MBW. Support from external bodies to bolster professional behaviour and to provide guidance and discipline was needed. Due to the institutions' general focus on the private sector, it was rarely provided.

The committees formed by the RIBA in the inter-war period, the standing Salaried Members Committee and its 1937 successor, the Official Architects Committee, addressed some of these issues. The Institute published the *RIBA Scale of Annual Salaries* in 1930 in an attempt to mirror the *RIBA mandatory minimum fee scales* charged by private practices and imposed by the Institute from 1872 to 1982. But the committees (under pressure from organizations including AASTA) mainly concerned themselves with issues of employees' welfare while ignoring professional standards and behaviour. This reflected the view set out by AATSA in 1920 (while it was still the Architects' and Surveyors' Assistants' Professional Union – ASAPU):

> The Union is not concerned particularly with ARCHITECTURE as an Art or Profession. It desires to raise the general standard of the design, construction and craft detail of BUILDING, and to ensure to the worker who produces these a standard of living compatible with his technical acquirements and skill.[4]

Even Summerson (1942, p241) thought this was the right focus:

> What one hopes for is initiative from the profession itself (will the RIBA do it, or leave it to the AASTA?) to press for a standard of organisation and working conditions in architectural departments so that a sense of individual responsibility is retained, so that the slur on 'salaried' practice is wiped out once and for all.

The conflict between the RIBA and the AASTA was abated by the onset of the Second World War. As a result the RIBA's Code of Professional Conduct remained a document focused on the 'independent consulting architect' for many more decades.

War

The Second World War was a turning point for the professional classes in a way that the First had not been. Although the British came out relatively lightly compared to other participants in the conflict it was nonetheless a total war involving the entire population, with every man and all unmarried women between 18 and 60, either enrolled in the military or directed into war work. With the civil service doubling in size, Britain's economy became intensively planned, with state intervention in all aspects of life the norm (Perkin 1989, pp407–9). The great majority of current and future building professionals participated; some directly involved in the fighting, others in frontline support roles or working to deal with the civil damage in the UK (Thompson 1968, pp331–2). As a consequence, after the war there was a consistent narrative from young professionals about how they had shouldered significant responsibility, often at a very young age, an experience they were intent on putting to good use.

The institutions carried on as best as they were able during the war under the *Chartered and other Bodies (Temporary Provisions) Act 1939*. They maintained entrance courses and examinations, both at home and, under an arrangement with the Red Cross, in prisoner of war camps in Germany. But, despite attempts at maintaining business as usual, their main activity was directing war work: ensuring adequate air raid precautions and, from 1942 on, beginning to consider post-war reconstruction. The RIBA established a Reconstruction Committee and the ICE a Committee on Post War National Development (Mace 1986, p.xxiv; Watson 1988, p164), both of which were influential in shaping the peace and the works necessary for rebuilding the country.

The national desire to rebuild in the aftermath of the war, and to rebuild for the common good, brought professionals, their representative institutions, government and the public into near perfect alignment (Tubbs 1945, p56):

> Now the Second World War, by bringing a common danger, has helped to make each man and woman aware of the fact that he or she is a member of a community. More people are realizing that the work of every individual should bear some relationship to a common purpose.

As historian Tony Judt has noted (2005, p163):

> The feelings of the moment are nicely captured in Humphrey Jennings' contemporary film documentary of England in 1951, ... there is much emphasis upon science and progress, design and work.

It was a moment when many building professionals believed that they had found the strong role and purpose, both socially and economically, that they had been searching for. It was a vision of a properly deserved destiny that would sustain them for many years and decades to come, but it would be only one response to the peace.

Post-war idealism and corporatism

In the 1940s engineers, architects and surveyors returned to or joined the construction industry after a formative period spent in the military, confident with taking on responsibility and determined to deliver a better society. In historian Andrew Saint's words (1987, pp30–1):

They wanted to build for the many, not the few, and to do so co-operatively, modestly, justly, efficiently. Through a blend of national and local initiative and through the coming together of architecture, science and industry, there was a glimmer of hope for their ambition in Britain.

Or in David Kynaston's (2007, p29):

If for Keynesians, social reformers and educationalists the war provided unimagined opportunities for influencing the shape of the future, this was even more true for architects and town planners and their cheerleaders. In their case a momentum for fundamental change had been building inexorably between the wars, and now the heady mixture of destruction and reconstruction gave them their chance.

As a result young professionals joined the public sector in large numbers, sacrificing the potentially higher pay and social standing of private practice for the opportunity to be useful and to work on a grand, often visionary scale. Public sector offices rebuilt cities and towns scarred by the war, designed the new schools, hospitals and housing estates of the welfare state as well as factories for newly nationalized industries, public buildings and vast infrastructure projects. The apogee of this programme of works carried out in a period of extreme austerity was the Festival of Britain, held in London in 1951, and its central temple of the arts, the Royal Festival Hall. The Festival buildings were considered by Winston Churchill to be 'three-dimensional socialist propaganda' and he had them hastily cleared away, following his return to government in late 1951. It marked the end of a period of innocence for public sector professionals.

But if the post-war generation had a spiritual home it was not the professional institutions, which were seen to be retrograde and concerned with 'the usual business of getting a living and mongering styles' (Saint 1987, p31), but the experimental wartime organization, the Building Research Station (BRS), in Garston, just outside Watford, north of London. At the BRS practical research into building techniques was combined with technology and science in order to develop and test new and affordable ways to construct buildings in a period of austerity and constrained material resources. The BRS output a constant stream of information papers and rapidly became the definitive knowledge hub for the industry's professions at a time when the future appeared to be science-based, planned and if utilitarian, also fun.

Despite the outbreak of post-war idealism, for many others the traumas and catharsis of two world wars and the great depression that separated them had left an overwhelming desire for a stable and well-planned society and the professions found they fitted even more cleanly and unquestioningly into this calmer vision of society. The principle of the structured, corporatist society that emerged was one described by Philippe Schmitter (1974, pp93–4) as:

a system of interest representation in which the constituent units are organised into a limited number of singular, compulsory, non-competitive, hierarchically ordered and functionally differentiated categories, recognised or licensed (if not created) by the state and granted a deliberate representational monopoly within their respective categories in exchange for observing certain controls on their selection of leaders and articulation of demands and supports.

The vision for such an ordered, hierarchical and essentially class-based system was one in which each social or functional group, whether an employers' organization, trade union, voluntary body, religious congregation or social association, would appoint representatives with the responsibility to speak on their behalf to decision-makers in business and government. In return they would have a large degree of freedom of action within their sphere of activity and the right for their voice to be heard and interests respected. In retrospect such a vision seems utopian and in practice it broke down almost immediately as politics reverted to business as usual and as the twin socio-economic pressures of individualism and consumerism developed. But at the time it appeared both sensible and desirable. Critically many of the professional and other representative bodies keenly wanted to believe in the corporate narrative – that their anointed role was to represent both their discipline and membership in society and politics – and for some membership bodies, especially the British Medical Association at the founding of the NHS, it was undoubtedly true. The narrative outlined a definitive role for professional bodies, and none more so than those in the construction industry, giving them an important (and, in their eyes, well deserved) stake in society. The problem was, once they had adopted this view of themselves they were extremely reluctant to surrender it, however much it was at odds with the world at large.

During the post-war period the institutions eagerly sought to become part of the establishment and regular visitors to the corridors of power. Institution presidents began to be drawn from the public sector, alongside the usual significant names from private practice, they already had a degree of access to the prime minister and the government and it all helped make the narrative more real. They were routinely given knighthoods and, in some cases, peerages. There was every appearance of pulling together for the future of the country, whichever party was in power, and the professional institutions became very much part of what was labelled at the time 'the Establishment', defined in *The Spectator* by journalist Henry Fairlie (1955, p380):

> By *the Establishment*, I do not mean only the centres of official power – though they are certainly part of it – but rather the whole matrix of official and social relations within which power is exercised. The exercise of power in Britain (more specifically, in England) cannot be understood unless it is recognised that it is exercised socially.

The extent to which this was a reality is debatable, certainly the institutions liked to believe that they were more significant and connected than they probably were. In practice they crop up in very few contemporary histories of the period, and often not in a good light. For example, in Anthony Sampson's 1962 dissection of the establishment, *Anatomy of Britain*, only the Institution of Civil Engineers is mentioned, in their 'stately palazzo off Parliament Square', with the damning comment that in comparison with their Victorian forebears, 'they are far less noticed by parliament or the public' (p516).

The divergence of these two streams in the 1940s and 50s meant that at just the time when the institutions achieved a role as arbiter and interlocutor with government and were able to act as a representative voice for their disciplines, they not only lost touch with their most energetic and innovative potential new members, who found other channels and opportunities for their (possibly just as unreal) plans to change the world around them, but they also unwittingly dropped much of their foundational role as conduits of specialist and expert knowledge. It would not turn out to be good exchange.

Public sector professionalism

The post-war expansion of public sector professional employment, allied to the relative lack of opportunities in the private sector, led to a point in the mid-1950s when half of the registered architects in the UK were in salaried employment (Mace 1986, p.xxiv) with nine out of ten of these employees working in public offices. The shift towards the public sector so worried the RIBA in 1948 that it created a committee 'to consider the present and future position of architects in private practice'.[5] In retrospect, the 1950s represented the high water mark of public sector employment and by 1964 it had already fallen back from 45% to 39%. The numbers for RICS members in 1955 were very similar, with 54% of building related surveyors working in private practice, 41% in public employment (split 39/61% between central and local government) and 5% working for commercial (non-professional) firms (Killick 1958, pp662–3).

The professionalism and competence of those within the public sector wasn't in doubt internally and was initially driven by the combination of idealism and post-war spirit of possibility and achievement with a supply of work that was enough to keep everyone busy and purposeful. At the outset external opinion was sceptical of the quality of the work of public offices. J.M. Richards in the first edition of his popular Penguin book, *An Introduction to Modern Architecture* (1940, pp105–6), wrote that 'typical official architecture is that of the London County Council. It is most competent technically, but the Council has taken the safe route of adopting for nearly all its buildings a modified Georgian style'. However, in later editions of the same book he, along with much of the wider population, had tempered his view: 'Not all [local authorities] are very progressive in their ideas, but an important exception is the London County Council, notable for its willingness to experiment and for the high quality of its designs' (1962, p118). But what was still missing was a version of professionalism that defined how the architect, planner, engineer or surveyor should act when they weren't acting as an intermediary between commissioning client and contractor, but were already working directly for a public agency.

The battle over representation in and control over the RIBA opened in the late 1940s and raged at least until the early 1960s when a period of calm briefly preceded fresh arguments. As with all such long running campaigns the point of the original arguments became long lost in the general fog of battle between the opposing factions, but it remained in broad terms a struggle for dominance between public sector employees and private sector employers. It was a period bookended by the presidencies of two significant 'official architects'; Lancelot Keay (RIBA President 1946–8), the Chief Architect of Liverpool Council's Housing Department, and Donald Gibson (1908–91, RIBA President 1965–6), from 1949 to 1954 Coventry's City Architect and Planning Officer and at the time of his RIBA presidency Director General of Research and Development at the Ministry of Public Building and Works. Keay, despite being the first serving official architect to become president made it clear in his inaugural address in 1946 (cited by Day 1988, p11) that he was unwilling to risk any change to the *status quo* or to recognize that public sector architects operated under any different set of pressures and responsibilities:

> You have elected as your President one whose whole career has been spent in the service of various local authorities. Lest any should be apprehensive as to the advisability of such a selection, I hasten to give an assurance that it will be my earnest endeavour to maintain the dignity of the office and to preserve the traditions associated with it. ...
>
> This Institute, with the catholicity to be expected of it, does not differentiate between those who serve as private individuals and those who elect to work as a servant of the community. Its function is to ensure that all who are admitted to its ranks are qualified to discharge satisfactorily the responsibilities they accept. Any who would attempt to divide our ranks do disservice to the Institute, for its strength depends upon the closest co-operation of all its members.

Gibson had been one of a group of official architects, collectively labelled the 'upper house',[6] who had started pushing for better representation for public sector architects on the RIBA Council and committees in the late 1940s and having achieved that, became part of a well-organized caucus that steered the RIBA towards giving priority to the concerns of public architecture. The group issued their informal manifesto in a series of 11 essays in *The Architects' Journal* (Gibson *et al.* 1952a–e) discussing office organization, the relationship with the building industry, the origins of the figure of the public architect, the current condition of the state sector and hot topics such as patronage, salaries and grading, which were 'still a barrier to the production of good Public Architecture'.

The essays laid out the claim for 'the recognition of architecture as an essential public service. ... coupled with any proposals which may be thought advisable in order to increase the public architect's fitness for the tasks that lie ahead' (Gibson *et al.* 1952a, p146). They hailed 'the development of a new tradition in public architecture with the twin growth of local authorities to carry out new buildings and government departments to watch over them and co-ordinate their activities' and a 'new system [that] enables the architect to tackle the broader job of town building' (Gibson *et al.* 1952b, p208).

The authors of the manifesto envisaged that the way to achieve their aims was through the development of large, hierarchically structured architects' departments. These would overcome the 'popularly held idea among young architects that the small private office is the best, not only for the human conditions of work but also for the production of good architecture' and 'the loss of individuality, the fear that he has become an impersonal factor in a large and unwieldy machine' (Gibson *et al.* 1952c, p597). The traditional large office would be broken down into smaller groups, led by individual group leaders, and would respect and trust professional workers to maintain their own discipline and timekeeping – in particular by freeing them from the tyranny of 'signing-in'. Group leaders alone would be responsible for communicating with the Chief Architect. As the essay 'Group working and the large office' (Gibson *et al.* 1952c, p600) puts it:

> An organization of small or smallish groups or teams can do the job, provided the organization of the group, and its responsibilities, are carefully considered. It must have a good leader, and he must be given a generous measure of staff responsibility. The closest co-ordination must exist between the chief architect and the group leaders. Finally, the physical conditions must not be overlooked; cut out the one large drawing office, see that the group leader is right by his group, and that the chief is not too far away. Each architect, as a group member, must be treated as a responsible professional person, and, in being given greater opportunities and freedom he must also accept greater responsibilities and obligations.

The professional terrain has shifted in this explanation to an all-encompassing position. The redefined professional can now work as a responsible individual, both as part of a team and with a clear position in a hierarchy. The group members might not work directly for a partner or even the Chief Architect, but the sense of individual professional responsibility and obligation that was in clear danger of being lost is reclaimed. The aim was that 'by breaking the large office down into units that are comparable to the good private architect's office in size and volume of work, the individual architect can enjoy as great a degree of freedom as in a good private architect's office' (Gibson *et al.* 1952d, p788) and presumably the office could get a comparable quantity and quality of work from its professional employees.

What the authors didn't attempt to define was the scope and nature of professional responsibilities and obligations in public service, although, in a discussion of the architect's relationship with building control officers, they suggest that each architect should take personal responsibility for the quality of their designs and compliance with regulations (Gibson *et al.* 1952e, p428):

> The restrictive tyranny of building controls could be alleviated if they were concerned with specifying principles of design, and not with specifying the precise and detailed requirements of actual products and if architects were treated as fully responsible, on their own integrity as professional men, for ensuring that the buildings they designed conformed to Government controls and to all building regulations.

The group wanted to 'give the architect the responsibility and status due to him' (ibid.). To enable this and to fulfil their sense of duty to 'public service' the group believed they needed, not only to get the message out to the growing number of public sector offices, but also to take control of their professional body, the RIBA, itself.

The upper house met regularly to plot strategy in the basement 'pub' of the publishers of *The Architects' Journal*, The Bride of Denmark, and from 1956 in the offices of Robert Matthew Johnson Marshall (RMJM) to co-ordinate tactics with their shock troops, the 'lower house' (later known as the Chain Gang). According to Andrew Saint (1987, p246):

they achieved an extraordinary about-face in the RIBA from the late 1950s onwards. They helped win a better status for salaried architects in public service and railroaded the RIBA into carrying out its first serious research into the state of the architectural profession and its problems, culminating in 1962 with the publication of *The Architect and his Office.* In these years hardly a candidate promoted by the 'Chain-Gang' and its allies failed to win election.

The Architect and his Office

The RIBA had been running for well over 120 years before it felt the need to study how architects' practices functioned in reality. It had long assumed the model of practice described by Summerson, involving a handful of (male) partners and a relatively small staff working their way up from office junior to draughtsman before becoming qualified and possibly becoming a practice associate. There was an expectation that every young architect would want to set up their own practice at some point in their career and replicate the pattern for the next generation. As discussed this image of the architect had been out of date for a considerable time and a corrective was long overdue. But it was not this that triggered the Institute to undertake its first major research project; a project that would eventually lead to a body of work that would justify, for a period, its claim to be a 'learned society'.

Management theory had arrived from the United States after the Second World War, in particular the work of the Harvard human relationship school of Elton Mayo (1880–1949), Mary Parker Follett (1868–1933) and T.N. Whitehead (1891–1969), and several institutes of management were founded in Britain in the late 1940s as a result (Perkin 1989, p305). The focus of the work at Harvard was the underlying psychology of industrial efficiency and productivity and its key research finding was: 'group collaboration does not occur by accident; it must be planned for and developed' (Brown 1954, p85). By the 1950s the idea that work outputs could be vastly improved by the proper application of modern management theory had been popularized by writers such as Peter Drucker (1909–2005) in *The New Society* (1950) and *The Practice of Management* (1954) or James A.C. Brown (1911–64) in *The Social Psychology of Industry* (1954). Men like Stirrat Johnson-Marshall (1912–81), chief architect to the Ministry of Education (1948–56), and Robert Matthew (1906–75), chief architect of the LCC (1946–53), the *éminences grises* of the 'upper house' (representing the interests of public sector architects in the RIBA), who had both managed large teams, were keen to apply management theory, with its tools of surveys and scientific analysis, to the business of architecture.[7]

The Architect and his Office by Andrew Derbyshire (1923–2016), J. Austin-Smith (1918–99) and others (RIBA 1962) was an in depth study of 47 private sector, 12 local authority, five other government and five industrial/commercial offices ranging from small to very large (over 51 architectural staff). Subtitled *A survey of organisation, staffing, quality of service and productivity* it was a groundbreaking attempt to rigorously uncover what actually occurred in and around architects' offices. The report is generally written in calm measured language, but clearly there was a degree of shock at what was discovered.

The first point in the summary of findings reads: 'Most offices seemed to work in virtual isolation from one another, with little interchange of information or experience even between working groups within the same office.' It also reported the excessive 'amount of administrative work done by technical staff' and the 'large part of their design responsibilities in the technical field' that architects had devolved to others (although less so within large and local authority offices, who might carry in-house specialisms). In particular, the survey team simply did not find the level of professionalism they were expecting (1962, pp63–65, 185):

> Some offices were recruiting juniors without the minimum educational standards required for RIBA Probationers.
> Less than 2 per cent of time was spent on reference to technical and trade literature.
> Only seven of the visited offices, or 11 per cent of the total, achieved an all-round excellence of performance in management, technical efficiency and quality of design.

The main recommendations of *The Architect and his Office* focus on education and training, fees and earnings, and standards of service (management and competence), but equally they can be seen as a call for the profession to standardize its activities in a number of key areas of both practice and office management. There is an implicit demand that both private and public should come together over a shared view of what it is to run a professional practice and the standards of the outputs expected. This is reflected in a series of recommendations aimed at standardizing professional practices that are equally applicable to both sectors. They include (1962, pp13–15):

3. The education and training of architects should be planned as an integrated whole;
14. The RIBA should exercise greater control over the standard of service given by its Members, in return for the protection given them by the Scale of Fees;
19. The RIBA Management Handbook should be published as soon as possible;
23. Uniform methods of costing, overheads analysis and budgetary control should be developed and spread throughout the profession;
25. A study of the purpose and use of drawings should be put in hand;
27. The RIBA should gather and disseminate information and experience on user requirements for different building types; and
28. The profession should promote and encourage the application of standardization and industrialization to building.

In practice there were only two recommendations that applied only to the public sector:

15. Local Authority architects' departments should exchange information with each other about costs and productivity achieved. They should establish their own basis of comparison rather than making unrealistic cost comparisons with other types of practice.

16. The possibility of local authority Chief Architects being allowed to run their own departments within a given budget, without being held to parity in staffing with other departments of the authority should be explored.

Both of these suggest particular issues being worked through that couldn't be resolved before publication.

The publication of the study marked a point when peace broke out in the profession, as architects effectively agreed to work together to a shared agenda and there was a concerted effort to fulfil the RIBA's charter purpose of advancing architecture. It certainly marked a change of direction, as an effective standardization process commenced. The RICS launched their Building Cost Information Service (BCIS) in 1961, providing cost information on building, broken down element by element; and in 1963 (the year after *The Architect and his Office*), the promised *Handbook to Architectural Practice and Management* and the *RIBA Plan of Work* were published. The *Plan of Work*, in particular, became an important unifying output affecting the whole industry; used for explaining and defining the whole process of design and construction for many decades and put to use by consultants, clients, funders, contractors and regulators as the baseline schema for the ever messy process of getting building projects realized. A decade later would see the introduction by the RIBA of the *National Building Specification*. The battle for the heart and soul of the architecture profession that had been raging in various forms for over a century reached a consensus in the mid-1960s. The future was going to be ordered, information-driven and delivered efficiently and competently. Construction professionals would henceforth be useful, well connected and highly respected members of society.

The appointment of Robert Matthew and then Donald Gibson as successive RIBA presidents in 1962 and 1964 respectively, marked not only the success by the 'upper house' in achieving control of the RIBA, but also the high point of the RIBA's institutional influence and membership of the establishment. Both men were knighted for their contribution to architecture (although it should also be noted that their immediate predecessor, William Holford, was made a life peer in recognition of his contribution to town planning, and their successor, Lionel Brett, succeeded to the title of Viscount Esher two years before taking the presidency). Gibson in particular had the direct ear of Geoffrey Rippon, the then Minister of Public Building and Works (Walford 2009, p339).

Robert Matthew, according to his biographer, Miles Glendinning, 'finalised the RIBA's transformation from a 'moribund learned society' into an influential, modern institution' (2008, p310). Gibson, who only served as president for one year, rather than the usual two, could therefore take a broader view of the future, one informed by his experience in both public service and government. In his inaugural address he stressed how the architect was now 'recognised as playing a key role in [the] nation's future' and how the profession could 'serve the nation' through raising industrial productivity and improving the quantity and quality of their output. Gibson's vision was technocratic, looking forward to linear cities utilizing monorails, moving pavements and district heating systems as well as having, in his most prescient prediction, playing fields with 'plastic grass'. He saw the future dominated by large architectural practices, but worried at the inability of the public sector to attract the 'calibre of men' required for them to compete. His greatest enthusiasm was reserved for industrialized building, for the prefabricated modular CLASP system for school building and for the work of Alec Issigonis, the designer of the Morris Minor and the Mini for the British Motor Corporation (Walford 2009, pp344–6).

During the battles for control of the professional bodies, the public sector itself continued to consolidate its direct control over much of the work being done by the construction professions either by recruiting large in-house departments or through the use of favoured firms. Despite this the numbers of professionals in the public sector continued to decline both overall and as a proportion of the workforce.[8] The LCC's Architects' Department employed 1577 staff including 350 professional architects and trainees in 1952–3 under J.H. Forshaw (Lang 2014), while its successor, the Greater London Council's (GLC) Department of Architecture (1965–86) employed only 250 staff at its peak (*Building* 2002). In 1972 a new overarching autonomous public sector body, the Property Services Agency (PSA), appeared to take over many of the functions of the defunct Ministry of Works, while leaving other departments and local authorities with their own separate workloads. It is worth quoting at length from the speech of junior Environment Minister, Kenneth Marks, delivered in the House of Commons on the subject of the relocation of the PSA in March 1977 (House of Commons 1977):

> The Agency has the job of providing, equipping and maintaining a wide range of buildings and installations for Government Departments, and the Armed Services, as well as other bodies. It holds and manages much of the Government's civil estate, including Government offices and establishments all over the United Kingdom as well as the diplomatic estate abroad. It manages Ministry of Defence property on its behalf, both at home and overseas. … The clients it serves are mainly Government Departments, but it has certain other clients, the most important of which is the Post Office, for which it provides services on repayment.
>
> The staff of the Agency is about 50,000 strong, of whom about 30,000 are industrial workers, including about 7,000 locally engaged staff overseas. Of the 20,000 non-industrials, more than half are specialist staff – architects, civil, mechanical and electrical engineers, quantity surveyors, building surveyors, estate surveyors, technicians and drawing office staff.
>
> The Agency undertakes all types of construction work – from houses and barracks for the Services to offices, research facilities, airfields, dockyards and telephone exchanges for the Post Office. It has about 1,500 major new works projects in various stages of design, and about 1,000 under construction. This year, 1976–77, the Agency's expenditure on new works will be about £400 million. Its maintenance bill will be about £300 million.
>
> The Agency itself is the largest single client of the building industry in this country. While, as I have said, it employs a full range of professional staff of its own, it also makes very considerable use of private consultants. This year the bill for consultants' fees will be about £26 million.

This is a vision of a well-staffed command economy with professional staff directly serving the public interest. The professional ethos is one of service rather than duty and relies on the assumption that public sector employees have, in the words of a much later report (IPPR 2001, pp131), 'an inclination to act with integrity and in the interest of the user rather than the producer – that make them uniquely placed to deliver public services'.

This conflation of professionalism with the public service ethos may largely explain why the concerns of the Official Architects Group within the RIBA were primarily concerned with working conditions, rates of pay and the mechanics of team working within large hierarchical organizations. It was assumed that issues of professionalism would be addressed elsewhere, which may well have meant that they were not dealt with at all, and there was effectively a reversion to a technocratic professionalism. As John Palmer characterized the UK's public sector planners in his 1971 introduction to the British edition of Robert Goodman's *After the Planners* (Palmer 1972, p41), they:

> were technical professionals, relying on such skills as surveying, building, civil and public health engineering and law in contributing to the planning of cities, and therefore ill-equipped professionally to question the nature of the political role.

This view ignores the creativity within the public sector at the time, whether the nurturing of the pop-architecture protagonists Archigram within the GLC's Department of Architecture and Civic Design or the team brought together by Fred Roche at the Milton Keynes Development Corporation, of which it was said 'If there was a rule, you found a way to break it and you'd get the job done' (Kitchen and Hill 2007). As the 1970s ran their course, the original vision of public service gradually faded as its most effective apostles retired. Consequently Palmer's view became an ever closer approximation of reality.

With the majority of high profile building projects in the 1950s being designed by or for the public sector, a close association developed in the public mind between the modernism of the period, the state and the need for austerity and 'making-do'.[9] Public opinion initially supported the barebones nature of the new style, but expectations changed. The desire for greater comfort and opportunities for individual expression grew rapidly, supported by ever increasing wealth and social confidence. At the same time the poor quality of much of the built output was exposed. The resulting rejection of austerity, and with it modernism, proved too great a burden for an industry configured for constructing large blocks on cleared sites to bear. Support for public sector architecture collapsed. The change can be illustrated by two editorial comments from the *Coventry Evening Telegraph* discussing the rebuilding of the bombed city centre under the direction of its chief architect, Donald Gibson. In 1952 the paper claimed that 'the public are increasingly pleased with what is being done', but by the following year it was denouncing the latest addition to the skyline, Broadgate House: 'Such a monstrosity darkening half Broadgate, and obstructing the view and traffic from Broadgate down Hertford Street, could never have been conceived by a Coventrian, who loved Coventry' (quoted in Kynaston 2009, pp274–5).

Progress was undoubtedly being made. The number of new homes completed annually in the UK reached 425 830 in 1968, a substantial increase from the low 200 000s constructed in the immediate post war years (Department for Communities and Local Government 2017).[10] By the mid-1960s these were being built to the new Parker–Morris space standards set out in the Ministry of Housing's *Design Bulletin 6 – Space in the Home* (1963). Minimum sizes for homes for different occupancies were mandated along with requirements for 'conveniences', including internal flushing toilets, heating systems and adequate storage. In the report published by the Parker–Morris Committee, *Homes for Today and Tomorrow* (1961, p4) a research-led approach to design was outlined:

> This report is not about rooms so much as about the activities that people want to pursue in their homes … the problem of design starts with a clear recognition of these various activities and their relative importance in social, family and individual lives, and goes on to assess the conditions necessary for their pursuit in terms of space, atmosphere, efficiency, comfort, furniture and equipment.

Large numbers of urgently needed schools, hospitals, transport, leisure and other facilities for the public were being built and used.

Yet the promise that the new built environment, combining the latest scientific thinking and knowledge with a helping of instructive art, would be responsive and supportive of social progress was not adequately fulfilled. Professionals frequently seemed to be missing when problems arose with buildings and continued repeating the same solutions instead of learning lessons from lived experience – how the buildings actually operate and whether they support occupants' needs. Andrew Saint (1987, p190) notes that such was the urgency of work in the 1950s that the Architects and Building (A&B) Branch within the Ministry of Education 'never had a picture of what the mass of teachers, let alone children, thought about the post-war schools, or what effect if any they were having on the ideas and psychology of children'. The job security that idealistic construction professionals had sought within the public sector in order to do their job effectively was also becoming a liability. In the opinion of the contemporary academic and social reformer Richard Titmuss (1958, p231):

> We have almost reached a stage … where it would be more appropriate … to speak of 'The Pressure Group State'; expressing a shift from contract to status; from open social rights to concealed professional syndicalism; from a multiplicity of allegiances to an undivided loyalty.

Or in the more recent analysis of Frank Duffy and Andrew Rabeneck (2013, p118):

> This secure setting made possible a great deal of painstaking and highly relevant design research. However the same security may also have fostered complacency, inefficiency and over-manning, and prepared for what in hindsight can be seen as an inevitable backlash.

Public and policy

The differences in outlook between designers of buildings and cities and 'public opinion', characterized and satirized by the cartoons of Saul Steinberg in the US and Osbert Lancaster in the UK, and by the films of Jacques Tati in France, barely scraped the extent of disillusion generated by the gulf between the promise and the reality of modern living environments. Flats in tower blocks may have possessed the modern conveniences (indoor toilets, fitted kitchens and central heating) which so excited the first occupants, but in due course the majority of such buildings became unloved and problematic places to live in, with their poorly realized environs often left neglected. This was not through any lack of goodwill on the part of architects, engineers, planners or public sector agencies – who produced sometimes excellent projects. The theory may have been right, but the practice, including lack of adequate consultation and feedback mechanisms, fell far short of its goals. The 'buy-in' from the users was negligible. Commentators, writers and activists stepped in to condemn modernity and progress and to propose alternative solutions. This is not the place to go over such familiar ground at length; Paul Harrison, echoing American writer Jane Jacobs's work of the 1960s, describes the widespread discontent that was shared by many (1983, pp204–5):

> The ideals of council housing … were noble: to take families out of those tenements, with shared or absent baths or toilets; to remove them from exploitation and insecurity; to provide decent, sound houses, each with their own bath, toilet and kitchen; to give tenants a humane and benevolent landlord without an interest in profit, offering secure tenures, rents they could afford, exemplary service. There may have been a time in the 1950s perhaps, when these dreams were roughly approximated by reality. In the inner cities of the early 1980s, they ring extremely hollow. For many of yesterday's pristine estates have become today's slums, some of them only fit for demolition, many more requiring massive and expensive repairs and remodelling to make them anywhere near habitable.

The irony is that at the peak of the institutional triumph in penetrating the inner sanctum of the establishment the outputs of the most prominent professionals in construction had managed comprehensively to lose the support of almost all of society at large. The general feeling against new buildings and towns was then reinforced by clear examples of professional failure, arrogance and criminality. Notorious examples include the gas explosion and subsequent collapse of multiple floors in the (public sector) 22 storey Ronan Point block of flats in 1968; the bribery of public officials by the architect, John Poulson, exposed in 1972; and the 34 floor plates deliberately left vacant in the (private sector) Centre Point tower in the heart of London from 1966 to 1975 during a period of housing shortage. This generated a commonly accepted narrative that architects, planners and other building professionals were working for the developers and other perceived wreckers of cities and not for the benefit of the public. However isolated the examples of malfeasance and however many building professionals campaigned for the conservation and preservation of places and buildings, huge damage was done to the professional standing of those working in construction. Ultimately the attrition of the period left the professional bodies too weak to fight back when the state decided to dismantle long-held professional privileges.

Competition

It would be wrong to characterize the 1960s predominantly or solely as a period of failing housing provision and the decline of city centres. The far greater issue pre-occupying politicians and public in the period was the industrial conflict between workers and employers with multiple strikes in sector after sector. This led to the publication in 1969 of a Government White Paper 'In Place of Strife' and culminated in the change of government at the General Election in 1970 from the Labour Party to the Conservatives. The two central causes of disputes in this period were 'fair' wages and the rights of certain categories of workers to particular jobs (Turner *et al.* 1967). As the social historian Harold Perkin (1989, p465) wrote, 'The remarkable thing about these objectives is how closely they paralleled those of the salaried professions', and it is perhaps no surprise that the practice of professional protectionism became entangled with the government's intent to bring the trade unions to heel and to root out demarcation of function in all areas of industry.

The first moves were made in 1967 when Douglas Jay, the President of the Board of Trade, announced in Parliament (Hansard 1967, pp779–80):

> Over the past year it has been increasingly suggested that there may be restrictive practices in the professions which are contrary to the public interest. I have therefore decided to ask the Monopolies Commission to investigate this matter and to make a general report to me under the extended powers in Section 5 of the Monopolies and Mergers Act, 1965. The Commission will be asked to consider such questions as restrictions on entry into a profession, the practice of charging standard fees or commissions for professional services, the terms or conditions on which professional services are to be supplied, and restrictions on advertising.

As this announcement suggests, the scope of the investigation was broad and the implications for the professions far reaching. Mandatory (or even recommended) fee scales, a common feature for many professions including solicitors and accountants as well as quantity surveyors, structural engineers and architects, were condemned in the Monopolies Commission report (1970, p78):

> In general, we regard a collective obligation not to compete in price, or a restriction collectively imposed which discourages such competition, as being one of the most effective restraints on competition. The introduction of price competition in the supply of a professional service where it is not at present permitted is likely to be the most effective single stimulant to greater efficiency and to innovation and variety of service and price that could be applied to that profession.

The original principal was that professionals should compete on merit and quality of service rather than price and that it was unethical to reduce the amount of time and resources required to do a job properly or attempt to undercut the fees of fellow professionals. Principle 3 of a new Code of Conduct adopted by the RIBA in 1976 stated unambiguously that 'A member shall rely only on ability and achievement as the basis for his advancement'; but in the face of a series of reports from the National Board for Prices and Incomes, the Monopolies and Mergers Commission (MMC) and The Office for Fair Trading (OFT) amongst others, the principle of mandatory professional fees, then recommended fee scales and finally even indicative fee guidance were outlawed. The RICS gradually withdrew their authorized schedule of professional charges following a 1977 MMC report specifically on Surveyors' Services. The report, though equivocal on fee scales, which they determined should 'not be binding', generally found that associations should 'permit their members freely to quote a fee in competition with other suppliers and so as not to prevent competition for business on the basis of fees' (1977). The RICS abandoned its final fee scale in 2000. Architects had their own MMC report on scale fees in 1977, but continued to challenge the idea that their Institute shouldn't publish any form of indicative guidance until 2003 when the OFT ruled finally against their permissibility.

While the MMC was investigating professional fees in the construction industry it was also looking into the rules of the accountants', solicitors' and barristers' bodies against advertising and determined that they too 'operated against the public interest' (Director General of Fair Trading 2001, p128). In response the codes of practice of other professions were rapidly amended to allow advertising under certain circumstances. The 1976 RIBA code read:

> A member shall not solicit either a commission or engagement for himself or his business for a client or employer, but he may make his or his practice's availability and experience known by giving information which in substance and presentation is factual.

In 1981 a more open version of the code was adopted:

> A member may make his availability and experience known by means of direct approaches to individuals and organisations.

A range of new methods of promotion also became permissable, including:

- Publication and exhibition of his work,
- Direct approaches to individuals and organisations,
- By entries in any directory and
- By advertisement in any printed publication.

By the 1997 edition of the RIBA's code of conduct all mention of advertising and promotion had disappeared entirely.

One by one the old professional conventions of practice were dismantled in the face of governments intent on liberalizing the economy, including the restrictions on professionals acting as or taking a direct financial interest in projects or contractors (see Chapter 8). Despite the high proportion of professionals working in the public sector it is notable that all these measures were taken against the, at the time, much weaker private sector. That was all about to change with the election in 1979 of a new Conservative government under Margaret Thatcher who, in Eric Evans' words (2013, p70):

was suspicious of professionals. She was irritated by the easy, apparently effortless expertise many possessed and she considered the higher professionals a pampered elite.

Under the Thatcher government of the 1980s public sector offices and services were dismantled and wherever possible sold off. Compulsory competitive tendering introduced by The Local Government Act 1988 would shortly include the professional services of surveying, legal services and architecture (Clarke 1994, p126) resulting in the at risk in-house departments being disposed of *en masse* in the early 1990s, while they still had a market value. The Greater London Council (GLC) was abolished in the face of determined political and legal opposition in 1986, together with its building and architectural departments, and the PSA, under a cloud of corruption allegations from the 1980s, was split up and privatized in 1990–3. Well before the end of the century the role of the Chief Architect, Surveyor or Engineer at the head of a sizeable department and directly responsible to the local authority had all but disappeared. By 2008 only about 8% of registered architects were working in the public sector, a figure that is still in decline (Chappell 2016, p62).

In less than a century members of the construction professions had gone through a thorough and intense immersion within the public sector and by the end had reverted to being a predominantly private sector phenomenon again. The experience changed the notion of professionalism within the sector and stripped it of much of its perceived duty to the greater public good. Commercial interest took the lead, at least for a while, as professional firms reinvented themselves foremost as businesses rather than arbiters broking conflicting interests or champions of cherished values. Building professionals have learnt from their experience within large multi-faceted public sector organizations to operate within corporate structures, whether led by professionals or not, and sometimes with remote ownership and very varied aims. In general this has been healthy for the industry but has meant abandoning the idea of the professional sector as a third force, free to set the agenda for the development of its individual disciplines and to act as intermediary between the forces of the market and the wider world of both society and the local and global environment.

Eliot Freidson (2001, p218) puts this counter view more forcefully in expressing his concern for 'the soul of professionalism':

On a broader level, the ideological assertion of economism that the primary purpose of work is to maximise personal gain should be declared as a frontal assault on professionalism. It should be declared unethical for professionals, whether self-employed or employed, to aim at maximising gain at the expense of the quality of their work and the broadest possible distribution of its benefits.

The next chapter will examine the changes in profession education that took place simultaneously with the developments in professional organization described above.

Notes

1 Forster's death was attributed to the 'harassing fatigues and anxieties of official duties' (Halliday 1999, p54).

2 RIBA, Memorial sent to Municipal Councils, District Councils and County Councils, 17 November 1904, cited in Mace 1986, p.xxxiii.

3 W. Hamlyn, *Official Architecture*, December 1937, quoted in Walford 2009, p25.

4 ASAPU, Pamphlet, June 1920, RIBA Library – MRC MSS.78/BT/10/1/1, cited by Walford 2009, p13.

5 *Architects' Journal*, 23rd November 1950, p414, cited by Walford 2009, p189.

6 The 'upper house' was a group of senior public sector architects brought together by the latest incarnation of the AASTA, the Association of Building Technicians (ABT): Robert Matthew, the LCC's chief architect, Percy Johnson-Marshall of the LCC, his brother, Stirrat Johnson-Marshall, chief architect of the Ministry of Education, Donald Gibson of Coventy City Council and Robert Gardner-Medwin of the Scottish Department of Health (Saint 1987), p245).

7 At roughly the same time (1958), the architect Leslie Martin brought practitioners and educators together to discuss what capabilities architects needed and how this could be provided through reform of higher education (discussed further in Chapter 7). The discussions from this educational conference also raised questions about the nature of architectural practice itself and helped to set the stage for *The Architect and his Office*.

8 The number of architects registered in the UK in 1964 was 20 911 of whom 39% (8155) were in the public sector. In 1990 overall numbers had risen to 32 004, but the proportion in public service had fallen to 22% (7040) (RIBA 1962, p4).

9 There were only a handful of exceptions to the link between major developments and modernism in the 1950s, for example Coventry Cathedral (1955–62) and the Vickers (Millbank) Tower (1958–62) in London.

10 For comparison the permanent dwellings completed in the UK in 2016 were 170 870 (Department for Communities and Local Government 2017, Table 241).

7

Openings

Once the institutions had taken the step, however reluctantly, of accepting examinations as the means of controlling the quality of new entrants to the professions and enabling them to guarantee the competence of their members, the transfer of professional training from practitioners to universities and colleges became an inevitable process. The move to a formally educated and degree-bearing professional workforce continued slowly and relentlessly, and despite all external events, throughout the 20th century. It affected almost all career choices and, remarkably, has yet to entirely run its course.

The change from practitioner-institution control to a series of more equal institution-university relationships meant a significant power shift away from the professions as a whole that was unenthusiastically accepted. In particular it meant that the institutions lost the ability to vet and restrict who would be allowed to start training as a construction professional. Even if they kept the sanction over the final entry to membership itself and were in practice able to demand ever higher standards of training and curriculum content it meant that membership became open to a much wider demographic as traditional barriers to entry fell away. For the first time women managed to enter the professions as well as those from the working class and an increasingly diverse range of ethnic and national backgrounds.

The institutions were not alone in finding the transition from the hierarchical structure of the late 19th century into the world of mass culture of the 20th difficult. Historian Eric Hobsbawm (1994, p339) summarized the hopes and fears of the people at the end of century as follows:

> The era seemed, for most people in the west, to come closer than any other before to the promise of the century. To its liberal promise, by material improvement, education and culture: to its revolutionary promise, by the emergence, the massed strength and the prospect of the inevitable future triumph of the new labour and socialist movements. For some.

It was a promise only to some in the professions, mainly the younger generation. To others it represented the serious threat to the *status quo* that they had striven to uphold and were dependent on. Education and entry to the professions was a significant battleground in widening access and it would be hard fought over in the years and decades ahead.

Engineers

The turn of the century appeared to be a point of momentous change. The telephone, patented in 1877 and commercially introduced in 1878–9, was rapidly expanding its reach (from two per thousand people in London in 1895 to 22 in 1911).[1] Petrol driven cars first appeared on the roads in 1894–5 and the first mass-produced Ford models built on assembly lines appeared in Britain in 1911. The Lumière Brothers showed their first film in Paris in 1895 and London in 1896, purpose built cinemas followed rapidly and in 1910 newsreels started showing in cinemas worldwide, with the Daily Bioscope in London screening them continuously from its opening in 1909. There would have been no mistaking that the advent of the 20th century presaged unprecedented change in design and construction just as in society and the economy. It was also evident that the technical innovations were coming from elsewhere, from France, Germany and, above all, the United States, but that Britain was underperforming.

Engineering in Britain emerged into the 20th century unprepared for the future. The gentlemanly and anti-intellectual traditions of the industry had prevented any substantial investment in education and training since the emergence of the profession. Even the Samuelson Commission, set up explicitly to address concern that English industry was failing in the face of foreign competition, concluded its report with bland statements of complacency (Samuelson 1884, Part IV, pp506–11):

> Great as the progress of foreign countries, and keen as is their rivalry with us in many important branches, we have no hesitation in stating our conviction, which we believe to be shared by Continental manufacturers themselves that, taking the state of the arts of construction and the staple manufactures as a whole, our people still maintain their position at the head of the industrial world.

At the end of the 19th century there were only four university engineering schools in England and Wales, with a further four in Scotland, collectively producing less than 200 graduates a year. In comparison, Prussia had 13 institutions teaching engineering in 1902. In the United States, in addition to the three specialist Institutes of Technology (Massachusetts, California and Case in Ohio), engineering was taught at almost all universities. Several new universities with engineering courses were formed in England in the early years of the 20th century: the universities of Birmingham (1900), Victoria (Manchester) (1903), Liverpool (1903), Leeds (1904), Sheffield (1905) and Bristol (1909). Some of these were adapted from existing engineering training initiatives, but most were new foundations. The UK was attempting to modernize and produce engineering expertise, but the number of students remained low. In 1908 there were 1 632 students in England and Wales, 20% of them under 20 years old, compared to over 10,000 in Germany and 18 000 in the United States (Albu 1980).

The ICE contributed to the general sense that action was necessary. The Institution held two conferences on the 'Education and Training of Engineers' (1903 and 1911). These consolidated the view that training should be a mixture of college study and practical experience, but otherwise achieved little of note.

The main concern of both events was the importance of practical experience in the training of British engineers, and the 1911 session developed the idea of 'training under agreement', which, for the first time, attempted to control the quality of in-work training (Watson 1988, p160).

The experience of the First World War was more salutary. It revealed systemic weaknesses in the ability of British engineering and provoked fears of post-war German competition. The Government assessed that the institutions were failing to produce the number of engineers required to drive and support the economic wellbeing of country and empire and in 1920 the UK Board of Trade stepped in with the aim of increasing recruitment. For this purpose it developed new qualification standards: the Ordinary National and Higher National Certificates (ONC, HNC). The ONC required three years of part-time study for 16–17 year olds at a technical college and the HNC two more, or 2 + 1 years if studied full-time (Percy 1945, p8). The engineering institutions were, in turn, obliged to recognize the HNC as exempting candidates from their qualifying examinations, sacrificing the quality of their membership for the quantity demanded by government.

Before the war the institutions made the college plus experience route for entering the profession the preferred option with the aim of raising standards to match those of the overseas competition, but in its aftermath the new system was compromised by imposition of much lower standards of entry. An attempt was made to disguise the differences in professional standards in 1922 when the ICE created a new catch-all title for its full members: 'Chartered Engineer'. However, the move only succeeded in devaluing the worth of the new title and in 1927 the Balfour Committee on Trade and Industry confirmed the lack of demand by industry for qualified staff (1927, p186), as did the Percy Report in 1944 (see below).

The available numbers are imprecise but indicate that approximately 700 students were awarded engineering degrees in 1939 against nearly twice that many achieving an HNC. During the Second World War, following an intensive effort to increase the numbers of trained engineers, the numbers in 1943 rose to (Percy 1945, p8):

- Universities etc.   1,250
- Technical colleges   1,438
- Institutions & private training   300.

For comparison, by 1927, 3500 engineering students were graduating each year from Germany's *Technischen Hochschulen*.

The British numbers are from a 1944 Committee Report (published in 1945), chaired by Lord Eustace Percy for the Ministry of Education 'to consider the needs of higher technical education'. The report states unequivocally 'The annual intake … is insufficient both in quality and quantity' (1945, p5). Percy's aim was to find ways of increasing the output of trained engineers while also raising the quality. His target was 3000 graduates per annum, to be maintained for at least ten years and spread equally over civil, mechanical and electrical engineering, not including overseas students, achieved from a 50/50 university/technical college split. Percy showed no appetite for radical solutions to the challenge – there were no plans for German or French *dirigisme* – although he wanted to expand post-graduate study and establish the equivalent of American-style Institutes of Technology. His recommendations, in tune with the times, were for greater central organization and planning, 'a substantial increase in teaching facilities and teaching staff' through the conversion of technical colleges into universities and the establishment of new specialist technology colleges, the provision of incentives to potential students in the form of bursaries and the awarding of degree qualifications by both universities and colleges (although this last proposal led to disagreement from some members of his committee panel, who thought it a step too far).

Percy's recommendations were of a piece with concurrent plans for the expansion of secondary education implemented by the 1944 Butler Education Act and other reports including that of Sir Alan Barlow in 1946 on 'Scientific Manpower'. Barlow envisaged a programme to expand university provision by 85% that would produce a doubling in output of scientists (Barlow 1946, pp7–8). In response the universities almost immediately started expanding and adapting to a much broader idea of higher and university education that included a raft of new technical subjects and courses. Despite this they couldn't keep up with the demand for places and by the late 1950s there was again a shortage of university places for those leaving school with relevant qualifications and the expectation of a university education. The numbers of students entering university grew from 0.8% of 18 year olds in 1900 to 1.7% in 1938 and 4.2% in 1959 (Robbins 1963, p16).

The institutions benefited considerably from the new supply of engineering graduates, with, for example, the ICE seeing the rate of growth of members in the post-war years rise from a steady 5–6% in the 1930s to 15% in 1952 and 22% in 1960. But the increase in the numbers qualifying from universities and colleges had taken place without their involvement. The conferences the institutions organized to discuss engineering education achieved little, with speakers at each event expressing disappointment that the recommendations of the previous one had been ignored. As late as 1972, and against all the evidence, the ICE president-elect, Roger Hetherington, announced at a conference in Glasgow that 'the Institution had not been idle, careful consideration was being given to the discussions and that results would soon be seen' (Watson 1988, p166).

By the early 1960s there were 28 universities and training colleges (counting all the London University colleges as one) as well as ten Colleges of Advanced Technology offering a mixture of Diplomas in Technology and degrees from London University. Of the 18-year-old age group, 4.5% were starting full-time degree courses with another 3.2% entering full-time higher education courses. These numbers were now significantly higher than in Germany (4% / 3%) and the Netherlands (3% / 3%),but were still well below those of France (7% / 2%) and Sweden (10% / 2%). Of the age group, 2.4% was reading science, engineering or agriculture and in this Britain was largely ahead of other countries, with only the Soviet Union (with 4%) in front. There was an overall split across higher education of:

- Humanities   28%
- Pure science   25%

- Medical subjects   15%
- Technology   15%
- Social studies   11%
- Education   4%
- Agriculture   2%

In Lord Robbins's groundbreaking 1963 Report on Higher Education (published the same year as Prime Minister Harold Wilson's 'white heat of technology' speech), he proposed that the number of students in higher education in Britain should be expanded from 216 000 in 1962/3, to 390 000 in 1972/3 and about 560 000 in 1980/1 (Robbins 1963). When his recommendations were accepted and implemented the universities and new polytechnics produced ever larger numbers of engineers.

With the expansion of the higher education system came new pressure on the engineering institutions to raise their standards. In particular, they needed to resolve the issue of entry qualifications and find means of recognizing the more highly educated cohort entering the profession. In 1964, with strong government encouragement, they were persuaded to collaborate through an Engineering Institutions' Joint Council (EIJC) to agree common standards. Lord Hailsham, as the UK government's Minister for Science, urged them to 'proceed with all reasonable speed to raise the academic standard for entry to the profession to that of a degree of a British university' (Chapman and Levy 2004, pp6–7, quoting Dr Wilfred Eastwood in 1981). The engineering institutions bowed to the inevitable and having recognized that 'the profession needed a new image' and 'the individual engineer had a yearning for greater status in the community', agreed to the EIJC recommendations to reorganize the designations in the industry so that all qualified engineers were either Chartered (CEng) (i.e. those with a degree or equivalent and a minimum three years of practical experience), or Technicians (TEng), with lower qualifications and a training largely based on practical experience.

A year later the EIJC, re-constituted as the Council of Engineering Institutions (CEI) (and from 1981 – the Engineering Council), was given the responsibility of running the registers for all Chartered and Technical (later Incorporated) Engineers, including setting the qualification standards right across the various engineering sectors. From 1970 the CEI started administering its own, independent examination system and individual institutional systems were phased out. By the end of 1973 all newly qualified chartered engineers had degrees in engineering and professionalism had taken another long-anticipated step forward.

As a result of the Robbins reforms engineering became embedded in the university system and obtaining an accredited masters degree in engineering the current default means of qualifying as a chartered engineer. The engineering institutions, with Engineering Council oversight, each run a variety of routes to chartered status with the norm being a three-part process of academic qualification, verified work experience and professional review interview. The last part may also include a membership examination (Institution of Structural Engineers) or preparation and submission of a report (ICE, CIBSE) – see Table 7.1.

Table 7.1 Qualification processes for three engineering institutions

	Institution of Civil Engineers (ICE)	Institution of Structural Engineers	Chartered Institution of Building Services Engineers (CIBSE)
Academic qualification	Accredited integrated MEng degree; or international qualification (Washington Accord/FEANI)		

Initial Professional Development (IPD)

	Institution of Civil Engineers (ICE)	Institution of Structural Engineers	Chartered Institution of Building Services Engineers (CIBSE)
Work experience	Structured training programme provided by an ICE approved employer under a supervising civil engineer (SCE); or	Accredited training scheme run by another organization (e.g. ICE); or	Details of employment (Curriculum Vitae); and
	Mentor-supported training self-organized under an ICE approved mentor	Individually managed training under appropriately experienced mentor/s	Organization Chart outlining position in company, level of responsibility and leadership of projects
IPD record	Submission report (up to 5000 words); and	Progress summary records and IDP quarterly reports;	Engineering Practice Report (EPR) (4000–5000 words)
	CPD records	Portfolio of work;	Development Action Plan
		Experience report; and	Sponsor verification
		IPD final reports (mentor-signed)	
Required Attributes / Objectives / Criteria (headlines)	**Attributes** 1. Knowledge and Understanding of Engineering 2. Technical and Practical Application of Engineering 3. Management and Leadership 4. Independent Judgement and Responsibility 5. Commercial Ability 6. Health, Safety and Welfare 7. Sustainable Development 8. Interpersonal Skills and Communication	**Core objectives** 1.1 Knowledge of the Institution and involvement in Institution affairs 1.2 Ability to demonstrate effective communication and interpersonal skills 2.1 Ability to produce viable structural solutions, within the scope of a design brief, taking account of structural stability, durability, aesthetics, sustainability and cost 2.2 Ability to carry out analysis and design of structural forms 2.3 Ability to specify and co-ordinate the use of materials	**Competence criteria** Al Maintain and extend a sound theoretical approach in enabling the introduction and exploitation of new and advancing technology A2 Engage in the creative and innovative development of engineering technology and continuous improvement systems B1 Identify potential projects and opportunities B2 Conduct appropriate research, and undertake design and development of engineering solutions B3 Manage implementation of design solutions, and evaluate their effectiveness C1 Plan for effective project implementation

(continued)

Table 7.1 (Cont.)

	Institution of Civil Engineers (ICE)	Institution of Structural Engineers	Chartered Institution of Building Services Engineers (CIBSE)
	9. Professional Commitment	2.4 Knowledge of relevant environmental, societal, sustainability and economic issues, and associated legislation	C2 Plan, budget, organize, direct and control tasks, people and resources
		2.5 Experience in construction techniques	C3 Lead teams and develop staff to meet changing technical and managerial needs
		3.1 Experience in management skills for programming and control	C4 Bring about continuous improvement through quality management
		3.2 Appreciation of the law and statutory legislation	D1 Communicate in English with others at all levels
		3.3 Experience in health and safety requirements and legislation	D2 Present and discuss proposals
		3.4 Appreciation of commercial **and** financial constraints	D3 Demonstrate personal and social skills
		3.5 Knowledge of procurement routes and forms of contract	E1 Comply with relevant codes of conduct
		3.6 Knowledge of quality systems	E2 Manage and apply safe systems of work
			E3 Undertake engineering activities in a way that contributes to sustainable development
			E4 Carry out and record Continuing Professional Development (CPD) necessary to maintain and enhance competence in own area of practice
			E5 Exercise responsibilities in an ethical manner

Professional Review Interview (PRI)

	With: 'experienced civil engineering professionals'	1 hour with: '2 trained reviewers'	1 hour (inc. 15–20 minute presentation) with 2 interviewers
	Written exercise: Assessment of written skills	Examination	Assessment by CIBSE Registration Panel

Qualification and Title

	CEng MICE	MIStructE CEng	MCIBSE CEng
	Chartered Civil Engineer	Chartered Structural Engineer	Chartered Building Services Engineer

Sources: ice.org, istructe.org and cibse.org

Surveyors

As the engineers moved towards becoming an all-graduate profession, the surveyors felt no urgency about changing from their traditional methods. The qualifying examination instituted in 1891 was considered an adequate test of competence, even if it wasn't regarded as enough to ensure membership of the Institution. Candidates for Fellowship were also required to provide 'proof of responsible experience', a backdoor means of ensuring that they were acceptable to existing members (Rogers 1883, p335, cited in Thompson 1968, p196):

> It is a misapprehension to suppose that a Candidate who has passed ... the Fellowship examination is entitled on reaching the minimum age ... to be transferred to the highest class of membership. It is apparently not clearly understood that the Candidate for transfer must fully satisfy the Council that he has had such experience in practical and responsible work as the public are entitled to look for in a person describing himself as a Fellow of the Institution.

The RICS council had also been keen to emphasize that 'professional education should be left to voluntary efforts, and if organized, should be organized by such technical schools as might chance to come into existence and chance to offer appropriate courses' (*Council Minutes*, V. 12th December 1904, quoted in Thompson 1968, p215).

The views of William Sturge on university education, as expressed in his paper 'The Education of the Surveyor' (already extensively quoted from in Chapter 5), appeared to have held sway in the profession until at least the 1960s (1868, pp50–1):

> Several of the most precious years of the young man's life are spent in an education by no means specially adapted to his future profession, years which he can ill spare. If he acquires the manners and tastes of gentlemen, he may also acquire the desultory and expensive habits, if not the vices, of too many of his associates; and instead of reading he may waste his time in frivolity and dissipation. The tastes and habits he will form will probably render the drudgery of a surveyor's office peculiarly distasteful to him. I arrive at the conclusion that the balance is against an university education for the surveyor.

Reporting in 1960, the Wells Committee Report on the Educational Policy of the RICS finally expressed the hope that 'the traditional method of attending evening classes (or depending on correspondence courses) will become less dominant than it has been in the past'. The numbers bear out the preponderance of the traditional approach: in 1928 just 1% of surveyors held a vocational degree, in 1955, 10% and 1963, 12%. With the profession rapidly growing in size over the same period this meant that the numbers applying for entry outside the university route increased from 2400 in 1939 to 10 000 in 1967 (Thompson 1968, p227).

The Institution had tweaked its qualifying examinations a number of times in the early part of the century but didn't have a systematic review until 1939, when the Healing Report led to the rationalizing of the three examination levels and a requirement for a period of practical experience prior to the Final Examination of which, 'two years at least must have been spent in the office of a surveyor approved by the Council'. Healing was followed by Watson in 1950 who recommended the development of full-time courses in surveying in technical colleges with their graduates receiving exemption from the Intermediate examination. In the event this measure was not substantial enough to tempt students in any great numbers, although it did lead to the availability of courses ready for the reforms recommended in the Wells Report. Its Chairman, Henry Weston Wells (1911–71), was the son of William Henry Wells (1871–1933) who had been instrumental in setting up the College of Estate Management as the President of the Charter Auctioneers' and Estate Agents Institute in a three-way collaboration with the Land Agents Society and, in a lesser role, the RICS. The College was born out of a takeover of Parry, Blake and Parry (see Chapter 5) in 1919 and inherited their commanding role in preparing candidates for the qualifying examination. It ran a mixture of correspond-ence courses and evening classes with just a handful of full-time students.[2] In the post-war years, with the ability to award London University degrees and encouraged by the Watson Report, the College expanded the number of full-time students from 50 in 1945 to 450 in 1967, although it continued, as it does to this day, to provide the majority of its tuition through distance learning.

It was a system of recruitment that took the bulk of its intake from those leaving full-time education aged 15 or 16. Fledgling surveyors, having decided for whatever reason against further schooling, were expected to continue training in their own time and at their own expense whilst working full-time. The system had maintained the old apprenticeship-based structure, even if it had the trappings of college education and examinations, and it was rapidly becoming unattractive to new entrants deemed suitable. The Crowther Report for the Ministry of Education (1959, p202) had noted:

> many professions which have been accustomed to recruit boys of real aca-demic ability at 16 will find that they can no longer do so at that age, because pupils of the required ability will not give up the chance of a Sixth Form education,

and surveying was far from immune to this trend. With educational opportun-ities expanding elsewhere during the post-war period it was both becoming more difficult to find applicants of sufficient quality for the needs of the pro-fession and the profession's leaders were becoming concerned that long-term damage would be caused by having to 'accept poorer intellectual material' (Wells 1960).

As a result the Wells Committee turned the concerns expressed in previous reviews about the lack of surveying experience on their head, recommending that in addition to the required practical knowledge of surveying 'no person shall be eligible to become a chartered surveyor unless he possesses the necessary academic qualifications', while still precluding 'any possibility of making Advance [18+] level passes compulsory'. Wells recognized that 'it is a matter of concern that, compared with other professions, surveying has not made a greater impact on university life and thought in general' and looked to the profession developing a 'post-graduate research activity on university lines' (Wells 1960). Fortunately the profession, unwilling or unable to act by itself, would not have long to wait for the opportunity for the university system to engage with it.

As with the engineers it was the Robbins Report of 1963 and the subsequent opening up of the university and polytechnic system that transformed the opportunities for surveyors to enter the profession as a result a full-time education and for the discipline to develop academic research opportunities. As a result the next RICS education review chaired by Hugh Eve (1906–69), like H.W. Wells from a dynasty of surveyors and also a past-President of the College of Estate Management, in 1967, was able to impose a far more radical position on the profession, requiring an entry standard of two A-levels, effectively requiring schooling until 18 plus, ensuring that surveying was taught and studied in the new polytechnics, with 'centres of excellence' for surveying and envisaging that 'in time the normal method of training for the professional examinations or for exempting qualifications will be by way of full time, or sandwich and study' (Eve 1967).

The reforms of the 1960s led, as Wells and Eve predicted, to a very different professional workforce, with the balance of new entrants rapidly shifting to being largely college-educated. In the wake of Eve's recommendations 13 'Centres of Excellence' were established, monitored by the Council for National Academic Awards (CNAA) and by 1978 the number of students taking all stages of the Institution's qualifying examinations had dropped to 4991 from 8639 in 1967 (Schultz 2010). By 1994 the Institution was almost entirely relying on exempt courses to qualify applicants and had effectively abandoned its own examining role (Plimmer 2003, p3).

The change in educational approach allowed the Institution to significantly upgrade its in-house final entrance procedure in 1973 when it introduced the Test of Professional Competence (TPC) to replace the 'proof of responsible experience'. This later became the Assessment of Professional Competence (APC) and remains the final hurdle to be crossed to achieve full Professional Chartered Status (see Table 7.2).

The APC tests a surveyor's competence with different versions of each of the RICS's 22 areas of practice (grouped into eight land, eight property and six construction areas) and each capable of being further interpreted to relate to particular specialisms and geographical contexts. Construction's six disciplines are: Building control, Building surveying, Built infrastructure, Project management, Taxation allowance and Quantity Surveying and Construction. Pathways through the APC vary depending on level of previous experience and whether the candidate is aiming at Associate or Chartered status. The process of taking the APC involves taking an online ethics module; a period (either one or two years depending on length of experience) of structured training, during which the candidate keeps a log-book recording their training; a document that summarizes their experience and that demonstrates acquisition of a series of pre-defined competences (mandatory, core and a selection from an 'optional' list); a case study of a project they have worked on; and a

Table 7.2 Competencies for RICS quantity surveying and construction APC

Mandatory competencies
- Conduct rules, ethics and professional practice
- Client care
- Communication and negotiation
- Health and safety Level 1
- Accounting principles and procedures
- Business planning
- Conflict avoidance, management and dispute resolution procedures
- Data management
- Sustainability
- Team working

Core competencies
- Commercial management of construction or Design economics and cost planning
- Contract practice
- Construction technology and environmental services
- Procurement and tendering
- Project financial control and reporting
- Quantification and costing of construction works

Optional competencies
- Building information modelling (BIM) management
- Capital allowances
- Commercial management of construction or Design economics and cost planning (whichever is not selected as a core competency)
- Contract administration
- Corporate recovery and insolvency
- Due diligence
- Insurance
- Programming and planning
- Project evaluation
- Risk management
- Conflict avoidance, management and dispute resolution procedures or Sustainability.

Source: RICS (2017a, p9).

final assessment interview with an RICS appointed assessment committee (RICS, 2017b, p19):

> The interview will last approximately one hour and is designed to determine whether you:
>
> 1. can express yourself clearly in an oral presentation and interview
> 2. can demonstrate, in support of your written submissions, your understanding of the knowledge gained and competencies achieved during your training
> 3. have an acceptable understanding of the role and responsibilities of a chartered surveyor

4. can apply your professional and technical skills to benefit those who employ your services.

The APC is intentionally tough, but also keeps control of entry into the profession firmly with the RICS. The various accredited degree courses that provide the route into the APC are subordinate to the extensive curriculum set for the APC itself. The dominance of the APC has also meant that the Institution has been able to attach a high degree of importance to ethics and professional standards. It is these aspects, rather than technical skill, that have become the cornerstone of the qualification process.

Increasingly ambitious, by the late 1990s the RICS had decided to stake its claim to a global role and launched its 'Agenda for Change' in 1998 under the presidency of Richard Lay. This envisaged significant and controversial structural changes to the Institution but also a strategy involving 'elevating the standing of chartered surveyors'. The APC may have been enough to satisfy existing members that new entrants were sufficiently qualified but for the envisaged external audience it was felt that the Institution lacked sufficiently clear and transparent entrance criteria. As part of the Agenda for Change a new RICS Education Task Force was established to enable a graduate entry system that could be delivered by across the world by a range of partner organizations. The Task Force reported back in 1999 with 'Investing in Change', a vision for 2010 (RICS 2008, p6), summarized as:

- a clear attractive image of the career opportunities offered by the RICS qualification at professional and technical levels;
- access to RICS membership for the best graduates in all disciplines;
- strong partnership between RICS and a limited number of recognized centres of academic excellence throughout the world characterized by:

    i.   highly competitive entry to courses at undergraduate and postgraduate levels;
    ii. an appropriate range of curricula at undergraduate and postgraduate level;
    iii. excellent teaching faculties working closely with practice;
    iv. international standards of research by focused teams;
    v.   increased freedom for selected universities to develop courses and methods of delivery at undergraduate, postgraduate and post qualification levels; and
    vi. attractive courses promoted by RICS.

To achieve this the Institution has had to relinquish jurisdiction over both courses and content, as it notes: 'RICS has devolved much of what was previous controlled centrally to a series of individual partnerships with its approved universities' (RICS 2008, p6). It travelled from a nearly non-existent academic base in 1960 to a high performing educational infrastructure operating across the world in half a century. New members are all degree-holders or above and the Institution has speculated about increasing this from a Bachelor's to a Master's degree. It has been a very rapid journey indeed.

Architects

The support of the RIBA for the new architecture schools that began to be established in the 1890s meant that the profession entered the 20th century with a firm base within the expanding university and art school system, even though there was not yet a standard approach to architectural education. In 1902 the RIBA offered probationers who obtained first class certificates from Liverpool School of Architecture exemption from the RIBA's Intermediate examination, and in 1906 the same exemption was granted to students who passed the four-year course at the Architectural Association in London (comprising two years of evening classes and two years of the day-time course).

The RIBA's involvement with these courses led in 1904 to the Institute establishing a 16-strong Board of Architectural Education (BAE) to maintain a watching brief over the schools and to develop the RIBA's policy on education. This meant from the outset that the Institute developed a strong relationship with the schools and was in a position both to impose a standard syllabus and dictate a four-year course length, including two years spent in an architect's office. The BAE's role was to act as an advisory body to the schools and to organize visiting inspectors to regularly assess and report back on each school. It was a strong and directive system that expressed a desire to allow schools freedom of action, but in practice preferred uniformity. To its credit the RIBA made sure that the BAE included a range of voices, including four signatories to the 1891 letter to *The Times* attacking the idea of examinations and diplomas. Others, including representatives from the Royal Academy and the architecture schools themselves, acted as advisors.

The grip of the BAE on architectural education ensured that non-recognized schools rapidly came into line and submitted to the Board's criticism of 'their isolation and want of co-operation'. The BAE's influence spread to include schools across the British Dominions (Kaye 1960, p158). By 1931 27 schools were recognized for exemption from either the Intermediate or both Intermediate and Final examinations, including one school in Australia, three in Canada, two in South Africa and one in India. They were inspected for their compliance with the syllabus and other edicts laid down by the BAE. As the system expanded the Board developed additional roles, including responsibility for the Examinations themselves in 1910 as well as prizes and scholarships. In 1925 the RIBA Visiting Board was established under the BAE's aegis along with five other committees devoted to running what had become an elaborate system of control (Mace 1986, p115).

Prior to the BAE's founding, the only established school of architecture,[3] the Liverpool School of Architecture and Applied Art under its first director, Frederick Simpson (1855–1928, Director: 1894–1904), had already adopted the *Beaux-Arts* model of training, imported from France – via America. The decision to embrace *Beaux-Arts* ideas may have been influenced by the city's strong trans-Atlantic trade links but it was also the result of the on-going confrontation regarding the style of the city's new Anglican Cathedral. The Church of England, under Ecclesiological influence, had imposed gothic but the city and university preferred a classical design (Crouch 2002, p38). Sides having been taken, there was no doubt which way the school of architecture would develop (Simpson 1895, p14, cited in Richmond 2001, p28):

They [the French] still have a vernacular style in which their architects and workmen are trained, and although it may not be a perfect one, it is surely better than 'the babel of tongues' which exists in England at the moment.

The *Beaux-Arts* approach, which would be maintained and heavily promoted by Simpson's successor, Charles Reilly (1874–1948, Director: 1904–33), included the system of *ateliers* run by individual 'masters' originally pioneered in Paris. This approach appealed strongly to the BAE members who saw it as a means of tempering the criticism of the Institute's examination system and its difficulty in dealing with issues of design and artistic value. The *atelier* system was half way between pupillage and academic training and allowed for unanimity across the Board. The American version of the *Beaux-Arts* system would become the model for architecture training in Britain from then on. This was critical to the RIBA's purpose of dispensing with the old, discredited pupillage system, even if the gradual progress on this front was not fast enough for some (Webb 1924, p586):

> The competition with the old system of pupillage and the hold which this somewhat haphazard system of training still retains in every part of the country except London and Liverpool is acting as a heavy drag upon the schools.

Confident that Britain had managed to catch up with international practice in architectural education and had something to show that it could be proud of, the RIBA participated in a Franco-British conference on architectural education in Paris in 1920 and then organized a remarkable six day conference in London in 1924, the First International Congress on Architectural Education, to coincide with the British Empire Exhibition being held that year at Wembley. Delegates from 20 countries gathered to discuss architectural education in the past, present and future. Forty-nine architecture schools from 12 countries exhibited their students' work at venues across London.

Held in a period of optimism just before the great crash, speakers described their systems in detail, generally referencing the debt owed to the *Beaux-Arts* in France before going on to describe their own priorities, which tended towards mastery of design backed by a good grasp of construction techniques. Discussions were held about the importance of art and presentation techniques, for and against construction knowledge and experience and the best moment to introduce the problematic renaissance style to students. Only on the third day of the Congress was the topic of professional purpose in architectural education raised. This came at the end of a letter from William Lethaby (1857–1931), Professor of Design at the Royal College of Art, read out at the event (RIBA 1925, p74):

> I wish I could suggest a motive strong enough to be a counter-attraction to the distinctions and prizes that are to be won by designing draughtsmanship 'architecture' in the air. All I can point to is the reason of the thing and public service; the architect who can improve ordinary buildings – for example, the small dwelling house – will be a benefactor to his country.

Lethaby's quiet point was ignored by most subsequent speakers but was addressed in a contribution from Arthur Beresford Pite (1861–1934), Professor of Architecture at Cambridge, who hesitantly noted that a group of architects had been meeting 'to consider enunciating an ideal for architectural education by creating a path for thought in architectural progress' (p80). It was only a small thread in a much broader discussion, but it suggested that there was already an undercurrent of dissatisfaction amongst students that would only become public in the following decade as the idea of architecture with a social purpose took hold (see Chapter 6).

The confidence of the RIBA at this time was such that it believed that it could take command over the entire profession in Britain and beyond. In 1925 the Institute drafted a clause in an agreement with the Society of Architects baldly stating its intention that 'the Royal Institute shall forthwith Promote and use its best endeavours to carry through a Registration Bill until it becomes the Act of Parliament' (Kaye 1960, p152). Registration was intended to close the profession and restrict the use of the title 'architect' to those who had qualified through the RIBA's system of recognized schools and examinations. Progress through this system would become the only route to join the legally protected register, which the RIBA also intended to manage. The Bill was supported by members of Parliament when it was introduced into the legislative process in 1927 on the grounds that it would protect the public against unqualified practitioners, encourage the production of beautiful architecture and because, curiously, it would strengthen the case of registration for dockworkers. Opposition focused on the obstacles it would put in the way of working class children entering the profession. The Parliamentary Bill failed to make its way through the select committee process, was shelved and only re-emerged, modified by MPs, in 1931 when the Standing Committee charged with its progress through Parliament decided (Kaye 1960, pp152–3):

> no single body should be in charge of the Register, and that the Registration body should not be a solely professional body.

It was a devastating result for the RIBA. Not only did the Act, when it was passed, establish a separate Architects' Registration Council of the United Kingdom (ARCUK) it also gave the Institute limited representation on the 75-strong Council alongside a long list of other organizations, including newly established societies and associated professional bodies.[4] ARCUK immediately convened an alternative Board of Architectural Education and introduced a raft of rules and regulations affecting architects (*Architects (Registration) Act 1931*). The RIBA had unwittingly managed to create a rival body, lost control of the status of its own examinations, seeded distrust throughout the industry and had failed to close access to the profession. The wording of the Act was flawed, only the term 'registered architect' was protected and there was nothing to stop unregistered persons from describing themselves as architects. The Act would need to be revisited and revised, and in 1938 a new Act was passed to protect the title 'Architect' (with the exception of naval, landscape and golf-course architects) (Mace 1986, p42). Having ridden so high in the 1920s, by the end of the 1930s the Institute had become deeply associated with intrigue and unscrupulousness. It would do little for professionalism and only complicated the already drawn out process of accreditation for newly qualified architects. The division between professional and registration bodies continues uncomfortably to this day.

In the debate in Parliament on the revised Bill in December 1937 the RIBA was shown little mercy, with the Institute being accused of disreputable conduct and of using any weapon to belittle people who opposed it (*House of Commons Official Report*, 17th December 1937).[5]

'The Bill is a sham. It has been brought before the House with the insinuation that it will do something to clean up building design in the countryside. It does nothing of the sort. It is an undignified attempt to try and smooth down the quarrels which exist in the profession.' I. Orr-Ewing MP

'I see absolutely no reason why this House should give the members of this association the right to put in gaol people who compete with them.' J. Wedgwood MP

Clearly the Institute, and with it the professions generally, did not command the respect that they had anticipated and considered their due. Although the Bill was passed into law there was no support, either then or later, for making the use of an architect mandatory when erecting a building. In part this may have reflected a question raised by several speakers during the debate: was the level of qualification under debate sufficient to warrant even the protection of title? (House of Commons Official Report, 17th December 1937):

'The facts which have been supplied to me show that in the examinations of the Royal Institute of British Architects, out of 215 successful candidates in 1936, 128 trained themselves by attending evening schools, while the remainder, 87, studied at home or through correspondence classes.' George Hicks MP

Despite demanding examinations and the growing role of universities in architectural education the profession had not managed to define itself as being recognizably and adequately educated for its role. This was hardly surprising. Although pupillage and apprenticeship had declined by 1943, the number of non-recognized courses and schools that fed into the RIBA's examination system had expanded vastly and, as the figures cited in the House of Commons in 1937 illustrate, this formed a substantial part of recruitment to the RIBA. The number of corporate members of the Institute rose from 8218 in 1938 to 10 706 in 1948 and 18 175 in 1957. By that time there were 3764 students at 17 recognized schools preparing for the Intermediate and Final examinations and a nearly equivalent number of 3342 at 41 non-recognized schools, together with a further unknown number preparing via correspondence and other courses. It wasn't easy for students not receiving course exemptions to pass the RIBA's examinations, which only had a 40% pass rate, but 90% of candidates, after repeated attempts, managed to do so (Martin 1958).

In April 1958 an RIBA Conference on Architectural Education was held at Magdalen College Oxford to address what had become a pressing issue for the profession. It was the first such event organized by the Institute since the 1924 International Congress. Despite its eventual significance and impact on architectural training the Oxford Conference was a small event with only 50 RIBA members and a handful of overseas visitors attending. According to the report prepared after the event by the conference Chairman, the architect Sir Leslie Martin (1908–2000), the 'ultimate purpose' was 'that the profession should attempt to improve its standards of competence at all levels', and after three days' discussion a series of recommendations was agreed (Martin 1958, p8):

1. The Conference unanimously agreed that the present minimum standard of entry into training (5 passes at 'O' level) is far too low and urged that this level should be raised to a minimum of 2 passes at 'A' level. The Conference agreed that courses based on Testimonies of Study and the RIBA External Examinations are restricting to the development of a full training for the architect and that these courses should be progressively abolished.
2. Ultimately, all Schools capable of providing the high standard of training envisaged for the architect should be 'recognised' and situated in Universities or Institutions where courses of comparable standard can be conducted.
3. Courses followed by students intending to qualify as architects should be either full-time or, on an experimental basis, combined or sandwich courses in which periods of training in a school alternate with periods of training in an office.
4. It may be that these raised standards of education for the architect will make desirable other forms of training not leading to an architectural qualification, but which will provide an opportunity for transfer if the necessary educational standard is obtained.
5. The Conference regards post-graduate work as an essential part of architectural education. It endorses the policy of developing post-graduate courses which will enlarge the range of specialised knowledge, and will advance the standards of teaching and practice.

The reform of architectural education that was put in train by the Oxford Conference moved at such a pace that within ten years 90% of students studying architecture were enrolled on degree level courses and the standard combination of five years spent in higher education and two years of practical experience was firmly established (Mace 1986, p117). The reforms would be further reinforced after implementation of the changes to higher education institutions that followed the Robbins Report in 1963, allowing the new universities and colleges to absorb previously independent schools.

By the mid-1960s the education of engineers and architects had come roughly to the same place, but by very different routes. Their separate journeys had been triggered by the examinations that had been established by their respective institutions at the end of the 19th century but while engineering had been in the hands of academics before eventually needing to be brought to order by government fiat, architecture had remained largely under the control of the profession and only became a predominantly academic subject much later in the day. This, as much as the differences in subject matter, accounted for the very different approaches that the disciplines adopted as their standard *modus operandi*. Engineering at university level developed in this period as an academic subject allied to research, but failed to engage adequately with industry and produce the number of graduates required, forcing government to step in. Architecture maintained its links with practice, but failed to either offer its students adequate training or engage in research, an activity left to external bodies like the Building Research Station and the Ministry for Education (see Chapter 6). Even within a university environment, architectural education found a way in the studio/*atelier* system to set itself apart and maintain a detached exceptionalism that kept its spirit of creativity and self-belief intact, even while it omitted to look after its knowledge base.

In this context it was no surprise that the Oxford Conference, having been presented with a paper by Richard Llewelyn-Davies (1912–81), 'Deeper Knowledge: Better Design', added 'enlarging the range of specialised knowledge' to its list of urgent concerns. Leslie Martin, in summing up, paraphrased the paper in his conference report (Martin 1958):

> Knowledge is the raw material for design. It is not a substitute for architectural imagination: but it is necessary for the effective exercise of imagination and skill in design. Inadequate knowledge handicaps and trammels the architect, limits the achievements of even the most creative and depresses the general level of design.

The paper was a call to restructure both training and practice, to ditch the old *Beaux Arts* approach, which he claimed 'few will nowaday defend' in favour of a *Bauhaus* approach, and to link architecture not only with industrial production but also the social sciences. He wanted to make university architecture departments centres of research with 'research architects', who also needed to be working with practising architects, contributing to teaching along the lines of medical education (Llewelyn-Davies 1958):

> We shall need a change of emphasis in the training of an architectural student. This must now have the object of giving him a broad grasp of the whole field of knowledge, and of teaching those attitudes and methods of work, already developed in the sciences, whereby the details of a subject can be fairly quickly learnt, so long as its essential principles have been understood.

Within two years Llewelyn-Davies was appointed Professor of Architecture at the Bartlett School of Architecture, University College London and would implement his programme, 'Education of an Architect' (Llewelyn-Davies 1960), at one of London's most influential schools. As promised the programme rigorously promoted a science-based research agenda to consciously develop an objective 'body of knowledge'. In doing so he alienated a body of architects who felt that a substantial part of the art of architecture had gone missing. Architect Richard MacCormac (1938–2014) recalled that in the intellectual environment of the Bartlett at this time 'there had to be a piece of objective research – a PhD to ensure a factual basis to underpin such a proposition – rather than a resort to human experience, to what architecture and the arts are about' (MacCormac 2005, p51). The emphasis on research-led education did not last long.

In retrospect there was little that should have concerned MacCormac, as he surely knew when he wrote his 2005 essay, as the system rapidly settled down to a mixture of encouraging practical knowledge, combined with artistic expression, that spread rapidly across the UK and beyond. The intention of integrating research into the curriculum and creating research-literate architects was not realized. The agenda was relegated to a relatively few PhD students and a handful of 'research units', if it was continued at all. The *Beaux Arts* ideas of design and composition may have been abandoned in favour of a mild version of the *Bauhaus* approach but its method of studio teaching led by an *atelier* master, having been temporarily weakened, gradually regained its strength and became almost universal practice. The Oxford Conference had been highly successful in both modernizing architectural education and establishing a normative standard, a standard that continued to be policed by RIBA visiting boards, later joined by ARCUK/ARB. If it failed to inculcate a thoroughgoing research culture in schools of architecture it had at least introduced the idea; an idea that would gradually take root.

Although they have become firmly entrenched as the *de facto* system of architectural education over the past 60 years the changes brought in by the Oxford Conference reforms are still not universally welcomed. Architecture and universities do not straightforwardly fit together and as Allen Cunningham noted in a paper in discussing 'Oxford's dangerous legacies' (1999):

> The discouragement of part-time education and the 'burden of pupillage' eliminated any obligation for practitioners to serve as masters to guide aspirants across the threshold 'when play becomes real' – as do lawyers, doctors, musicians, sports people and fine artists – and therefore placed the onus upon academics to produce 'complete' architects, a fatuous expectation which endures.

Oxford ensured that the universities and polytechnics would take over responsibility for delivering architecture training, although the subject was never turned into a fully academic discipline concerned with both the expansion and inculcation of knowledge. Only a few schools, predominantly the Architectural Association and others located in Schools of Art, remained outside the system. In practice the universities, as they all eventually became, also took full control over the examinations and professional qualifications, leaving the RIBA and ARB with the more strategic roles of setting the curriculum and validating individual schools.

After Oxford the standard length of architectural training was formalized into three distinct parts: Part 1 – a three year 'degree' course, Part 2 – an advanced 'Diploma' course of two years and Part 3 – a minimum of two years practical experience in an architect's office with final qualifying examination in professional practice at the end. Variations on this basic pattern, in the form of sandwich and part-time courses, were also developed. All three of these elements are run by educational, mainly university, bodies with the qualifications being universally recognized and with students free to transfer between them on completion of each part.

At Part 3 candidates are assessed on:

- 24 months of practical experience together with an official record of experience
- Their professional career and CV
- A written case study
- A written examination

- A final oral examination.

Passing the examination allows successful candidates both to register with ARCUK/ARB and become a Chartered Member of the RIBA.

Although architecture had developed a fairly robust system for qualification in 1985 along with a handful of other professions including nurses, dentists, veterinary surgeons, midwifes, pharmacists and doctors, architects became subject to a European Economic Community (EEC) Council directive on the mutual recognition of qualifications in architecture (EEC 1985, L223/18) that added another layer of regulation and control. The Directive specified that training should comprise a total of at least four years of full-time study or six years part-time. The course length was upgraded most recently in 2013 to:

a) a total of at least five years of full-time study at a university or a comparable teaching institution, leading to successful completion of a university-level examination; or

b) not less than four years of full-time study at a university or a comparable teaching institution leading to successful completion of a university-level examination, accompanied by a certificate attesting to the completion of two years of professional traineeship.

The content of an architect's training is also specified (EU 2013, L354/157–8):

The study shall maintain a balance between theoretical and practical aspects of architectural training and shall guarantee at least the acquisition of the following knowledge, skills and competences:

a) the ability to create architectural designs that satisfy both aesthetic and technical requirements;

b) adequate knowledge of the history and theories of architecture and the related arts, technologies and human sciences;

c) knowledge of the fine arts as an influence on the quality of architectural design;

d) adequate knowledge of urban design, planning and the skills involved in the planning process;

e) understanding of the relationship between people and buildings, and between buildings and their environment, and of the need to relate buildings and the spaces between them to human needs and scale;

f) understanding of the profession of architect and the role of the architect in society, in particular in preparing briefs that take account of social factors;

g) understanding of the methods of investigation and preparation of the brief for a design project;

h) understanding of the structural design, and constructional and engineering problems associated with building design;

i) adequate knowledge of physical problems and technologies and of the function of buildings so as to provide them with internal conditions of comfort and protection against the climate, in the framework of sustainable development;

j) the necessary design skills to meet building users' requirements within the constraints imposed by cost factors and building regulations;

k) adequate knowledge of the industries, organisations, regulations and procedures involved in translating design concepts into buildings and integrating plans into overall planning.

Although the UK's training system is compliant with the EU definition of training as an architect it has allowed architects from elsewhere in the EU with a minimum of five years training before qualification also to be recognized as architects by the ARB. This is likely to change again as and when the UK leaves the EU. While recognizing this alternative route to qualification both the ARB and RIBA have continued to work with the full three-part system and the RIBA requires the seven years minimum qualification period before granting Chartered Architect status and membership of the Institute.

Although the ARB General Criteria specify that students at the end of Part 2 of training will have an 'understanding of the profession of architecture and the role of the architect in society, in particular in preparing briefs that take account of social factors', most aspects of training for professional practice are dealt with within Part 3, where the ARB demands a longer list of 'understandings', each with a substantial list of sub-clauses (ARB 2010):

PC1 Professionalism
PC2 Clients, users and delivery of services
PC3 Legal framework and processes
PC4 Practice and management
PC5 Building procurement.

The PC1 Professionalism criterion has the following sub-clauses:

1.1   Professional ethics
1.2   The architect's obligation to society and the protection of the environment
1.3   Professional regulation, conduct and discipline
1.4   Institutional membership, benefits, obligations and codes of conduct
1.5   Attributes of integrity, impartiality, reliability and courtesy
1.6   Time management, recording, planning and review
1.7   Effective communication, presentation, confirmation and recording
1.8   Flexibility, adaptability and the principles of negotiation
1.9   Autonomous working and taking responsibility within a practice context
1.10 Continuing professional development.

The Oxford Conference firmly moved architectural training to degree status, with the majority of architects gaining a triple qualification from the three-part process including two academic qualifications. The EU criteria (see above) specify full-time study at a university or a comparable teaching institution. Yet the UK maintains a route around this in the form of the RIBA Studio, the successor to the RIBA Examination for office-based candidates. This allows independent trainees working full-time in architectural practice to gain Part 1 and Part 2 qualifications. The Part 3 qualification is also available outside the university system. The numbers taking the RIBA Studio route in any year are small but architecture remains the last of the traditional construction professions not to be a fully degree-holding profession.

Diversity

The transfer of responsibility for education from practices and institutions to universities and colleges, especially when state funded, encouraged a much wider set of entrants to seek access into the professions including women, ethnic minorities and many students from families and communities who had never before sent anyone to university or considered a professional career was possible. A route through education with selection largely based on merit (rather than on perceived compatibility with the existing culture) allowed them to start the journey towards professional status, to develop confidence, proceed with persistence (often against the prejudices of existing professionals) and to break in to one of the closed worlds of architects, engineers and surveyors.

The involvement, particularly of women, in professional roles in construction has a long history with individuals acting as architects and engineers and to a lesser extent, surveyors, from an early date. Feminist histories have retrieved the names of pioneer women architects such as Katherine Briçonnet (c1494–1626) who oversaw the building of the Chateau of Chenonceau[6] and Elizabeth Wilbraham (1632–1705) who has two country houses, Wotton House and Weston Hall, somewhat uncertainly, attributed to her (Millar 2010; Harwood 2014a). The 18th century Cumbrian architect, Sarah Losh, was discussed in Chapter 5. Early engineers include the bridge designer Sarah Guppy (1770–1852) and electrical engineer Hertha Ayrton (1854–1923) who had studied mathematics and physics at Cambridge University, while engineering and surveying share Alice Perry (1885–1969) who took an engineering degree at Royal University Galway and briefly became County Surveyor in Galway West (IEI 2000). But despite these individuals female construction professionals remained extremely rare even after a number of pioneering women managed to gain entry around the beginning of the 20th century and particularly in the aftermath of the First World War.

Two sisters, Ethel and Betty Charles (1871–1962 and 1869–1932), were the first and second women members of the RIBA. They had both done three years apprenticeship in the London offices of Ernest George and Harold Peto, but although barred from studying at the Architectural Association, they had managed to enrol on the university extension course of the Bartlett School of Architecture at UCL, graduating with distinctions. In June 1898 Ethel then took and passed the RIBA examination for associate membership (Walker n.d.; women in architecture 2012). At that point the RIBA took notice and 'the candidature of Miss Charles for the Associateship gave rise to a long discussion' at their December Council meeting. In practice a small number of Council members tried to block her election, claiming in the words of W. Hilton Nash that 'it would be prejudicial to the interests of the Institute to elect a lady member', but the far greater number of members agreed with H. Heathcote Statham when he said (*Journal of the Royal Institute of British Architects* 1898, p78):

The Council might have elected a lady member any time these 50 years. The new departure was that for the first time a lady had had the spirit and ability to go through the examinations. ... Considering the general public feeling now, it would seem that the profession should be open to women, and it would be the greatest possible mistake for the Institute to oppose the election.

The RIBA Council elected Ethel Charles to membership with 51 for and 16 against. Bessie Charles was elected to membership two years later in 1900. In 1905 Ethel Charles was awarded the RIBA's Silver Medal for Architecture and also won an international competition for a church in Germany, but the two sisters never managed to win any large-scale projects to implement and they went into practice together in Cornwall and worked largely on domestic projects. Between them the Charleses had made a momentous breakthrough, but as the historian Lynne Walker records, following their admission to the RIBA 'the number of women architects actually fell significantly, from nineteen in England and Wales in 1891 to six in 1901 and seven in 1911' (Walker n.d., p3). While other schools of architecture started admitting women, with the Glasgow School of Art around 1905 and University of Manchester in 1909, it would take the exigencies of a wartime economy to start making a change to women's real employment opportunities.

Women were actively encouraged during the 1914–18 war to fill civilian jobs while men were drafted into the armed forces, producing 'a revolution in the employment of women outside the household', if a temporary one once the men were demobilized (Hobsbawm 1995, p44). In 1918, in recognition of their contribution, various groupings of women were given the vote in Britain for the first time,[7] and the following year under pressure from the Labour Party the Government enacted the Sex Disqualification (Removal) Act 1919 to clear away various other restrictions. The Act had specific clauses opening up the civil service, judiciary (including the ability to serve on a jury) and universities to women, but its main provision was plainly worded, and sweeping (HM Government 1919):

A person shall not be disqualified by sex or marriage from the exercise of any public function, or from being appointed to or holding any civil or judicial office or post, or from entering or assuming or carrying on any civil profession or vocation, or for admission to any incorporated society (whether incorporated by Royal Charter or otherwise).

In practice the Act was not tested in court until a case concerned with a racehorse trainer in 1966 (Bennion 1979). The Institutions were then obliged by law to allow women into their ranks. They didn't act with any alacrity.

The RICS had been rebuffing potential female members since at least 1899 when the Birkbeck Institution requested permission for a Miss Beatrice Stapleton to enter their examinations. It did the same in 1915 to an application from Miss M.V. Smith, in both instances claiming that the constitution wouldn't allow it (even through it presented no actual barrier). In 1921, in response to the 1919 Act, the RICS passed an unnecessary amendment to its by-laws formally opening membership to both sexes (Thompson 1968, p139). A year later the institution qualified its first female Professional Associate, Irene Barclay (1894–1989). She was made a fellow in 1931. Barclay had taken a degree in history and a diploma in social science at Bedford College, University of London, with the latter taking her into housing management. Emboldened by the 1919 Act she took the evening classes, alongside her friend Evelyn Perry (?–1976), to qualify as a surveyor. Perry was elected to the RICS the following year (and became a fellow in 1937). The two women set up in partnership together as Barclay and Perry in Somers Town, London, with Irene Barclay continuing to work with numerous London housing associations until her retirement in 1972 (RICS 2017d; Barclay 1976, p15; de Silva 2017, p24).

The engineers, following their wartime experiences, were more organized and in 1919, reacting to the pressure to relinquish their newfound responsibilities, a group including Katherine and Rachel Parsons, Verana Holmes and Margaret Partridge with Caroline Haslett as their first secretary, founded the Women's Engineering Society (WES) (Hatfield 2005, p4):

> to promote the study and practice of engineering among women; and, secondly to enable technical women to meet and to facilitate the exchange of ideas respecting the interests, training and employment of technical women and the publication and communication of information on such subjects.

Despite this early lead the ICE took until 1927 to elect its first female member, Dorothy Donaldson Buchanan (1899–1985), a bridge designer trained at Edinburgh University who worked on the Sydney Harbour, Tyne and Lambeth bridges. It took until 1957 for them to elect their first woman fellow, Mary Ferguson (1914–97), also a graduate of Edinburgh and a bridge designer, 18 years after she joined the Institution as a corporate member in 1939 (ICE 2017b).

Women architects fared slightly better with a number of schools actively supporting them. The Architectural Association accepted women students from 1917 and despite dismissive comments from the Head of School that they would find their abilities 'more particularly in decorative and domestic architecture rather than the planning of buildings 10 to 12 stories high' it rapidly became, in Lynne Walker's view '<u>the</u> place for young women to train' (Walker n.d., p6). By 1937 architecture schools like Liverpool, with the publication by Charles Reilly of an article 'Architecture as a Profession for Men and Women' in the *Journal of Careers*, were explicitly marketing themselves to women (Richmond 2001, p37). In 1928 a major breakthrough occurred when Elizabeth Scott (1898–1972), just graduated from the AA, won the international architectural competition to design and realize the replacement for the burnt-out Shakespeare Memorial Theatre in Stratford-upon-Avon. Her success, reported by the press as 'Men rivals of two nations beaten' and 'Unknown girl's leap to fame' was 'seen as a victory for all women and as evidence of their ability to obtain large scale public commissions, breaking the stereotype of women as solely domestic architects' (Walker n.d., pp7–9).

Despite such progress the RIBA, like its fellow institutions, was struggling to move with the times. Only in 1931 did it elect its first female fellow, Gillian Harrison (1898–1974) and in 1932 it agreed to a Women Members Committee. Gertrude Leverkus (1899–1976), who had graduated from UCL in a year group of 500 men as the only women and became an RIBA member in 1922, chaired the Committee for most of its 11-year life, during which it dealt with topics including: slum clearance, housing standards, town planning, the presence of women on Local Authority housing committees, lower salaries of women architects and the Progress of Women Exhibition held in 1936 (Mace 1986, p285; Historic England 2017). Their numbers remained relatively few, but their influence was growing strongly, as noted by *The Architects' Journal*'s diarist 'Astragal' who commented in 1934 that two prominent members of the RIBA, Elizabeth Denby (1894–1965) and Judith Ledeboer (1901–90) 'wield more influence – and get more work done – than any six pompous and prating males' (cited in Walker n.d., p22).

All three institutions proceeded gingerly, accepting women when they fought their way through a hostile system to become qualified, but generally making it clear through their behaviour and language that theirs' were male professions that didn't unnecessarily encourage female participation. In 1957 there were still only five female members and one graduate member of the ICE (Watson 1988, p130). As a result although language and attitudes gradually changed there was not one point at which the proportions of women in the three professions suddenly took off. Progress has been maintained at a steadily growing pace despite the social revolutions and equal opportunities legislation of the 20th century, rarely declining and equally taking an inordinate amount of time to approach anything like 50% (see Figure 7.1).

The institutions have clearly found the issue difficult to deal with over the decades. In the absence of institutional initiatives, independent groups developed to champion women in the professions. These included WES, Women into Science and Engineering (WISE) established in 1984 in response to the *Finniston Report* and the Association of Women in Property (WiP) founded in 1987. An alternative means was through in-house groups such as the RIBA's Women Members Committee and the RICS Lioness Club established in 1980 (later the Women Surveyors Association). Despite the level of public discussion about gender equality, especially following the passing of the Sex Discrimination Act in 1975 by the UK government and academic studies, is it difficult to see any significant activity from the institutions until the late 1990s and early 2000s when they began to commission studies such as the RICS's *Surveying the Glass Ceiling* (Ellison 1999) and the RIBA's *Why Do Women Leave Architecture?* (Manley *et al.* 2003). These reports were intended to provide a way out of a situation that was acknowledged to be very poor to extremely bad. In 2009 a report for the Construction Industry Council (de Graft-Johnson *et al.* 2009, p2) concluded:

- The Construction Industry has a long way to go to achieve diversity
- Some professional institutes are not committed
- There is a lack of available information on diversity.

All three institutions gradually increased the proportion of their female membership until eventually in the 21st century (and in their established order of activity) they each elected their first women presidents. Jean Venables, a water engineer and specialist in flood risk management, became the 144th President of the ICE in November 2008. Ruth Reed, an architect with a practice

Figure 7.1
Percentage of female professional members (published institutional data)

Sources: The data for the graph has been scraped from numerous sources, preferably annual reports or from official spokespersons reporting on the latest statistics. These include: ARB – Annual Reports 2005–17; EngineeringUK reports 2003/04–2017 (note: the data reporting is inconsistent and only some of the reports contain relevant gender breakdowns); RIBA – The annual trustee reports and from RIBA research reports (note: the RIBA is very reluctant to share membership numbers or information, except in the most general terms); ICE – There is almost no relevant published information from the ICE. One data point comes from its 2015 Annual Report and Accounts and the other from de Graft-Johnson et al. (2009); and RICS – again the Institution does not publish relevant information directly but it does occur in its diversity reports in Greed (1999), de Graft-Johnson et al. (2009) and in Howard (2017). The institutions' tendency to opaqueness does not match the transparency that they urge on others and is far from good practice.

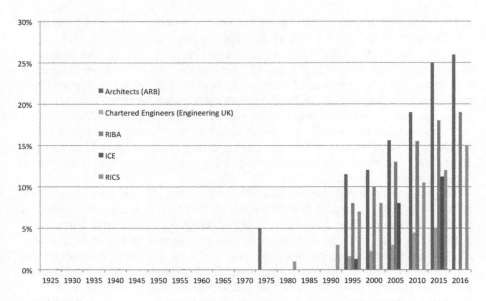

in mid-Wales, followed as 73rd President of the RIBA from 2009 to 2011 and Louise Brooke-Smith, a Birmingham-based planning and development surveyor, as the 133rd President of the RICS in 2014. Both the RIBA and RICS have since followed up this breakthrough with two and one, respectively, subsequent women presidents.

The arrival of women in senior leadership positions in the institutions also triggered a renewed round of activity aimed at encouraging greater access to the professions for a broader section of society. The RIBA appointed a future president, Jane Duncan, as their diversity champion in 2013 and launched its Role Model campaign in 2015 (Young 2015). The RICS launched its Diversity and Inclusion campaign in 2015 (RICS 2015a) following it with their Inclusive Employer Quality Mark (IEQM) the next year (RICS 2016c). The ICE's contribution was perhaps more searching, with a study, Disruptive Diversity, commissioned from the materials engineer and founder of National Women in Engineering Day, Dawn Bonfield (Bonfield 2015). Again this was followed by initiatives including the ICE Diversity and Inclusivity Plan, which promoted the *Ten Step Plan* developed from the Royal Academy of Engineering *Diversity Concordat* by WISE (RAEng 2013).

In 2016 the numbers of female registered architects in the UK was 26% of the total, while the RICS reported that 15% of their worldwide membership were women and the Engineering Council reported a further increase in the number of female engineering registrations with women in the combined engineering grades, 79% of whom are Chartered Engineers, rise from 4.88% in 2015 to 5.22% in 2016 (ARB 2017a; Howard 2017; Engineering Council 2017a).

If women have had a difficult time entering the construction professions it has been far more straightforward than other sectors including black, Asian and minority ethnic (BAME), disabled and lesbian, bisexual, gay and transgender (LGBT) groups who have had very low level representation in the professions.

The Society of Black Architects (SOBA) was founded in 1990 and was instrumental in the professions beginning to address the representation of minority groups in their ranks. The RIBA started collecting data on the ethnic origins of its members in its annual *Architects' Employment & Earnings Survey* from 1991 (although not in the years 1995–2000) and other institutions including the RTPI started collecting figures from around 1993 (Barnes *et al.* 2002). In this spirit the 1994 Latham Report, *Constructing the Team*, included the forthright recommendation (21.2) that (Latham 1994, p72):

> Equal opportunities must also be vigorously pursued by the industry, with encouragement from Government. The CIC, CIEC and CLG should produce co-ordinated action plans to promote equal opportunities within the industry and to widen the recruitment base.

Arguably, it was the murder in London in April 1993 of the 18 year old aspirant architect, Stephen Lawrence, and the subsequent work of his parents, that did most to make the professions aware of the position of black and other minority entrants and their opportunities as architects, engineers and surveyors. The Stephen Lawrence Charitable Trust founded in 1998 brought the discussion into the mainstream and in the same year the RIBA under its then President, Marco Goldschmeid, instituted the Stephen Lawrence Prize for projects costing less than £1 million, in his memory.

As with the opportunities for women in the early 20th century it was the education system that facilitated BAME entrants into the professions with, according to UCAS statistics in 2001, entrants to Architecture, Planning and Building courses being 1% mixed race and 9% from minority ethnic group, with another 7% who did not declare their ethnicity. These figures are tempered by other statistics collected from the same period that there was a significant dropout rate and that professional registration may have been as low as 4% (Barnes *et al.* 2002, pp7–8). More recent research has indicated that the increasing diversity of the British student body has encouraged the BAME intake but in employment experiences are reported still to be problematic and discriminatory (CABE/Centre for Ethnic Minority Studies 2005).

As with the gender split in the profession the information on the ethnic diversity in the industry is very partial. The ICE, RIBA and RICS all collect information but choose not to make it widely available. The best information is still from 2001 when 6.6% of architects were from ethnic minorities (although the RIBA was only reporting 2% at the time), 1% of surveyors and 2.9% of civil engineers, although as the research used sampling data and the samples were small the numbers are subject to high levels of variation (CABE/Centre for Ethnic Minority Studies 2005, p22). A subsequent study indicates that the proportion of BAME architects may have risen as high as 8% (Fulcher 2012). The 2009 CIC study noted that the RICS's global membership had an 8% minority ethnic component mainly composed of Chinese members, but otherwise notes the lack of available information (see above).

Information on inclusion of the disabled in professions is hard to find, as is the equivalent data on sexual orientation. The institutions have addressed the issues in their inclusion and diversity publications, but without providing any precise information. The RIBA commissioned a report 'Disabled

Architects: unlocking the potential for practice' (Manley *et al.* 2011) and has covered both disabled and LGBT issues in its Role Model project but greater inclusivity within the professions appears to be a goal that is still not robustly engaged with.

The 20th and 21st centuries have demonstrated an immense change in the backgrounds of new entrants to the professions and although there are still barriers to be overcome they are broadly open to all-comers. At the same time the entry requirements have been raised significantly so that almost all professionals now carry degrees or higher qualifications, in stark contrast to the position at the start of the 20th century. Twenty-first century professionals undoubtedly are more expert in their field, but find themselves competing in a world where expertise is more widely available, with and without professional baggage. The expectation can only be that professionals will have to continue raising their standards, both in their education and expertise if they are to maintain their advantage and social distinction into the future.

Notes

1   *Weltwirtschaftliches Archiv*, 1913, I/ii, p143, cited in Hobsbawm (1994, p347).
2   'In 1925 there were 870 postal students, 214 evening students, and 31 full time students. Moreover it seems there was very little genuine full-time work over the whole academic session, since many of the full timers were postal or evening students attending for one term only' (Thompson 1968, p22).
3   In 1904 when the RIBA Board of Architectural Education was established the Architectural Association was still in transition from being a mutually organized to a staff-led school.
4   Those entitled to membership of the ARCUK Council under the 1931 Act included the RIBA, the Incorporated Association of Architects and Surveyors (IAAS), the Faculty of Architects and Surveyors, the Architectural Association (AA), the Association of Architects, Surveyors and Technical Assistants (AAST), the Ulster Society of Architects, the Board of Education, the Minister of Health, the Commissioners of Works, the Department of Health for Scotland, the Governor of Northern Ireland, the Chartered Surveyors' Institution, the Institution of Structural Engineers, the Institution of Municipal and County Engineers, the Society of Engineers, the Institute of Builders, the National Federation of Building Trades Employers and the National Federation of Building Trades Operatives.
5   'When the council was formed, it became a hotbed of intrigue, and instead of concerning itself with architects it concerned itself with the glorification and predominance of the Royal Institute of British Architects' (Robert Tasker MP, from Parliamentary Debates, House of Commons Official Report, 17th December 1937).
6   The motto over the door of the Chateau reads 'S'il vient à point, me souviendra' (If it is built, I will be remembered).
7   Women over 30 who were, or were married to, a member of the Local Government Register, were property owners or graduates voting in a University constituency were granted the vote in 1918. The universal franchise for all those over 21 was not granted until the Representation of the People (Equal Franchise) Act 1928.

8
Transition

Breakdown of trust

In early 1998 a funeral director, Deborah Bambroffe, in Hyde, Greater Manchester became concerned about the number of unexpected deaths of elderly women being handled by her family firm. The women had all died under the care of or been found dead at home by their general practitioner, Dr Harold Shipman. She expressed her concerns to a local doctor, Linda Reynolds, who compared the numbers to that of her own, neighbouring practice and discovered that Shipman had signed cremation forms at over three times their rate over the previous three months. Dr Reynolds reported her detailed findings to the local Coroner, who in turn informed the Police. An investigation followed and concluded that there was no substance in Dr Reynolds' concerns (Smith 2003, pp7–9).

When, just over two months later, Mrs Kathleen Grundy, another of Shipman's patients, unexpectedly died having supposedly recently rewritten her will for the benefit of her doctor, the alarm was raised again. This time the police investigated with more vigour and on the 7th September 1998 Shipman was arrested and charged with murder, attempting to obtain property by deception and forgery. Fourteen further charges of murder were added before his trial a year later, at which the doctor was convicted on all counts and sentenced to 15 terms of life imprisonment and, for the forgery, a concurrent term of four years' imprisonment. Only after the guilty verdict did the General Medical Council (GMC) feel it was appropriate to act and their Preliminary Proceedings Committee suspended him from practice. On the 11th February 2000 the Professional Conduct Committee of the GMC erased Shipman's name from the medical register (Smith 2002, pp13–16).

The subsequent inquiry found that Shipman had killed at least 215 of his patients over a period of more than 20 years, usually with a lethal injection of diamorphine, with another 45 very probable victims and possibly another 38 on whom inadequate evidence was available. Three of Shipman's victims, including Kathleen Grundy, were murdered in the interval between the police closing the first investigation in April 1998 and his re-arrest that September (Smith 2002, pp12–3).

There was an immediate impact on the medical profession, with a raft of measures put in place to 'ensure the quality of care, to focus healthcare organisations on continuous quality improvement, and to ensure that seriously deficient clinical performance is rapidly identified and dealt with' (Home

Secretary and Secretary of State for Health 2007, p3). In the light of the Shipman case and some other notable health scandals in the late 1990s, particularly that of the unacceptably high mortality rate for children's cardiac surgery at the Bristol Royal Infirmary, there was a widely reported view that the public trust in doctors had been badly damaged. The Chair of the Royal College of General Practitioners in a May 2000 editorial asked (Pringle 2000, p355):

> If a patient cannot trust their GP not to deliberately harm them then how can they trust their doctor not to avoid accidental harm? Or covertly to deny them access to effective treatment on the grounds of, for example, cost?

Likewise public and media confidence in the GMC collapsed and a 2004 editorial in the *British Medical Journal* declared: 'it has broken its contract with the public – to protect patients in exchange for the privilege of self regulation' (Smith 2004).

The medical profession and its representative and disciplinary bodies were sent reeling, but the investigation and subsequent findings were also understood as a moment of truth for all the other professions in the UK. The lessons drawn from Shipman's behaviour and the institutional failure to deal with it were applied across the board to professional relationships, and the trust between professionals, their regulatory bodies and the public was routinely questioned and routinely found wanting. A narrative of mistrust in professionals and professionalism became a feature of the final decades of the 20th century, a narrative given exemplary shape by the Shipman case. If the unspoken compact between professionals and society was to survive, it was believed that it had to be reshaped and redefined, and a new, better-understood version forged. Reform of professional practice and oversight was declared to be a necessity.

A reasonably representative version of this was the RICS's Strategic Foresight 2030 study, *Just Imagine!*, authored by John Ratcliffe (2011). As a 'provocative thought' at the start of the report he wrote:

> there is a widespread perception that the traditional professions are under siege. Their authority and status, their exclusive access to specialised knowledge, and their right to regulate their own affairs are all seriously being challenged. No longer able to claim special privileges as disinterested, altruistic occupational groups acting detachedly in the public interest, professions are finding their traditional values and loyalties eroded.

The issue was also addressed in a timely and questioning way in the 2002 Reith Lectures, *A Question of Trust*, delivered by the philosopher and ethicist, Onora O'Neill (2002a, p11):

> A standard account of the supposed 'crisis of public trust' is that the public rightly no longer trusts professionals and public servants because they are less trustworthy. But is this true? A look at past news reports would show that there has always been some failure and some abuse of trust; other cases may never have seen the light of day. Since we never know how much untrustworthy action is undetected, we can hardly generalise. Growing

mistrust would be a reasonable response to growing untrustworthiness: but the evidence that people or institutions are less trustworthy is elusive.

O'Neill emphasized the fragility of trust and difficulty rebuilding it once it has been dissipated, but also challenged the effectiveness and value of the alternative, the new culture of accountability, proposed and frequently imposed in its place (2002a, p13).

In theory the new culture of accountability and audit makes professionals and institutions more accountable to the public. This is supposedly done by publishing targets and levels of attainment in league tables, and by establishing complaint procedures by which members of the public can seek redress for any professional or institutional failures. But underlying this ostensible aim of accountability to the public the real requirements are for accountability to regulators, to departments of government, to funders, to legal standards. The new forms of accountability impose forms of central control – quite often indeed a range of different and mutually inconsistent forms of central control.

O'Neill diagnosed that the culture of accountability was not the remedy for a loss of trust and suggested 'we need to think less about accountability through micro-management and central control, and more about good governance, less about transparency and more about limiting deception' (O'Neill 2002c, p11). Nonetheless accountability through innumerable reporting mechanisms – whether key performance indicators (KPIs) as recommended by the Egan Task Force (1998, p15), BREEAM ratings or Design Quality Indicator (DQI) scores – became, at least for a period, key to the construction industry's operations, just as exam league tables and Ofsted ratings for schools or the Care Quality Commission rated the health and social care systems. Meanwhile better governance in the sector began to be tackled through the development of a new public sector agency, the Commission for Architecture and the Built Environment (CABE) – see further below.

Market forces

A study co-ordinated in 2002–3 by the present author for the RIBA-CABE sponsored Building Futures think tank and published as *The professionals' choice: the future of the built environment professions* (Foxell 2003) addressed many of the concerns about professionalism in the sector. In particular it noted the unease expressed by the many professionals and external experts consulted, about the way the industry was likely to develop. The study took the form of five scenarios written by a range of authors looking forward to 2023, together with a context-setting essay and a conclusion paper. This latter paper, written by William Davies and John Knell of The Work Foundation, identified two broad areas of major concern for the professions that had emerged from the scenarios – 'the need for better risk management, and the need to facilitate better forms of cultural expression' (p132). With the benefit of hindsight their insight was prescient. In the years that followed, professional firms largely chose to follow the path of risk mitigation and management. In contrast the institutions developed narratives explaining and giving a public face to their individual and separate disciplines. They looked for ways to increase their media exposure, to attract the attention of politicians and other policy makers and to bring in new members. The divergence would leave the two sides dangerously far apart in the years to come.

The major issue that the study examined was the anticipated impact of market forces, as unleashed by the proponents of privatization and limited government intervention in the final two decades of the 20th century. This had already resulted in heightened competition and a reorientation of corporate behaviour around shareholder interests. Davies and Knell saw in this a threat to free and honest speech in and around the institutions from the forces of commercialism and short-termism (p18). But above all the scenarios raised concerns that managerialism would be the chosen panacea for concerns over public accountability. If the professions wanted to maintain a reputation for expertise, fair dealing, social betterment and ethical responsibility then the institutions needed to respond creatively, take ownership of the issue and 'seek to play a role in target setting, and accept that they need to demonstrate competence clearly to the public, to the government and to investors'.

The Building Futures report (p157) warned that the institutions:

> will lose freedom in the process, and their work will change. Yet the ethical purpose of managerialism – creating transparency and trust – is not at odds with that of the professions. The problem is not the goal, but the *naiveté* of certain methods. Combatting this *naiveté*, and opening up new spheres of autonomous accountability, is the challenge.

In the event the institutions chose not to assert an already diminished authority to engage with the challenge. The market moved ahead by itself and established its version as the solution.

The recession of 1991 and 92 was particularly harsh on design and construction and many industry companies and consultancies collapsed or were stripped back to a small core of partners or directors. Latham (1994, p7) – see below – recorded that 'by 1993 construction output was still some 39% below its 1990 peak, whereas for manufacturing the dip was 3% and for services any lost growth has now been regained'. The impact on public sector bodies tasked with managing the public estate and research programme was equally adverse. The GLC had already been abolished (in 1986), but the National Economic Development Office (NEDO) was closed down in 1992, with its subsidiary, the Property Services Agency, the main offspring of the once mighty Ministry of Public Buildings and Works, sold off in 1993. The Building Research Establishment (BRE – the renamed Building Research Station) lasted a little longer, but it was eventually privatized in 1997.

The solution that emerged in the construction industry to these issues in the period around the millennium was one that would largely ignore the existing professional structures. It took the form of a small number of major multi-disciplinary professional service firms (PSFs),[1] each large enough to believe they could act independently. The downturn in work in the early 1990s encouraged a number of far-sighted companies to reinvent themselves and invest in better-protected and more predictable ways to secure work, at a time when many other companies in the industry collapsed. Similar moves were also made in other sectors, including law, medicine and accountancy, with small and medium sized specialist professional companies merging and morphing into large multi-service firms. This process was actively encouraged by government, which as Gerard Hanlon (1999, p121) has noted was 'engaged in trying to redefine professionalism so that it becomes more commercially aware, budget focused, managerial, entrepreneurial'. This development has been tracked by the sociologist, Julia Evetts, who, over a series of papers from 2003 on (2003, 2012a, 2012b) recorded the change in professional work resulting from the 'commodification of professional service work relations' (2012b, p22).

The new breed of consultancy, the PSFs that emerged from the recession (alongside a few companies that survived the recession by working overseas) grew haphazardly. Firms merged or acquired one another; famous names were absorbed by large, and sometimes anonymous, companies; and organizations were purchased cheaply from the public sector. Each move was carefully appraised with an eye for securing new work streams, but the outcome was corporate growth and eventually corporate gigantism. Although tending to maintain their roots in one area of the industry, the new organizations rapidly became multi-disciplinary operations, keen to offer strategic advice to clients in emulation of the big accountancy firms as well as the practical design, engineering or surveying work that was the original basis of their workload. Some firms were prepared to provide almost any service that they could hire staff to deliver, from consultancy to site security. This led to confusion about how they differentiated their professional services from other commercial work.

The rise of the PSF would reshape what it was to be a professional and redefine the term professionalism in new 'organizational' ways. Individual companies moved towards greater self-reliance and exercised more active control over their employees. They developed and maintained their own internal codes of behaviour, protocols and standards rather than relying on those emerging from the professional institutions. The rationale for this was clear to the multinational and multi-disciplinary PSFs. They wanted to maintain direct control and, in any case, their professional staff, spread across numerous jurisdictions, often outnumbered the membership of many of the traditional institutions. The shift in power inevitably left the institutions less relevant and with diminished influence.

The underlying neoliberal economic narrative of the Thatcher ascendancy was that the power of the market could not be gainsaid and attempting to challenge it or even to moderate it was a fool's errand. When the New Labour government was elected in 1997, after nearly 18 years of Conservative rule, the repeated promises of economic stability and prudence, followed by the assurance that that there would be 'no return to boom and bust' left the prevailing market ideology determinedly in place. Throughout the period almost everything that could be was exposed to market forces: professional services were seen as being no different from any other service.

One view saw the early-1990s recession as an opportunity to clear away restrictive, adversarial and inefficient work practices and to break down many of the barriers in the industry. The alternative (minority) view was that it was an opportunity to restore a collective ethos of professionalism and higher standards to an industry in danger of racing to the bottom in its pursuit of survival and competitive advantage. In a climate of reduced government oversight and control in many spheres (including the built environment), professionalism was argued as a necessary force to protect the public good and provide stewardship against short-term market forces. Although much aired, this latter view never achieved any significant purchase.

Industry reform

In July 1993 the UK government appointed the former conservative MP, Michael Latham (1942–2017), to prepare a report on procurement and contractual arrangements in the construction industry, matters at the heart of public sector concern during a time of a serious downturn in construction and public finances. He published it a year later as a review jointly owned by government and industry having chosen to dig much further into what *Building* magazine described as 'the grim reality' (Gardiner 2014), and set out an agenda for modernizing the industry as a whole. Although the focus of the review was notionally on the relationship between contractors and their sub-contractors, it covered, as his chosen title, *Constructing the Team* suggested, the whole supply chain, including clients, consultants and builders. The report made clear they were all expected to change their behaviour forthwith.

Latham took the opportunity to propose a far-reaching plan for how construction should and could work that would bring both efficiency and quality to the process of designing and building. A plan that would, in his view, cut out the waste generated by the antagonistic practices and constant litigation used by firms to boost their otherwise meagre profits. His central vision was of a client-led industry that would work through collaboration and co-operation, avoiding disputes and resolving them, if necessary, through swift and effective mechanisms. His intention was to create 'win-win' solutions through the application of teamwork, supply chain integration and a new process known as project partnering (Latham 1994, p23):

> There must be integration of the work of designers and specialists. A design team for building work may include an architect, structural engineer, electrical services engineer, heating and ventilating services engineer, public health engineering consultant, landscape architect and interior designer. Some or all may have further specialists working either with them or for them. Installers – contractors, subcontractors and sub subcontractors – are also likely to have design responsibilities.

The report didn't discuss the challenges of achieving this in terms of professionalism but was scathing of both the public sector's devotion to competitive fee bidding, effectively requiring the design team to work for the minimum possible fee, despite the official line of 'value for money – not lowest price', and the recorded response of professional consultants to the need to cut costs, at any price. To this end the report quotes research carried out by the Association of Consulting Engineers (ACE), which found that in order to reduce their fee tender, firms of engineering consultants were prepared to:

● give less consideration to design alternatives (73%);
● give less consideration to checking and reviewing designs (31%);

- allow for higher risks of design errors occurring (40%);
- produce simpler designs to minimise the commitment of resources to a task (74%);
- allow for capital costs of construction and operation to be higher as a result (60%);
- increase the number of claims for additional fees (84%);
- risk the frequency of problems on site being higher (33%);
- lower the frequency of visits to site (49%);
- pay less attention to environmental concerns (29%);
- pay less attention to health and safety both in design and on the site (12%);
- resist client changes to designs (67%);
- have less trust between client and consulting engineer (69%);
- spend less resources on training graduates and technicians (79%);
- spend less resources on Continuing Professional Development training and courses (77%);
- spend less time on the writing of professional papers (75%)
- devote less time to professional activities (56%);
- bid low to maintain the cash flow or (on occasion) to test the market (94%);
- bid low with the intention of doing less than in the enquiry (35%);
- bid low with the intention of making up fees with claim for variations (61%).[2]

Almost all these actions would, unless discussed in detail with the client, amount to unprofessional behaviour; many would still be unprofessional even with the client's consent. Latham also cites similar comments from a Royal Incorporation of Architects in Scotland (RIAS) report, *Value or Cost* (McKean 1994). Clearly under pressure from a very hard market, professional standards were extremely vulnerable to commercial pressure.

Opportunity costs were high for accessing the work streams that slowly returned to the market, especially those emerging from the public sector. The sector was, above all, seeking reassurance and a safety-first approach along with the promise, if not always the delivery, of cost savings. This favoured established firms with the track record that gave them the required credentials to satisfy nervous procurement officials. It was an approach that provided few opportunities for small start-up companies, but it attracted the interest of overseas companies who established offices, predominantly in London, or bought existing UK firms to take advantage of the promised pipelines of work. All these firms were necessarily ready and prepared to work under the new and experimental forms of procurement and contract: design and build, construction management, public-private partnerships and other similar arrangements that gave project control to financiers and contractors. In practice this meant that professional firms were repositioned in the supply chain, away from a direct engagement with the project client to become second tier suppliers to the main contractor. What relatively few ever got the opportunity to do was to operate in the integrated supply teams on partnering projects that Latham and his successor, John Egan, in his 1998 report, *Rethinking Construction*, recommended. The market never warmed to long-term arrangements when it could happily operate with short-term contractual relationships.

The rise of PSFs

The UK government in the early 2000s, with an overwhelming interest in cost-control, determined that the public sector should, in future, source its building needs from an ever smaller number of larger firms (Gershon 2004, p14). Ideally these could provide a single point of responsibility for many different services and disciplines. This was a perfect environment for the growth of the increasingly confident PSFs, especially given their ability and willingness to speak the language of commerce. As many of the earlier consultancy practices specializing in commercial work were now defunct or amalgamated, the PSFs became the preferred providers of services to development companies and overseas governments. Commissioning bodies, whether in the public or private sectors, reacted favourably to the structure and reach of the often internationally based PSFs and, in a reflection of the old corporate adage that 'no-one ever got fired for buying IBM' (e.g. Porter 2008, p27), felt reassured by the scale and apparent security offered by these new companies.

By 2005/6 the business-minded PSFs had expanded their market share in the sector to the extent that 2% of UK construction professional firms (approximately 1150 out of a total of 58 000 companies) generated between 70% and 80% of all fee income (Davis Langdon 2007). This compared to 3% (approximately 700 out of 23 500) earning 60% in 2001/2 (Davis Langdon 2003). They were also employing increasing numbers of staff. In 2013 John Connaughton and Jim Meikle compared the top 20 UK-based construction PSFs in 1995 and 2011, and found that the number of qualified staff employed by PSFs had more than doubled (from 11 527 to 25 528 – down from a high of 32 011 in 2008), although they also noted that percentage fee levels possibly halved over the same period. Their study also examined the transition in the structure and ownership of the firms named. In 1995, seven of the top 20 companies were publicly traded; 16 years later this number was 12, with half of them registered overseas.

In a relatively short period (approximately 1980 to 2000) the composition of the commissioning organizations employing professional firms changed from one of dominance by the public sector to being largely private sector led, although many of the projects were still publicly funded. This was a direct result of the use of new procurement methods, whether design and build (D&B) or public-private partnership (PPP) approaches. The large public sector professional offices, having largely been dissolved or sold, were replaced by organizations, the PSFs, which could provide built environment services on an on-demand basis. Cost, although part of the declared rationale for the shift from public to private, was less of an issue than a desire for downsizing the public sector and the mantra that project risk could be effectively transferred to commercial organizations that were adept at dealing with it. For the private sector this created opportunities for those companies willing to carry and manage project risk and offer a level of accountability for the complexity of construction projects. As a result the public sector lost much of its ability to run projects and became further reliant on the PSFs. A spiral of dependency developed with the public sector relying on PSFs to provide them with the client-side services necessary to put in place the complex procurement and legal arrangements associated with large single-point contracts. Governance and oversight as well as delivery capacity were all supplied by PSFs, giving them an extraordinary business advantage over any other companies.

PSFs rapidly became indispensable in the market for building services, both defining and providing the services that were required for any project and working, predominantly with one another, to bring them to fruition. They developed control of the necessary market intelligence (see discussion on knowledge management below), the labour force capable of being deployed at short notice to any location and a command of the confidence of project funders and corporate boards. To an extent they represented an unstoppable force, for the time being at least.

As construction-based PSFs grew larger their service offers necessarily became more diverse. For example: the largest UK PSF, the multi-disciplinary consultant Atkins, grew from 1927 qualified staff in the UK in 1995 (as WS Atkins) to 6340 in 2007 (and 16 824 total staff worldwide)[3] offering design services in everything from Access & Inclusive Design to Wellbriefing with Infrastructure Engineering and much else in-between. With growth and development also came changes in ownership and corporate structure. Atkins itself was established in 1938 as a civil and structural engineering company by William Atkins (1902–89). For over 40 years the firm was a partnership, then a private limited company, before being listed on the London Stock Exchange as a public limited company (plc) in 1996 (Atkins 2013). In July 2017 Atkins was acquired by the Montreal-based engineering and construction company, SNC-Lavalin. With the acquisition of Atkins the company 'will have over 50,000 employees and annual revenues of approximately C$12 billion' (£7 billion) (SNC-Lavalin 2017).

PSFs emerged as a major force and opportunity within the industry at a time of economic crisis (although only four of them in 2016 exceeded the UK government definition for a mid-sized business).[4] They were far from a UK-based phenomenon; similar restructuring of professional services, if with different cultural emphases and content, was occurring across the world. There was a well-understood need for reinvention, both economically and technologically, and the collapse of long-standing structures followed by the rapid recovery and expansion of the private sector at the end of the century, combined with the potential of ICT for new ways of working, created ideal conditions for the new business-led model of the PSF to thrive. But how did such firms develop from the low base that existed at the end of the 1990s recession? Where did the PSFs come from?

The old ideal of the partner run and owned office described by John Summerson in his essay 'Bread & butter and architecture' (see Chapter 6) continued to operate and maintain its appeal to those keen on individual control and creativity or the ethos of the *atelier*, but with a few exceptions, small to medium-sized firms,[5] in whatever numbers, were not a viable replacement for the large public sector organizations swept away by the pro-market reforms of the 1980s and 90s. It might have been possible to revive the so-called 'commercial' practices of the 1980s but, if anything, these firms had suffered more than most in the 1990s recession, especially those who had overstretched themselves or mistimed a public listing. They were far from an attractive proposition, except possibly for a fire-sale takeover. The alternative model that developed was to build on and rapidly expand a range of medium-to-large sized traditional firms that had not only survived the recession but, critically, were run by a second or later generation of partners and directors.

This was a tough competitive world and not all companies succeeded at making the transition. Some famous and long-standing construction names disappeared in the shake up, often being absorbed elsewhere, as the number of larger sized firms reduced year-on-year to be replaced by fewer and ever larger PSFs. As the market opened up and became more international they were joined by a number of, often overseas, firms with a more opaque nomenclature and structure, including AECOM, Arcadis and CH2M, an opening that itself resulted in more mergers and acquisitions and the increased presence of multi-national firms with a global reach, using their own internal standards to maintain their brand and only accountable to shareholders and the whims of the market.

Inevitably as such companies grew and became more diverse they developed into partially self-sufficient organizations, increasingly detached both from their origins and the professional structures originally designed to support, inform and discipline them. This represented a challenge both to them and, most significantly, to the professional bodies. Large companies needed fewer of the services provided by the institutions, although they still relied on them in many ways, including for the oversight of education and qualifications; but the expense of paying a large number of professional subscriptions was, inevitably, seen as a burden. The professional institutions, originally set up to support large numbers of smaller and traditionally organized companies, also became increasingly reliant on the membership subscriptions coming from the large PSFs who employed the majority of their existing and potential membership, but found it difficult to address their needs alongside those of their most vocal and active members who tended to be based in single-discipline SMEs. The response of the institutions and others will be discussed in the next chapter.

Redefining professionalism

The rise of a new type of professional firm, the PSF, was aided during the 1990s by the simultaneous, and much commented on, 'professionalization of work', where workers in numerous different disciplines, whether childcare, information technology or even the Army ('the professionals'), were encouraged to revalue and redefine themselves as professionals and see their work and career as carrying a 'professional' badge of quality.

Valèrie Fournier, in a 1999 paper, quotes the five core values of a large British service company (British Telecom) (p295):

- We put our customers first
- We are professional
- We respect each other
- We work as one team
- We are committed to continuous improvement.

She notes that 'the image of professionalism was not used to refer to certain groups of employees possessing specific skills and knowledge (such as say accountants or engineers) but to index a certain form of conduct or work ethics' and 'the disciplinary logic of professionalism is deployed to new organ-isational domains to profess "appropriate" forms of conduct when employees' behaviour cannot be regulated (at least so economically) through direct control'.

The transition from traditional 'occupational' types of professionalism to 'organisational' forms has been chronicled by Julia Evetts, following Fournier and others. In a 2012 paper she wrote: 'The discourse of professionalism is taken over, reconstructed and used as an instrument of managerial control in organizations, where professionals are employed', but at the same time organizations were still keen to hang onto the well-established image of the professional in society (2012b, p11):

> The image was of the doctor, lawyer and clergyman, who were independent gentlemen, and could be trusted as a result of their competence and experi-ence to provide altruistic advice within a community of mutually dependent middle and upper class clients. The legacy of this image, whether in fact or fiction, has provided a powerful incentive for many aspiring occupational groups through the 20th century and helps to explain the appeal of profes-sionalism as a managerial tool.

In this view organizations were taking those parts of the professional system that they favoured: standardization of work, trust, a carefully delineated duty to clients and a hierarchical structure, and commodifying them, while ignoring or jettisoning other aspects, including: individual control over the way work is carried out, personal judgement, adherence to ethical codes, stewardship, responsibility for the public good and professional collegiality and interdepend-ence. In particular professional knowledge, instead of being seen as something that should be shared across the wider community of practice, was identified as having corporate value and was therefore to be kept private and protected. Professional work was being monetized and converted into 'service products to be marketed, price-tagged and individually evaluated and remunerated' (Svensson and Evetts 2003, p22).

There were clearly compensations for staff working in PSFs, provided in exchange for submitting to organizational control and for the partial loss of autonomy. These included the pay package, but also the organizational support and opportunities to work in new areas and sectors that would have been out of their reach as single discipline practitioners. Inter-disciplinary boundaries inside large organizations, as those previously working in the public sector had discovered, can be crossed with relative ease. As successful PSFs developed their business activities in novel directions (including project management, urban development planning and tax advice) a greater range of career possi-bilities opened up for employees; including the potential for working with a far more diverse set of colleagues, in different locations and on new and challen-ging projects.

Individuals inside companies might still be organized in their familiar dis-ciplines and represented by the old professional institutions, but their work duties became subject to change, as did allegiances to traditional professional borders. PSFs were also able to recruit workers with new skillsets and respon-sibilities, with or without ties to the mainstream built environment institutions. The stability of the relationship between professionals, professional firms and

Table 8.1 Two different forms of professionalism in knowledge-based work

Occupational professionalism	Organizational professionalism
• Discourse constructed with professional groups	• Discourse of control used increasingly by managers in work organizations
• Collegial authority	• Rational-legal forms of authority
• Discretion and occupational control of the work	• Standardized procedures
• Practitioner trust by both clients and employers	• Hierarchical structures of authority and decision-making
• Controls operationalized by practitioners	• Managerialism
• Professional ethics monitored by institutions and associations	• Accountability and externalized forms of regulation, target-setting and performance review

Source: adapted from Evetts (2012a, p7).

professional institutes was at best being considerably loosened and at worst showing signs of breaking down. As Evetts points out: 'the increasing focus on marketing and selling expert solutions connects professionals more to their work organization than to their professional institutions and associations' (2012b, p22) – see also Table 8.1.

Corporate Social Responsibility

The rise of the new multi-disciplinary firms, selling professional services but not necessarily bound to individual professional bodies or the mores and traditions of 19th century professionalism, allowed them, at least theoretically, to evade society's expectations and the social compact that traded respect and status for impartiality, fair dealing and a contribution to the public good. They were no longer firms owned by their partners, although a number became owned by trusts acting for the benefit of their employees. Most incorporated as modern limited liability companies, run by boards acting on behalf of their investors and shareholders; companies not significantly different from other international commercial businesses and able to operate with the same freedoms.

The need to be seen in the same category as other companies by investors and owners has meant that most PSFs have chosen to embrace the practice loosely known as Corporate Social Responsibility (CSR) (Center for Ethical Business Cultures 2005, p16) as their prime means of dealing with issues of business ethics and reputation.[6] CSR has largely replaced a previous reliance on professional and community-based ethical standards and codes of conduct, further separating the PSFs from the institutions. Professional staff members are still required to comply with their institutional codes, although in practice, sheltered by their employers, they are unlikely to be called to account as individuals, and their obligations as employees and pressures from corporate management will be far more powerful drivers of their conduct and behaviour.

Table 8.2 Guidance on social responsibility

ISO 26000:2010	
Key principles	**Core subjects**
1. Accountability	1. Organizational governance
2. Transparency	2. Human rights
3. Ethical behaviour	3. Labour practices
4. Respect for stakeholder interests	4. The environment
5. Respect for the rule of law	5. Fair operating practices
6. Respect for international norms of behaviour	6. Consumer issues
	7. Community involvement and development
7. Respect for human rights	

Source: ISO 26000:2010.

CSR first emerged as a proposition in the 1970s and in the 1980s suddenly and very rapidly expanded, to the extent that CSR support became a business sector in its own right. An expectation developed that all firms and especially the larger, more public facing ones would publish, abide by and report on compliance with, often bespoke, codes of conduct and standards of behaviour. It was a voluntary process with few rules or guidelines and entirely reliant on self-reporting by companies. In the absence of any authoritative version a range of individuals, groups and organizations proposed CSR 'standards', including: the Sullivan Principles (1977), the CERES Principles (1989) leading towards the Global Reporting Initiative (GRI), the UK Corporate Governance Code (1992), the Caux Round Table Principles for Business (1994) and the UN Global Compact (2000) – amongst many others including ones that deal with specific issues such as employment practice or carbon emissions. There is no bar to signing up to multiple different schemes.

One of the most recent and widely accepted of these standards, and probably the most authoritative, is ISO 26000, developed by a process of lengthy international consultation and released in 2010. It is recognized, even if not necessarily implemented, by eight of the UK's top 20 leading PSFs. It defines seven key principles and seven core issues of social responsibility to be addressed by organizations using the standard (see Table 8.2).

The reasons why a company should adopt social responsibility standards are varied. The text accompanying ISO 26000 lists:

- its competitive advantage
- its reputation
- its ability to attract and retain workers or members, customers, clients or users
- the maintenance of employees' morale, commitment and productivity
- the view of investors, owners, donors, sponsors and the financial community
- its relationship with companies, governments, the media, suppliers, peers, customers and the community in which it operates.

These are all rationalizations in concord with management consultant Peter Drucker's (1984, p62) view that 'the proper "social responsibility" of business is to tame the dragon, that is to turn a social problem into economic opportunity and economic benefit, into productive capacity, into human competence, into well-paid jobs, and into wealth'. It is a distinctively different proposition from the civic idea of professionalism and its obligation to society.

The standard criticism of CSR is that it is an attempt by companies to disguise their true and amoral corporate nature. Certainly the reporting mechanisms that underpin it give rise to reasonable suspicion that it is more self-regarding and promoting than self-regulating. If there is an external force that encourages CSR promises to be kept it is more the threat to a company's reputation and the possible ensuing damage than any comeback on missing advertised standards. The codes of ethics and ISO checklists may be more effective at ensuring nothing goes wrong than making sure that whatever is done is done because it is the right thing to do. This, of course, can be equally true of professional codes of conduct. The question is whether a CSR policy is any more or less effective at achieving civic goods (in addition to reining in any poor behaviour) than professionalism?

An analysis of the public CSR commitments made by the UK's 20 largest built environment consultant firms (*Building* 2016) shows that they all work to a fairly standard list of concerns including sustainability, corruption, modern slavery and equal opportunities and use many common measures to alleviate them; including statements of visions and values, codes of conduct, community outreach work and quality standards. Indeed most topics are dealt with by the majority of firms – see Figure 8.1.

It should be noted that the analysis (carried out specifically for this book) is only of intentions. Firms may claim a commitment to, for example, sustainability, but their record may be no better or worse than any other company's, or they may choose to highlight only those aspects that will show themselves in a good light. Other commitments, for example to avoiding and reporting modern slavery and corruption, are necessary and legal obligations for larger companies and it is more surprising that they aren't claimed by 100% of the surveyed companies. It is to be hoped that having published their intentions, companies will regularly publish the outcomes in the cause of transparency (see ISO 26000), possibly even in a standardized format. There are signs this is happening in some areas, such as equal opportunities and carbon footprints, where legislation, competition and social pressure have forced change. But progress in this direction is gradual, even when disclosure appears to benefit a company by, for example, highlighting their qualities as an employer.

As a consequence of a commitment to CSR, construction industry PSFs typically assert and apparently demand higher 'professional standards' than are required by any of the professional institutions that represent their staff. Similarly, their corporate mission statements feature many of the same keywords used in institutional codes and guidance, such as 'integrity', 'ethical' and 'collaboration'. This is also true for many other, non-professional service companies, and ethical positions are no longer the main preserve of charities or professional firms.

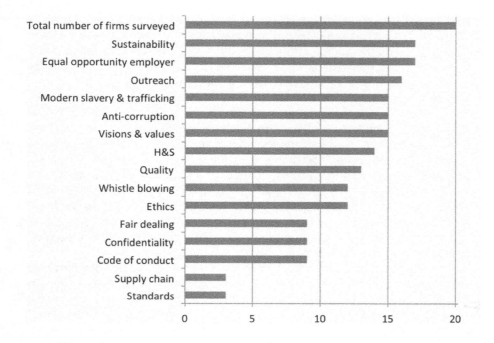

Figure 8.1
Commitments to CSR
Issues of the largest
20 Built Environment
Consultants

Note: Faithful + Gould is
included within Atkins.

Source: graph based
on data accessed from
the UK websites of the
following companies
on the 10th February
2016: Atkins, Mott
Macdonald, Aecom,
Arcadis, WSP Parsons
Brinkerhoff, Arup, Turner
& Townsend, CBRE,
Mace, JLL, Gleeds,
Waterman Group,
Gardiner & Theobold,
Ramboll UK, Foster
+ Partners, Deloitte
Real Estate, BDP, Buro
Happold Engineering,
Peter Brett Associates &
Hoare Lea.

The focused approach of the PSFs has redefined the narrative of professionalism in the industry in recent decades. Vision and values are highly prized and a great deal of effort is placed on communicating them through the complex structures used by firms to manage many thousands of staff in several separate offices. Peter Drummond, chief executive of the multidisciplinary firm of architects, designers and engineers, BDP between 2004 and 2013, explained this in an interview in 2008 (quoted in Powell 2008, p52):

> At BDP, we have tried to place our vision and plan for the next five years in the hearts and minds of the business and of all of our people. The Plan, as it is affectionately known, is the basis of financial business planning and decision taking, a clear statement about our aims and values, and a set of actions (known as strategic imperatives) for all the main operational parts of the firm.

As Drummond made clear, BDP's visions and values are inextricably tied to financial planning and operational considerations. For PSFs the values (professional or otherwise) belong to the firm and although they may have been developed by the original owners, partners or directors, they have been maintained and reconfigured by the processes of corporate planning and management. Corporate values are hence heavily influenced by the need to maintain competitiveness, achieve a continuous and predictable flow of work and provide a consistent level of growth. PSFs operate in a tough economic world in which all projects have to deliver satisfactory financial returns and improve their performance on a quarter-by-quarter basis in order to remain in business contention, attract the best staff and maintain the confidence of investors and clients. Although there are always tales of firms turning away lucrative work for good, ethical or professional reasons, including one in the book by Joe M. Powell, the source of the above interview, there is little sense that it is common in practice.[7] There is little time in such pressured circumstances for taking a long-term view or worrying about ethics and 'professionalism'.

In the long term, visions and values have to be backed by real outcomes. Claiming good environmental design needs to be evidenced by in-use monitoring of completed projects. The danger for companies comes when their espoused visions and values are shown to be hollow; that they are not following through on their commitments; or that the work of their junior staff, under intense pressure to perform, contradicts the good behaviour promised by senior management. Firms may be shown up by failing or inadequate designs, but they may equally be caught out when their claims to be equal opportunity employers are exposed by the make-up of their senior team or their good name is tarnished by some of the contracting methods or clients they agree to work with.

As with large public sector organizations in the past, the values of a firm may also restrict the autonomy of individual professionals and the degree of freedom of professional conscience and judgement that they are allowed. A professional should be able to act in 'the public interest' and in accordance with the code of conduct of their professional body, but this may not be possible if there is an in-house system, possibly even including an externally provided whistleblowing service, controlling their actions. To what extent can professionals maintain the breadth of vision required for them to maintain professional oversight and responsibility if they are only partial players carrying out a standardized task in a much larger organization? Company values may attempt to 'do the right thing', but without judgements being capable of assessment by an accountable external body, made up of professional peers, there is always a danger of moral hazard and at least the suspicion that corporate rather than public interests are being put first.

McWilliams and Siegel (2001, p117) define CSR as operating in situations where a firm goes beyond compliance and engages in 'actions that appear to further some social good, beyond the interests of the firm and that which is required by law'. This would appear nudge CSR into the territory of charity and charitable good works – something that is done after the main business is completed, using part of the surplus generated. Companies attempt to go beyond this by asserting their values and principles. But inevitably profession-alism and company values, although always potentially closely aligned, have different motivations and need to be differentiated.

Professional systems maintain external and peer-judged methods of inves-tigation, decision-making and discipline. Companies will always ultimately prioritize the interests of their shareholders over any other causes and the development of large firms selling professional services on purely commer-cial terms threatens to unpick the relationship between the professional and society. Connaugton and Meikle (2013, p106) speculate that such pressures may result in 'a form of de-professionalisation of services that reduces professionals' autonomy and independence in decision-making' and they raise the concern that the extent to which PSFs 'control and manage knowledge in their own interests would appear to bring them into conflict with professional institutions that seek to develop knowledge in the service of society *as well as* in the interests of their members'.

Practice models

The last 100 years have seen the nature of the predominant mode of professional practice in the built environment sector cycle through four very different organizational approaches:

- the partnership model – involving a relatively small firm, usually very small by today's standards, run and with responsibility taken directly by the owners
- the public sector organization – often very large but with a clear sense of purpose and lines of responsibility, if having a tendency towards bureaucratic sclerosis
- the professional service firm – large, sometimes gigantic, private sector companies frequently with a remote ownership (although in some cases an employee-owned trust) focused on financial outcomes and with a need to achieve continuous growth in order to remain viable
- a hybrid form – with the ownership remaining vested in the original, still active founders, who have transformed themselves into employee directors and their practice into a limited company.

Each mode of practice has not entirely been replaced by the succeeding version; there are still plenty of examples of all four types within the mixed economy of the sector, although relatively few public sector organizations have survived as service providers, rather than as regulatory or commissioning bodies.

The fourth, hybrid form of practice, has proved surprisingly successful and resilient within the industry. These companies, usually still bearing the name of their founder directors, trade on the personality and brand of one or two individuals, even though their throughput of work is clearly far greater than any individual can maintain direct, rather than critical, engagement with. They have aspects of both the partnership model and the large PSFs – there is even one of them, Foster + Partners, in the top 20 listings already discussed. Predominantly associated with design-led companies, mainly architects but with a number of structural engineers in their ranks, the idea of creative individuals leading by example and through capability, experience and judgement maintains its potency. Given the apparent success of this, relatively minor, cult of personality it is surprising that almost all the other PSFs have chosen a largely blank demeanour, preferring a corporate plainness to the charismatic leadership often used, even by the largest PLCs, to project company identity in the market place.

Hybrid firms may simply be in the first stages of the journey to becoming corporate PSFs, only biding their time before making the transition. But as their identity and offer to clients revolves around the contribution of the leading individuals, their ability to connect to clients and their personal control over the quality of the work they do, they may find it difficult to make a success of the transition. The role played by the firm's headline names is often an essential part of the narrative and brand of the firm, popular with clients looking to achieve a 'signature' product and a personal endorsement from a glamorous individual.

At the extreme end of this tendency is a group of firms, mainly headed by architects and product designers who trade, on an international basis, on their celebrity and charismatic personas to the extent that they are identifiable by their first names alone: Norman, Renzo, Rem and Bjarke among them. The late Zaha Hadid was an exemplary member of this group whose death in 2016 presents an extraordinary challenge to the large firm she left behind in how to maintain the brand without its central dominant personality. To an extent this group has been able to move outside the smallish world of design and construction to command cultural recognition and support well beyond the industry, and they have become essential for certain types of grand projects that need high degrees of public fascination, in order to succeed. Such firms, which command the largest share of the broader media coverage of the industry, may not always be the best examples to discuss more general issues of professionalism, yet they are and are likely to remain the focus of both the public's and students' view of what constitutes professional practice in the UK.

The institutions

The development of alternative powerful models for providing consultancy services in the property and construction industries, with their own apparently strong set of 'professional' values, does not, of course, mean that there has been a simultaneous abandonment of the traditional professional model. A need still exists for the work done by the institutions that individual firms, however large, do not seriously attempt. This includes oversight of educational and training systems, control of entry into the professions, development and maintenance of professional standards, representing individuals throughout their life in the industry and providing peer recognition and honours through qualifications and award programmes. But all these things can be and are carried out by other independent bodies, and some even in-house by the larger PSFs. It is possible to conceive of a professional world without professional institutions, if that is ever seen to be desirable.

Equally there are many activities performed by both institutions and PSFs; sometimes in collaboration, at other times independently, but often in competition. These include research, knowledge and technology development, policy engagement and lobbying government bodies. Arguably the PSFs are gradually dominating all these areas of activity with their ability to commit greater resources and also because of the business advantage and rapid payback that can be achieved. In comparison the institutions' efforts can often be under-resourced and lack follow-through. If anything meaningful is to be achieved then all parties need to act in concert.

Technology

There is another development that has possibly had an even greater impact on professional work in the sector than changes to employment and organizational structure: the widespread adoption of computer aided design (CAD) and the use of information technology (IT – later information and communications technology – ICT). The revolution started with the use of the first drawing packages in the aerospace and automotive industries in the mid-1960s, but it was only when the first desktop machines with graphics capability were developed in the 1980s and drawing software became both useful and affordable that the industry began to take notice and various pioneers started to abandon drawing boards for mice, keyboards and screens.

By the millennium CAD was standard in construction professionals' offices and had swept away almost all the drawing boards and special apparatus of design studios along with the need for well-lit studio spaces. They were replaced by the computer equipment and environment common to almost every other white-collar office. The exceptionalism and romance that the idea of the *atelier* fostered went and with it much of the sense that the nature of the work was different to any other service-orientated business.

The promise of a seamless journey from design to manufacture to assembly has been particularly alluring; a vision of a frictionless integrated construction industry overcoming all the current barriers to collaboration and providing computer enabled design with customer choice, pre-programmed consent, instant cost control, resource management, automated construction and much more all factored in. It was a vision outlined by Will Hughes in his contribution to the 20-year horizon scenarios in the study, *The professionals' choice* (2003, p94):

> Choose the building type, choose the amount of space, choose from a range of configurations, move a few customisable elements around with a cursor, input the details of special spaces and functional areas, specify the location and the degree of flexibility permitted, check out the bank balance, enter the details of where the money is coming from and how it will be paid, then click 'go' and the design and construction process starts. Within 10 seconds, the trucks start rolling from their distribution headquarters towards a site, and the building is commenced.

If that future hasn't arrived just yet the world of building information modelling (BIM) has. First mooted in the mid-1980s the idea of a digital representation of a building or other facility, encoded with high levels of information and intended for use throughout the life-cycle of a project from inception to demolition and including the critical management of the facility during its day-to-day use, has been widely embraced for use on larger projects. The UK government has required 'fully collaborative 3D BIM (with all project and asset information, documentation and data being electronic) as a minimum' for all its projects since April 2016 to Level 2 as described in the official standard, PAS1192:2[8] (HM Government Cabinet Office 2011, p14). Against this the National BIM Report 2016 found that only 54% of the 1000+ firms who responded to their survey were using BIM at all, and of these 37% were using 3D information models (if not ones that included all the building information), 29% were using the PAS standard and 17% had passed on the model to those 'responsible for the continued management of the building' (Malleson 2016). Perhaps because the transition to BIM is costly and requires a change in working methods, it has mainly been taken up by those firms with workloads large and consistent enough to allow them to hire specialist staff and to achieve the throughput to make it cost effective. The survey found (p37):

near universal agreement that the adoption of BIM requires significant changes in how design is carried out; more than 90% of both users and non-users of BIM agree that it requires changes in workflow practices and procedure. Adopting BIM is not an easy thing.

It also found that amongst those who had yet to adopt BIM 55% feared 'if we don't adopt BIM, we'll get left behind', suggesting a strong pull factor whether or not it was a cost effective way of working.

The transition to BIM has already created organizational change in the industry (Whyte and Hartmann 2017) and has the potential to cause a rift between those firms large enough to make it cost effective and the very long tail of small to medium enterprises (SMEs) in the industry which neither need, nor can afford, the complexity of systems designed for managing large and multi-authored projects. Although the threatened ICT divide is not at the same point of rupture as that between PSFs and single-discipline consultants, in both cases the increasing use of very different tools and ways of working at opposite ends of the industry is threatening to create new rifts within the professions that will be ever more difficult to bridge as they become deeper.

Procurement

The third development that has changed the structure of relationships between professionals and the construction industry over the past 30 years is the transformation of procurement and financing mechanisms. This occurred principally, if not exclusively, for work carried out for the public sector but it had its roots in private sector experiments of the 1980s with the invention and use of project securitization, construction management and fast track construction. Securitization offered a means of assembling large funding packages for projects at relatively short notice. The challenge was then to build such projects at a pace and profit to satisfy the investors, and methods that promised savings of '8 to 12 per cent in dollars' and '25 to 45 per cent in time' (Foxhall 1972, p11) were beyond temptation. These experiments also meshed with a search for alternative ways to procure buildings that improved both the reliability of the design and construction process and minimized risk to the client. Some experiments focused on reducing the time from inception to completion, others on outturn costs and a few on the performance of the end product. What they all had in common was the proposition that the designers, contractors and suppliers should work together from an early stage as an integrated team under the direction of a project/construction manager.

Despite an on-off record of success with these experiments at locations including Broadgate and Canary Wharf in London (Artandilian and Campbell 1987), the private sector gained a reputation for dealing imaginatively and effectively with project risk. It was a moment when political doctrine decreed that the market, and therefore the private sector, could do no wrong. The UK government turned to this as 'the solution' for major infrastructure and building projects. The belief in private sector efficiency, together with its ability to raise almost unlimited amounts of 'finance', was key for the development of funding and development innovations such as the Private Finance Initiative introduced in 1992 by the then UK Chancellor of the Exchequer, Norman Lamont, with the explicit rationale of 'bringing in private finance for public capital projects' (IPPR 2001, p77) and the subsequent range of similar mechanisms generically described as public-private partnerships (PPP). Such funding arrangements also had the welcome advantage of allowing capital expenditure to be kept off the public sector's borrowing sheet – a necessary fiction for maintaining international confidence.

The idea behind PPPs was to turn over the production of major capital projects to those prepared to finance and develop them against the guarantee of long-term leaseback and maintenance agreements. The public exchequer was intended to gain from drawing on the incisive private sector ability to cut out time, waste and duplication and thereby provide better value than traditional procurement could offer. As it turned out in practice, the same contractors who had previously worked the traditional system to their advantage took command of the new one, organized the capital investment, employed consultants directly themselves and became the single point of contact for both the client and the users of the building.

This is necessarily a highly abbreviated description of many separate procurement initiatives that were developed and used extensively from approximately 1995 to 2015 to invest in the public estate. They were essentially financial mechanisms designed to bring a plentiful supply of private money to bear on public projects without too much concern for the long-term consequences. Although frequently defended, or mocked, for being off-balance sheet they were a means of spending tomorrow's money today, or, in other words, a form of loan. The impact on the construction professions was two-fold. First, they brought a large amount of work to the industry. Second, they reversed the traditional client-consultant-contractor relationship. No longer were many consultants working for traditional clients, the owners and users of buildings. Instead they were working for the builders of projects, whom they had been used to telling what to do. There was little choice in the matter, as the Minister of Health, Alan Milburn, declared at the time, PFI was 'the only game in town' (BMJ 2000).

The change of employer from owner/user to contractor crystallized for many professionals their loss of status, power and control in an industry they had once sought to lead and to run in line with their worldview. They believed they no longer commanded the respect in the process that they thought was due to them following a lengthy education and qualification process. The benefits of the professional compact were also less tangible when, even on publicly funded projects with an ostensible social purpose, the focus was resolutely on the bottom line. Many argued that this had been a long time coming and the arrogance of the professional project, the 'conspiracies against the laity' in George Bernard-Shaw's (1906) phrase, had been finally punctured.

Conclusion

The construction industry professions had managed to avoid the damage of the Shipman affair to professional reputations by choosing to leave the professional high ground without any attempt at mounting a defence. Instead a radical change occurred with a reconfiguration of part of the industry into large professional service firms, espousing organizational over occupational or vocational professionalism. For this group of firms it became enough to do the job efficiently and adequately. There was no concern as to whether the work was of wider benefit. The changes to the nature of the work brought in by ICT and changes to procurement methods fitted well with this transition, suiting larger organizations and obscuring traditional professional mores in new clothes.

Public policy since the 1980s has been intent on making professionals, and especially those working in the built environment, more responsive to market forces. The combination of economic pressures, the insistence on reducing the numbers of potential bidders for projects and the effect of technological change has gradually been splintering the communality and collegiality of the professions and impeding mobility of individuals within the industry. Professional institutions have seemed powerless to prevent such tectonic shifts and perhaps wisely have not made any serious attempts to do so. Paradoxically only government has made an attempt at defying the tide, perhaps in order to soften the impact of its own policies.

This hasn't removed the need for all forms of professional companies (whether vocational, occupational or organizational) to build and maintain social trust and for there to be a perceived penalty for those that transgress even unspoken rules. The claim to be acting within the law is frequently not adequate to gain trust nor to assuage anger at actions deemed to be unthinking or unfair. The work of professional institutions in holding their members to account may never have been seen as adequate, but at least it was there. As professional service firms grow larger, become parts of a larger businesses with other interests and are controlled by an ever more remote ownership the suspicion of conflicts of interests and lack of accountability will expand with them. This is a challenge for all sorts of professional firms, whether lawyers, financial advisers or accountants and they have created mechanisms, such as internal firewalls, external oversight and, of course, CSR, that purport to maintain their integrity in response. Whether such approaches will work in the long term has yet to be seen.

Professional services need to be independent, understood to be free of bias and have the all-important commitment to the greater public good. Institutions need to rebuild their definitions and policing of these qualities in order to defend the fragile state of professionalism in the construction industry. It is the institutions that the next chapter addresses.

Notes

1 Also, more clumsily, known as Built Environment Professional Services (BEPS) companies.
2 Adapted from ACE evidence (May 1994) cited in Latham (1994, pp44–5).
3 Building (2008), Top 250 Consultants 2008, 10th October2008. The 2016 figures are 4260 qualified staff in 57 UK offices and a total of 18 052 staff worldwide in 190 offices – see Building (2016).
4 The UK government's definition for a mid-sized business is companies with a turnover between £5 and £500 million. Four companies had annual UK fees over £500m: Mace (£1559m), Aecom (£600m), Atkins (£944m) and Mott MacDonald (£517m) (Building 2016).
5 Small to medium sized enterprises (SMEs) – classified as firms with less than 250 employees and a turnover of less than £5million.

6 For example Boggan 2001, 'We Blew It': Nike admits to mistakes over child labor', or Gordimer 1997, 'In Nigeria, the price for oil is blood' on Royal Dutch Shell's links to human rights abuses and the execution of the campaigner Ken Saro-Wiwa in the Niger Delta.

7 See also the evidence sessions for the Edge Commission of Inquiry on the Future of Professionalism, www.edgedebate.com/?p=2209

8 PAS 1192–2:2013: Specification for information management for the capital/delivery phase of construction projects using building information modelling.

9

Institutions

Official interest in the quality of the built environment in England has waxed and waned in recent decades as governments, and particularly ministers, have come and gone. In 1992 the last vestige of the Office of Works, the Property Services Agency (PSA), was sold to the contractor and building materials company Tarmac (which became Carillion) and for the first time in many centuries the UK government was without a formal architectural advisor, although in practice the former Director General of Design Services at the PSA and past-President of the RIBA, Bryan Jefferson (1928–2014), continued to advise the Department for National Heritage (later the Department for Culture, Media and Sport) until his retirement in 2001 at the age of 73 (Rhys Jones 2014). The devolved nations of Scotland and Wales would prove to be very enthusiastic about the quality of architecture in their countries once they had gained their own administrations in 1999 (Scottish Executive 1999; Design Commission for Wales 2014).

The Thatcher government wasn't blind to the power of the built environment as a policy lever but looked to private sector solutions, including development corporations and enterprise zones, as ways of tackling deprivation and rioting in inner cities. Other built environment measures included the selling of council homes to their tenants with the intention of creating a property-owning (and conservative-voting) electorate. What it wasn't greatly interested in was the quality of design. This changed in 1994, when John Gummer, as Secretary of State for the Environment under Thatcher's successor, John Major, revived the spluttering flame of official concern for design quality in a policy discussion document, *Quality in Town and Country* (DoE 1994, p2):

> We are concerned with the quality of England's entire built environment. Much of what we say applies to all buildings, including those in the wider countryside. Our main focus, however, centres on urban areas, as it is here where most people live and work, where most development takes place, and where the greatest opportunities for improvement lie.

This renewed interest by Government in the built environment as a solution for the challenging problem of the hollowing out and alienation of cities turned into a positive passion for the subject with the change of government from Conservative to Labour in 1997. The arrival of John Prescott as Deputy Prime Minister, in charge of a merged mega-ministry, the Department for the

Environment, Transport and the Regions, led to the launch in 1998 of the Urban Task Force with a membership drawn from the built environment professions, academia, pressure groups and government and chaired by the architect Richard Rogers. It was given a mission to (1999, p1):

> identify causes of urban decline in England and recommend practical solutions to bring people back into our cities, towns and urban neighbourhoods. It will establish a new vision for urban regeneration founded on the principles of design excellence, social well-being and environmental responsibility within a viable economic and legislative framework.

The Task Force's report, *Towards an Urban Renaissance*, was published in June 1999 with the rallying cry that it was 'Time for Change', a change that required 'a wide range of people … to implement the urban vision – planners, designers, architects, landscape architects, engineers, environmental scientists, surveyors, developers, project managers, housing and neighbourhood managers' (p158). The call had gone out from government, albeit only through a consultative report, for building professionals interested in serving the public interest to step forward. It was an argument that might have been effectively made by the professional institutes if they hadn't been thoroughly intimidated by government and hadn't taken to fighting their own, largely internal battles. Individual professionals were only too glad to have their contribution and value recognized again. They had been invited back into the corridors of government and were very pleased to accept.

CABE

Professional response was enthusiastic and the enthusiasm was mirrored in a round of new government interventions, including the development of a policy on architecture for Scotland starting in 1999, the introduction of the Commission for Architecture and the Built Environment (CABE) (1999–2011) in England and the establishment of the Design Commission for Wales (DCFW) in 2002, alongside many other smaller organizations across the UK. CABE in particular became an organization that in a brief 12-year life did an extraordinary amount of work on improving the quality and the mechanisms that underpinned design decision-making in England. In the process it inevitably took over swathes of the territory that had once been the fief of the professional institutions and did so with an energy and determination that would leave them running to catch up.

CABE was born from a much smaller predecessor body, the Royal Fine Arts Commission (RFAC), taking over its funding, some of the staff and the sponsoring government department, the Department for Culture, Media and Sport; but it rapidly made its own way as a new player in the world of the built environment. CABE defined its remit (see Table 9.1) to be 'the champion for architecture in England' and to 'promote high standards in the design of buildings and the spaces between them' (CABE 2001) with a commission board of 13 including five architects, a quantity surveyor, an engineer, an historian, a journalist and several prominent figures from the arts world, chaired by the developer Stuart Lipton – essentially a professional grouping.

Table 9.1 CABE's objectives, 2000

- To enable, enthuse and educate government, local authorities and the private client, making good architecture central to the nation's vision of itself.
- To facilitate and co-ordinate the design and construction process, including architect selection, brief writing, construction on time, quality, cost and environmental issues.
- To champion good new architecture in all environments.
- To address the spaces between buildings and promote a high-quality public realm.
- To address the social dimension of architecture, enhancing the quality of life of the ordinary person by helping to lift the visual and functional standard of the built environment.
- To harness the popular desire for high-quality environments and help ensure that questions of value as well as cost inform public policy making.
- To stimulate and encourage all those entrusted with building in this country and to act as an ally of those committed to quality in the built environment.
- To address deficiencies and promote good practice in the construction industry generally and in particular areas such as the design and procurement of public buildings and social and speculative housing.

CABE's plan, as expressed in its 2001 vision, was 'to inject architecture into the bloodstream of the nation', and to do this it acted very much like one of the early professional institutions, recruiting what it referred to as 'the CABE family', roughly equivalent to the active membership of a developing organization, and a team of enthusiastic staff to carry out its programme of work. The main difference being that it was government funded and had the direct ear of policy makers; a model closer to the Napoleonic style of professional formation than the essentially independent Anglo-American structure.

A recent study of the organization (Carmona *et al.* 2017, p91) has established that:

A central pillar of CABE's strategy was to address professional 'silo' cultures. Deep divisions had long been a feature of relationships between the different professional specialisms in the United Kingdom, with each group failing to fully appreciate the role and significance of the others. ... CABE therefore faced the somewhat daunting task of promoting a broader culture change, both in the professions and among the politicians and in the population at large.

Of the four key activities traditionally undertaken by professional institutions:

i) providing a distinct body of knowledge;
ii) defining and maintaining a threshold for entry and standards of behaviour through self-regulation;
iii) aiding mutual association; and
iv) serving the public interest,

CABE rigorously pursued i) and iv) and concerned itself with parts of ii); displacing institutional activity in the process but mainly filling gaps and revealing decades of institutional inactivity and the thinness of the various excuses put forward for it.

CABE's research and publication programme produced large numbers of freely available documents covering many topics, including the value of the built environment, guidance for clients, designers, planners and policy makers and audits of various building types alongside numerous case studies and educational materials. At the end of its life CABE claimed on its website to have made '309 publications available online' (CABE 2011) and:

> delivered over 80 informative and purposeful research studies, devised to address gaps in industry knowledge, with the intent of improving the understanding of the relationship between the physical quality of cities, their spaces and buildings and the quality of life of their users.

The subject of diversity, or lack of it, within the industry also attracted the campaigning instincts of CABE. One of its core values (2001) was 'if we increase the diversity of the construction industry in terms of ethnicity, gender, disability and age profile, we will increase the skills base and improve the quality of buildings and places', and it pursued this with research and recommendations on action for education, recruitment practices and leadership in the industry (CABE/Centre for Ethnic Minority Studies 2005).

Serving the public interest was at the core of the organization from the outset – expressed in its tagline 'improving people's lives through better buildings, spaces and places'.

Core public funding for CABE ended in March 2011 as a result of spending decisions taken by the new coalition government, and the organization was merged into the charity, The Design Council. In its new guise it has concentrated on offering paid-for services, particularly design (peer) review, to cities and local authorities, but it is working to a much more constrained remit and is no longer seeking to occupy the same ground as the professional institutions. The original organization, having shown what was possible with a modest budget and substantial will, disappeared, begging the question whether the institutions or any other organizations will step in to fill this void.

Well before the invention of CABE and its Scottish and Welsh cousins, most built environment institutions had already pulled back from fully supporting all four of the pillars of professionalism listed above. They were concentrating on activities where there was member demand and, if possible, allowing their commercial arms to take over the provision of chargeable services in their stead.

Body of knowledge

As discussed in Chapters 3 and 6, a body of knowledge is a critical aspect of a professional institution's activities and it is a thread that has run through much of the discussion in this book. The most significant recent proponent of a knowledge-based professionalism has been Frank Duffy (1940–), who through his work at DEGW (a research and design consultancy where he was a founding partner), writings and time as RIBA President (1993–5) promoted the idea that there is (Duffy and Hutton 1998, p.xiv):

> an urgent need for architects to reaffirm the intellectual basis of their profession, to align it with other rapidly developing disciplines to make sure

that the design of the environment takes its proper place in a society based increasingly on the development and transmission of all kinds of knowledge.

In this Duffy was consciously following sociologist Ulrich Beck, with his stress on the importance of knowledge in professionally controlled organizations (see Chapter 1). But the institutions, in Duffy's view failed to embrace his challenge or appreciate the urgency, and writing with Andrew Rabeneck (2013, p118), he reviewed the state of play:

> As the state has been rolled back, architects and related professions bear a heavy responsibility for not having maintained and developed adequate knowledge bases that systematically connect client objectives to built outcomes, based on an open-ended and developing body of knowledge about buildings and their use.

Duffy advocated an active approach to gathering and harnessing knowledge from professional and building user experience; relaying lessons learnt back to the industry though case studies. The lack of such mechanisms, feeding back knowledge from completed projects in order to continually update professional expertise, remains a key weakness in the sector, but the institutions have continued to develop professional knowledge in other ways. Increasingly this is achieved through the repackaging and dissemination of information generated by others in the form of conferences, training events, journal articles and research awards. They maintain their libraries and provide a range of ways for members and others to obtain information from them, including information lines and web portals.[1] Several institutions run publishing houses that produce information that anyone can purchase or subscribe to.

In a number of instances such as the RICS's professional guidance and statements ('All practitioners must inform themselves of new and updated professional statements in order to remain professionally competent' (2017c)), the 'ICE Thinks' ('a conduit for everyone's ideas, views and opinions that builds into a think tank focused on the future challenges that we need to address right now' (ICE 2017a)), and CIBSE's technical memoranda (TMs), the institutions are directly researching and generating new knowledge for the benefit of their members. Such activity should be at the heart of knowledge organizations like professional institutions, even though it currently represents only a very small proportion of the knowledge required to practice or even to stay up to date.

The institutions have developed, with considerable investment, globally significant knowledge activities, including the National Building Specification (NBS) and the Building Cost Information Service (BCIS); both profitably owned and run by the RIBA and the RICS respectively. They represent a high-level knowledge resource for their members and an essential tool for practice, exactly the sort of knowledge that a profession needs to have at its command. In practice, of course, the knowledge is available on equal terms, and at the same cost, to whoever wants to purchase it, making the professional distinction significantly less relevant. Also, having become successful stand-alone business units within, yet isolated from, the institutions, these important knowledge centres have found it difficult to feed in to the necessarily more open knowledge culture of a community of professional practice. Somewhat more surprisingly their success has also not been replicated in the development of other similar intensive knowledge- and intelligence-based activities within the institutions, focusing, for example, on building energy performance or data on products and materials.

The development, maintenance and sharing of a distinct body of knowledge, while critical for the continued existence of a profession, represents a huge challenge for any organization with limited resources. It is expensive to both collect and keep up to date; difficult to make comprehensive and useful; and can duplicate the work of others who are more adept and agile at producing and disseminating it. Despite all of their early investment in substantial libraries during the 20th century the institutions gradually reduced their support for them as useful tools for practice, as opposed to historical research. Late in the century these expert knowledge bases were replaced or supplemented by their online equivalents. In 1998 the RIBA launched RIBAnet, an online self-help community for its members. The initiative was largely replaced by RIBA Knowledge Communities in 2008 and has since become a LinkedIn discussion group. The ICE and RICS, along with most other professional communities, run equivalent groups. All the institutions, to a greatly variable extent, provide knowledge resources online, including traditional documents in PDF form, filmed lectures and discussions, checklists and 'toolkits'. What appears to be absent is any significant use of digital innovation by the institutions to commission, curate and provide access to professional knowledge for their membership as part of their subscription package.

In practice the great majority of knowledge gathering and dissemination has long been ceded to other bodies: to the universities for detailed academic work; to research and technical organizations including BRE, BSRIA, CIRIA, product manufacturers and others for practical building intelligence and skills; and to official and historically focused organizations for much on-the-ground information on the existing built environment. As a result of the ICT revolution, knowledge applicable to the industry has never been greater or easier to find, even though the sheer quantity of information available makes the prospect of keeping up to date, synthesizing information and maintaining competence extremely challenging.

The availability of knowledge has meant that it is no longer domain specific and within the control of particular professions. Despite the existence of pay walls and exclusive membership areas of institutional websites, individuals no longer need to be members of professions in order to be able to gain the information base they need to practice. As a result many individual practitioners have chosen to and will increasingly compete from outside professional structures. This challenge to professional knowledge also occurs at a time when many companies and other organizations have developed private bodies of knowledge, retained and protected in-house with the purpose of giving them a commercial edge. Many PSFs now have their own information resources and knowledge tools, which gives them the freedom to ignore institutional knowledge sources if they so choose. This potentially increases the developing divide between large and small professional firms. Developments in knowledge accessibility and management have thus simultaneously hurt the professional institutions, potentially causing serious damage to one of their core activities. They urgently need to regroup and regain control over this key terrain (see Samuel 2018 among other analyses).

The problem facing the institutions is a presumption that privately controlled knowledge provides competitive advantage and should therefore be kept confidential and used predominantly for commercial purposes. This has compounded a historic reluctance among professionals to share current knowledge in the industry, even in anonymized form such as the CarbonBuzz programme. This has led to a dangerous paucity of specialist knowledge developed by and shared across the breadth of professional communities. Such knowledge is now more consistently being developed within PSFs or focused special interest research groups and then retained for its commercial value. This problem is compounded by the commercial confidentiality afforded to private sector partners and the invisible, but nonetheless significant, barriers between practice knowledge and academic research. There are areas of knowledge, particularly topics like energy usage, where analyses of some of the large data sets available to researchers are becoming publicly available and are beginning to influence practice. But overall professional knowledge in the industry, whether institutional or private, is atomized and reliant on a series of relatively small, walled-off data sets. The construction sector compares poorly with areas such as transport or cartography, where publicly available 'big data' has allowed citizen 'hacktivists' to produce up to date and socially usefully information.

The ownership and situation of professional knowledge has changed significantly in recent years. Openness and accessibility on one side and hoarding on the other has left the institutions little space to operate and capacity to flourish. Nonetheless there is a gap that they could fill through mandating the collection of knowledge from practitioners (cf Duffy and Rabeneck 2013; Foxell 2017), particularly about building performance in use, and, in return, providing an essential resource for practice. The work of CABE showed that it is possible to synthesize important industry issues into a series of straightforward and readable briefing documents; to undertake research into critical issues and publish the results; and to provide guidance, best practice information and case studies within a defined and limited budget.

Qualification and membership

The response of the institutions to recent challenges and the diversity of practice in the industry has been to become more service-orientated and intent on providing a popular product and experience for their membership. In order to make this work economically, institutions now pursue a growth model of ever expanding membership and services. This has many ramifications, not least in the way that the institutions run themselves, which will be discussed below. The challenge will be ensuring that the quality of individual members is maintained while their numbers continue to increase and, further into the future, how to respond when the pool of potential members runs dry.

Control over entry into a profession and the standards required to achieve it is a critical aspect of the professional system and is key to everything that follows. Such control can be exercised over education and training or at the point of qualification itself, or more likely both. The development of education in the UK, as described in Chapters 5 and 7 has led to a range of organizations being involved in the education process, from universities to semi-governmental registration bodies, narrowing down towards the final qualification stage. During the latter part of the 20th century the professional institutions played a strategic role in education: setting curriculum objectives, and providing visiting panels and course accreditation. This avoided involvement with individual students and other applicants, barring exceptional circumstances, until they seek chartered status. From that point the systems of entry into the different institutions vary greatly, from the acceptance of qualifications provided by the external bodies, including the all important Architects Registration Board (ARB) required for RIBA membership, to the full rigours of the RICS's in-house Assessment of Professional Competence (APC) process (see Chapter 7). Unfortunately, for a membership growth model, institutions do not have any direct control over the number of applicants coming forward, at least in the UK.

This has meant that the main target areas for expanding membership have inevitably been overseas, particularly in the major growth economies such as China and India where equivalent national professional bodies may be unable to offer as useful and transferable a qualification as one from a well-known international institution. For any institution the overseas market represents an immense potential pool of members, but in seeking to access them they find themselves competing with their international rivals. They will need to address serious challenges of how to provide relevant services across geographical, cultural and political divides.

Recruiting and providing services is much easier at home and expansion there too has not been neglected with offers of institutional affiliation being made to a wider and wider range of people. Different membership grades have been developed, accompanied by greater flexibility around who is qualified enough to start the membership process – see below. All of this nonetheless carries the danger that the quality threshold for membership will be seen to have dropped and that the potency of institutional brands may become diluted.

To date the overseas expansion of the institutions has appeared to be generally positive. It has allowed the institutions to become more 'international' and therefore relevant in a connected world. The RICS, in particular, recruited two of its recent presidents, See Lian Ong (2011–12) from Malaysia and Martin Brühl from Germany (2015–16), from its overseas membership. This in turn has created more opportunities for exporting professional services from the UK supported by the recognition of various professional 'superbrands'.[2] The RICS in particular has stated that 'the levels of skills and professionalism prevalent in the UK are in huge demand overseas, are marketed by UKTI and are setting the benchmark for professionalism in many countries worldwide' (Tompkins 2014, p2). Sean Tompkins, current CEO of the RICS, has stressed the importance of supporting this with an increased emphasis on standards and vigilance concerning professional ethics – 'At the RICS we spend about a third of our turnover specifically on developing standards, education, ethics and regulating the profession' (The Edge 2014b, p4).

Professional affiliation is not obligatory for practice in the UK and UK professional membership is certainly not a requirement elsewhere. This means institutions need to make membership as attractive as possible, find ways to encourage potential professionals to join and ensure that the process is as smooth as possible, especially for graduates of accredited courses. This has led all the institutions to develop student and graduate membership schemes. Other forms of membership are offered to those who are working in the same field who are without the necessary qualifications for full membership. This has led to Associate memberships of the RICS, RIBA and CIBSE, Technician membership of ICE and Non-Chartered membership of the CIOB. The membership rules for each of these are different but all expect some degree of experience or qualification, recognize these members as professionals and bind them to the institution's code of conduct and regulations. There is then a further category of members, who might be described as professional friends. These are Associate members of the ICE and Affiliate members of the RIBA and CIBSE. Qualifications are not necessarily required, no additional professional status is bestowed but compliance with codes of conduct is still required. It is highly unlikely that a member of the public would be able to differentiate between many of these alternative grades of membership.

Despite their desire for expansion the institutions have found it difficult to get their memberships to reflect the demographics of society (see discussion and numbers in Chapter 7). This represents an especially pressing difficulty in the UK given the problems that the industry is likely to be faced with: replacing an aging professional workforce; workers from overseas who are being encouraged by government to return home or not to come in future; and ensuring that future professionals in the industry are as diverse as the communities and building users they work with. For all the outreach work done by the institutions to inform and enthuse school children and students about the benefits of a career working in the industry and of the opportunities in different specific professional disciplines, the demographic profile of construction professionals has been slow to change. Professional structures have also found it difficult to accommodate individuals with changing and flexible career paths. The institutions have been hesitant in extending a welcome to those who have arrived from elsewhere in the industry.

Membership clearly retains its importance and value to individuals in the industry well beyond obtaining an institutional affiliation to display. Their role as membership bodies remains the institutions' most significant function: both for the members as sources of validation, collegiality and support and for the institutions in terms of influence and their own financial survival. But what does institutional membership offer in the early 21st century?

The institutions themselves make their pitch to new entrants with the promise that membership brings status and respect:

'Professional registration provides a benchmark through which the public, employers and their clients can have confidence and trust that registered engineers and technicians have met globally recognised professional standards' (Engineering Council 2017b).

'Joining RICS not only provides you with a prestigious professional qualification, it also offers you a genuine competitive advantage. An RICS professional qualification demonstrates that you work to the highest standard of excellence and integrity' (RICS 2017e).

'We provide the standards, training, support and recognition that put our members – in the UK and overseas – at the peak of their profession' (RIBA 2017).

'When you have ICE letters after your name it gives you the recognition and credibility to work in the built environment anywhere in the world' (ICE 2017b).

At the core of membership is the notion of a standard reached and verified at the point of qualification and then maintained and enhanced throughout a career. The credibility of this notion was much enhanced in the 1990s with the introduction of compulsory continuing professional development (CPD) across the industry, but it is still true that from the day of qualification (often organized and verified by a third party provider) an individual professional needs to have relatively little contact with their institution. So far there is little demand, or possibly need, for the regular retraining and competency training that doctors are required to go through. Nonetheless the claims that the institutions make err on the side of hyperbole. See Figure 9.1 for an overview of the services that institutions offer both their members and the wider public.

Conduct and behaviour

On joining members will be agreeing to abide by the Code of Conduct of their institution. For several disciplines this is effectively a double code. Engineers are subject to the code of the Engineering Council as well as their home institution. Architects need to comply with the ARB code and, if they become chartered members, the RIBA's.

The codes of conduct of the RIBA, RICS and ICE, along with eight other UK professional organizations plus that of the umbrella group, the International Ethics Standards coalition and the Society of Construction Law (see Table 9.2), are compared in Figure 9.2, sorted by theme and subject. Many of the organizations use the four broad categorizations of Integrity, Competence, Consideration for Others and Professional/Institutional matters to lay their codes out and the table follows the same arrangement, even if individual clauses have not always remained in the same categories as their organizations placed them. The level of detail in the codes varies considerably. In part this is due to the additional guidance notes and references to other documents used by some of them, but often it is a matter of style and confidence.

The codes cover very similar ground, sometimes with almost identical wording with the majority stressing issues like honesty, fairness, impartiality, client confidentiality, skill, care, health and safety, recording appointments in writing and non-discrimination as well as injunctions against bribery. Some topics on the other hand are only covered in one or two codes including:

- Plagiarism (ICE)
- Finance and dispute resolution (ARB and LI)
- Using best endeavours (RIBA).

Each of the nine chartered bodies included has a public interest obligation agreed with the Privy Council and written into its charter. Of these, five have expressly passed a requirement to act in the public interest on to their members via a clause in their code of conduct:

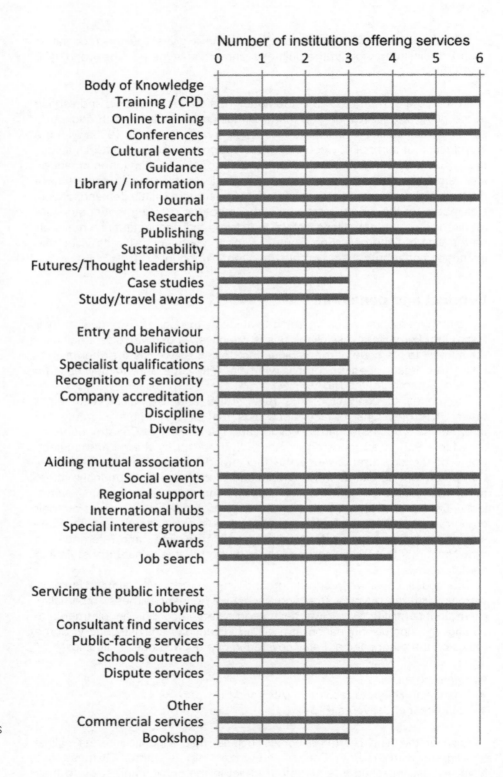

Number of institutions offering services

Body of Knowledge	
Training / CPD	6
Online training	5
Conferences	6
Cultural events	2
Guidance	5
Library / information	5
Journal	6
Research	5
Publishing	5
Sustainability	5
Futures/Thought leadership	5
Case studies	3
Study/travel awards	3
Entry and behaviour	
Qualification	6
Specialist qualifications	3
Recognition of seniority	4
Company accreditation	4
Discipline	5
Diversity	6
Aiding mutual association	
Social events	6
Regional support	6
International hubs	5
Special interest groups	5
Awards	6
Job search	4
Servicing the public interest	
Lobbying	6
Consultant find services	4
Public-facing services	2
Schools outreach	4
Dispute services	4
Other	
Commercial services	4
Bookshop	3

Figure 9.1
Range of services offered by six institutions in 2017

Source: based on a survey of the public websites of six institutions: CIBSE, CIOB, ICE, RIBA, RICS and RTPI.

Table 9.2 Society of Construction Law, Statement of Ethical Principles (2017)

Ethical conduct is compliance with the following ethical principles:
1. Honesty – act with honesty and avoid conduct likely to result, directly or indirectly, in the deception of others.
2. Fairness – do not seek to obtain a benefit which arises directly or indirectly from the unfair treatment of other people.
3. Fair reward – avoid acts which are likely to result in another party being deprived of a fair reward for their work.
4. Reliability – maintain up to date skills and provide services only within your area of competence.
5. Integrity – have regard for the interests of the public, particularly people who will make use of or obtain an interest in the project in the future.
6. Objectivity – identify any potential conflicts of interest and disclose the conflict to any person who would be adversely affected by it.
7. Accountability – provide information and warning of matters within your knowledge which are of potential detriment to others who may be adversely affected by them. Warning must be given in sufficient time to allow the taking of effective action to avoid detriment.

See www.scl.org.uk

- 'All members shall have full regard for the public interest' (ICE 2014, Rule 3)
- 'Acting consistently in the public interest when it comes to making decisions or providing advice' (RICS 2012, Standard 1)
- 'to safeguard the public interest' (CIBSE 2015, p1)
- 'Members shall … have full regard to the public interest' (CIOB 2015, Rule 1)
- 'have regard to the public interest' (The Institution of Structural Engineers 2014, Rule 2)

The others are more circumspect, with the Landscape Institute restricting the obligation to a 'regard' for 'the interests of those who may be reasonably expected to use or enjoy the products of their work' (2012, Standard 1) and the RIBA urging their members to 'balance differing and sometimes opposing demands (for example, the stakeholders' interests with the community's and the project's capital costs with its overall performance)' (2005a, Clause 2.1).

Keeping up to date and developing and maintaining skills has in recent years become a central feature of all institution memberships, even if not all of them spell it out in their codes of conduct (seven out of nine). They require that all their professional members participate in a relevant programme of continuing professional development (CPD), usually backed by a monitoring scheme and a discipline process for penalizing errant members. Typically an institution requires a minimum number of hours of CPD per annum: 20 for the RICS, 35 for the RIBA and 50 over two years for the RTPI. This obligation mixes formal sessions with more informal learning and needs to follow either a personal development plan (PDP), a CPD curriculum, or both. Members are expected to record their CPD and to submit their records to their institution when requested to do so.

CPD is promoted through a mixture of carrot – 'you need to keep your career moving upwards' (ICE 2016) – and stick – 'Failure to comply with the rules on CPD will be treated in the same way as any other rule breach under RICS Disciplinary Rules. The range of sanctions includes censure, fines, publication of disciplinary findings and expulsion from membership' (RICS 2016a).

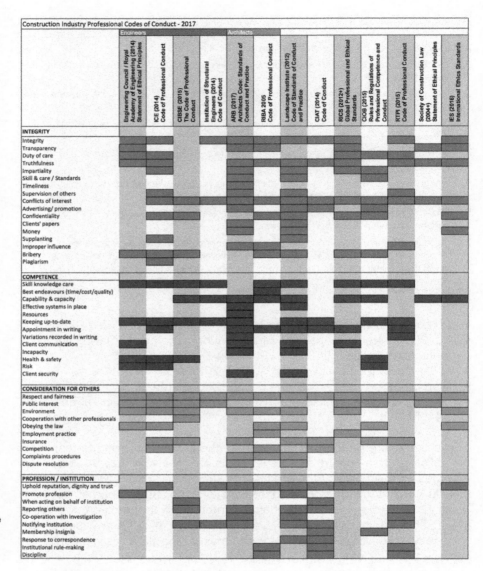

Figure 9.2
Professional codes of conduct comparison for 13 organizations, showing where they have explicitly included issues in their published code documents

In addition The Society of Construction Law has drafted and published a statement of ethical principles which 'apply to the work of all professionals working in the construction industry, whatever their original qualification or affiliation' (see Table 9.2).

The various codes of conduct have, in recent years, been the subject of detailed analysis and criticism. Jeremy Till (2009, p181) looking specifically at codes for architects has written:

> The point, whether in the ARB, RIBA, or AIA Code, is the same: behaving according to professional 'ethics' is not the same as behaving ethically. Indeed they might actually be Codes of Misconduct.

Hill *et al.* (2013, p13), writing from the point of view of surveying and land valuation, discuss the construction and property professions in general:

> The institutions seem unwilling to engage with the political and ethical dimensions robustly, as this appears to them to be somehow beyond the remit of their essentially technocratic skills and purpose.

The problem, as expounded by Till (pp181–2), is that there is a deep confusion between acting professionally and acting ethically:

> There are ways of acting professionally and there are ways of acting ethically: the two operate according to different parameters. The former, professional, life is prescribed by the various codes. These are overseen not by a sense of duty to society at large, but by service to the client and employer. They have no aspirations; they are there merely to draw a line across the minimum, extremely basic, standards. As minimum standards they are reasonably easy to fulfill; the problem is that they are often taken as the complete description of an architect's duties. ... They do not address the wider responsibility of the architect.

Till suggests, following Zygmunt Bauman and Emmanuel Levinas, that ethics should be 'defined simply and directly as "being-for the Other." To assume an ethical stance means to "assume responsibility for the Other" ' (2009, p173). However this definition threatens to conflate two aspects of conduct and behaviour that this book has been keen to differentiate between: ethics and the public interest. Ethics are here defined in the narrow sense as group behavioural codes – specifically, in this case, of construction and property professionals. Morality is a general and universal system, as defined by Audi (1999, pp284, 586), in contrast to ethics:

> **ethics**, the philosophical study of morality ... sometimes it is used more narrowly to mean the moral principles of a particular tradition, group or individual. ...
> **morality**, an informal public system applying to all rational persons, governing behaviour that affects others, having the lessening of evil or harm as its goal.

The importance of maintaining such distinctions is to achieve clarity about what is expected of professionals. There is no reason why professionals should be expected to behave better (i.e. with more goodness) than anyone else in a role requiring a degree of responsibility. Certain professionals may or may not choose to take a position in line with their conscience and so long as that does not conflict with the freedoms of others they are free to do so. But by becoming members of a chartered body, which passes its public interest obligations on to them (see above) they have a significant obligation to act in the greater public interest; that is, with morality. This may result in an obligation to design sustainably to protect the interests of future generations or to take decisions with even-handedness and fairness in order to uphold the mechanisms of civil society. These should not be interpreted as professionally 'ethical' acts, unless their profession has specifically agreed and instructed them to be so.

This reasoning leads to the principle that although effective professional codes of conduct need to be comprised of sets of both components: those intended to serve the public interest (morality) and those that maintain professional propriety and standards of service (ethics), there should be complete clarity between them. Of the four code elements discussed above, 'consideration for others' is in the ethical+moral category while 'professional/institutional matters' is only in the ethical. 'Integrity' and 'competence' serve both categories and should be disambiguated. Goodness has little to do with any of these, although, for the professions' sake, it is important that professionals shouldn't behave 'immorally' (in the common sense).

Discipline

Having attracted members with a persuasive mixture of status and services it becomes incumbent on the institutions to ensure those members then maintain their standards of expertise and performance, comply with the code of conduct and avoid bringing the profession and institution into disrepute. This requires having mechanisms and procedures that monitor, judge, discipline and occasionally expel errant members. Historically institutions have relied on complaints from clients and others to flag breaches in professional behaviour, although they've also paid attention to prosecutions involving their members. A guilty verdict in the courts may provide grounds for rescinding an individual's membership even if the case has nothing to do with their professional work. They do not generally actively investigate cases on their own behalf, or monitor insurance claims made against their members.

The one current exception to this is CPD compliance. The RICS disciplines its members when they don't fulfil their obligations and is the only institution in the sector to openly publish the findings of their judgements and any measures taken against members for breaches of their code. It is notable that the great majority of recorded disciplinary measures have been taken against practitioners working for themselves or for small businesses, even while they account for a low proportion of the work carried out by members of the RICS.

In practice the emergence of PSFs has moved a significant group of all institutions' members effectively beyond regulatory reach. CPD tends to be provided automatically for employees and almost all misdemeanours by large firms or their employees are handled internally or at least in private. From a complainant's point of view their concerns are best taken up with the company directly or their insurers, and if an issue needs to be taken further it is likely to end up in adjudication or in extremis in the courts, rather than with the institutions. The only recent case where a professional institution has disciplined a major PSF was when the RICS fined the Sweett Group PLC in August 2016 (RICS 2016b), after the company had been convicted in the Crown Court under the Bribery Act 2010.

Values and standards

Despite any domestic concern for the standard of professionals' performance, the UK institutions' reputation for inculcating and maintaining ethical conduct and high standards of behaviour is perceived to be strong when viewed from overseas. This is a significant market lever in many of the emerging markets of the world where local standards are relatively weak or lack sanction. As Will and Cathy Hughes (2013, p36) have noted in a paper on the relationship between professionalism and professional institutions:

> the current brand of professionalism that includes ethical codes of conduct buttressed by self-enforced sanctions is extremely popular in those jurisdictions. This has undoubtedly contributed to the surge in membership of institutions such as the RICS and CIOB in many developing countries where CSR has yet to emerge.

Hill *et al.* (2013, p13) concur when they report that the RICS has:

> found that one of its major attractions to individual property professions in other countries, particularly in South America, Africa, India, the Middle and Far East, is its regulation of educational standards and code of conduct, and its perceived independence of national political and business interests.

This view of the institutions as upholders of professional values and standards is an essential part of their appeal to both members and educators – who, possibly reluctantly, agree to operate to their standards – as well as to clients and the general public. The potency of institutional branding would appear to chiefly be of benefit to SMEs and smaller, single-discipline companies who can also use it as an imprimatur for themself. It is also valued by the PSFs, who actively choose to hire both qualified and chartered professionals for their staff and then pay their (tax-deductible) institutional subscriptions to ensure they keep up their memberships. This is likely to continue so long as the institutional brands are stronger than those of the PSFs themselves, but as the power of the PSFs continues to rise and that of the professional institutions remains fragile this cannot be indefinitely relied on.

The PSFs themselves continue to have an ambivalent relationship with the institutions. Although numerous linkages exist between the two there appears to have been a failure to embrace each other as firmly as the institutions and the public sector departmental offices did in the mid-20th century. One indicator is the composition of the institutional presidencies. In the years since 1993 the ICE has been the most successful at bringing in a range of backgrounds, with nine presidents out of 24 who were arguably from a PSF background, the RICS has had five, the CIOB one and the RIBA none from its 12 bi-annual presidencies.

This distance also affects the ability to enable and maintain a collegiate solidarity in each discipline, enabling the exchange of ideas and the provision of support when it is required, as well as the notion of a community of practice collectively working towards a shared goal. At present those communities are still approximately in place: they enable practitioners from different parts of the industry to come together to jointly tackle issues of concern, there is mutual recognition of qualifications and individual professionals can move relatively freely between different sectors. It is still even possible to grow a small firm into a major industry name. But as the PSFs increasingly operate in accordance with their own rules and use specialist and company-specific technology and ways of working, the existing fracture lines in the industry could become impassable. This might make it necessary for future entrants to choose at an early age which part of the industry they want to build their career in, with limited opportunity for movement across sectors later on.

Traditionally the industry has divided vertically into specialist silos. Generally it has proved difficult to cross these disciplinary boundaries. The current threat is different, today's fissures are horizontal, threatening to separate small from large, and practitioner run and owned from larger corporate entities. The institutions will need to work hard to keep their memberships together and all benefitting from a system of professions that works for everyone.

The division is also one that affects one of the cornerstones of professionalism, the principle of self-regulation. Traditionally professional firms owned and run by practitioners were directly connected to institutional structures based in the nation state of their main office. As PSFs have grown and adopted ownership structures that are remote, not necessarily professionally based and potentially international, self-regulation has become increasingly strained. Individual members may still subject to regulation but their employers, where responsibility properly lies, frequently cannot be. If judgement is to be made by professional peers how are the individuals who volunteer to join the institutions' discipline panels to grapple with corporate issues?

Association

Meeting to share information, experience and expertise is a critical aspect of professional organizations. Traditionally this will have been done at various face to face gatherings – business, educational, cultural and social – which will have contributed to the group's cohesion and collegiality, helped develop its knowledge function and increased the skills and confidence of members in the performance of their work. The downside was that this functioned best in big and busy cities and poorly in less populated and geographically challenging areas. In theory the advent of social media and other forms of virtual communication should have been hugely beneficial in bringing often widespread and sometimes isolated practitioners together.

Modern institutions have invested, with varying degrees of success, in communications technology in order to bring their membership communities together and provide them with news, information and access to professional activities wherever they are. The perception that most institutional events happen in and around their headquarter buildings is still largely true, but all organizations have become better at keeping members connected and updated on events and at providing them with platforms to share their concerns and to tap and share knowledge with their co-professionals. The potential is there, at least, but the anticipated interaction based around the membership community does not appear to have taken place.

Professional institutions have made relatively good use of new communications capacity to enable and expand traditional forms of professional relationships, especially the ability to communicate outwards from the centre, but they have yet to explore in any depth its ability to develop and foster new forms of professional exchange and organization. There is a degree of interactivity on official institution sites maintained on LinkedIn, Facebook and other social media sites that allow members to post opinions and questions, and start discussions, but they appear to be used predominantly for promotion rather than to enable two-way communication. Despite the potential there are few non-traditional ways of interacting with the life of the institutions, inputting policy ideas or helping with the decision-making processes of what are membership bodies. The professional institutions have largely chosen to be non-interactive, something they need to urgently remedy (Foxell 2017).

In contrast successful virtual groups have developed outside the institutions' immediate jurisdictions, through popular blogs, Twitter feeds and discussion groups on various platforms. Sites have developed (and disappeared again) run by numerous organizations, including charities, think tanks, pressure groups and commercial bodies as well as enthusiastic individuals. Such groups are less immediately exclusive than membership bodies but may have restrictions of their own, but, in general, everyone, professional and non-professional alike, has the same level of access to them.

One explanation for the absence of interactivity is that the professional institutions are gradually transforming themselves into professionally run service organizations efficiently providing a range of experiences, amenities and services to their membership and others rather than membership clubs for association and collective action. A small number of members may be elected or selected to a governing board that makes strategic decisions and sets targets, but the main actors are a strong chief executive and staff empowered to make decisions for themselves. An entrepreneurial executive may develop new and better services for both the benefit of the membership and the finances of the organization but such a control structure is at odds with principles of membership control and decision-making that were central to the institutions until the turn of the century. It is a transition that has been marked by a reduction in membership committees, increased staffing levels and the arrival of management consultants.

The outcome has been to downplay the mutual association aspect of institutions in favour of greater and more active provision of services, as well as improved financial stability. The institutions still have members in positions of authority: as presidents, vice-presidents, board members and trustees; and they could choose to revert to a more collegiate form of control if they wished. Members still sit on committees but it is more commonly in order to provide advice on workstreams run by staff than to discuss and develop projects to be undertaken by members themselves.

Local professional groups, outposts of the regional offices of the institutions, still work as traditional membership groups running their own activities with minimal central support. They are vestiges of a former way of running the professions and, although mutually supportive, locally based organizations may reappear in the future, for the moment it appears that virtual communities are adequate to fulfil the greater part of the need for member interaction.

Public interest

The transfer of the public interest obligation from chartered body onto the membership has already been noted in the discussion on codes of conduct, but the obligation to serve society positively remains with the individual institutions under each of their Royal Charters and should be central to their activities. The eight institutions discussed above have obligations as shown in Table 9.3.

Table 9.3 Institution Charter Objects concerning the public interest

Institution	Charter Objects
CIBSE (1976)	I. The promotion for the benefit of the public in general of the art, science and practice of such engineering services as are associated with the built environment and with industrial processes, such art, science and practice being hereinafter called 'building services engineering'; and II. The advancement of education and research in building services engineering, and the publication of the useful results of such research.
CIOB (1980)	a. The promotion for the public benefit of the science and practice of building and construction; and b. The advancement of public education in the said science and practice including all necessary research and the publication of the results of all such research.
ICE	To foster and promote the art and science of Civil Engineering.
The Institution of Structural Engineers (2005)	To promote for the public benefit the general advancement of the science and art of structural engineering in any or all of its branches and to facilitate the exchange of information and ideas relating to structural engineering amongst members of the Institution and otherwise.
Landscape Institute (2008)	To protect, conserve and enhance the natural and built environment for the benefit of the public by promoting the arts and sciences of Landscape Architecture (as such expression is hereinafter defined) and its several applications and for that purpose to foster and encourage the dissemination of knowledge relating to Landscape Architecture and the promotion of research and education therein, and in particular to establish, uphold and advance the standards of education, qualification, competence and conduct of those who practice Landscape Architecture as a profession, and to determine standards and criteria for education, training and experience.
RIBA (1971)	The advancement of Architecture and the promotion of the acquirement of the knowledge of the Arts and Sciences connected therewith.
RICS (1974)	To secure the advancement and facilitate the acquisition of that knowledge which constitutes the profession of a surveyor, namely, the arts, sciences and practice of: a. Determining the value of all descriptions of landed and house property and of the various interests therein and advising on direct and b. Indirect investment therein; c. Managing and developing estates and other business concerned with the management of landed property; d. Securing the optimal use of land and its associated resources to meet social and economic needs; e. Surveying the fabric of buildings and their services and advising on their condition, maintenance, alteration, improvement and design; f. Measuring and delineating the physical features of the Earth; g. Managing, developing and surveying mineral property; h. Determining the economic use of resources of the construction industry, and the financial appraisal, management and measurement of construction work; i. Selling (whether by auction or otherwise) buying or letting, as an agent, real or personal property or any interest therein and to maintain and promote the usefulness of the profession for the public advantage in the United Kingdom and in any other part of the world.

It is noticeable that the objects in all the charters focus on the overall benefits of the institutions to society and their role in acquiring and disseminating knowledge, avoiding any direct mention of their members. The public benefit or advantage is cited in six of the eight sets of objectives, while those of the ICE and RIBA, transposed and abridged from a period when the public good was still a nascent concern, deal only with the 'art and science' of civil engineering and architecture respectively.

In an era that stresses service delivery over public obligation it is unclear how the institutions currently understand their public service remit and in what way it influences their activities. All of them have, to an extent, a public facing role, although this can vary considerably. They organize lectures, hold exhibitions, run award schemes, work with politicians to promote their part of the industry and run workshops with schoolchildren to encourage them into the profession. They have a direct interest in maintaining and improving the standards in their discipline and ensuring the skill levels of their members are adequate to be able to deliver them. What they don't have is a clearly articulated narrative explaining their public interest obligations either to themselves, the public or even to the Privy Council, the body responsible for their charters.

The Edge Commission on the Future of Professionalism directly tackled this issue in the hearing sessions it held with the institutions in 2014. The Inquiry repeatedly tried to determine the public interest role of the institutions, but achieved very little elucidation. As the author of the Commission report, *Collaboration for Change* (2015, p52), Paul Morrell, noted:

> ambivalence about the precise definition of the public interest and institutions' expectations of themselves and their members becomes, at best, a cause for confusion and contention within the membership, and at worst grounds for an accusation of hypocrisy. ...
>
> It would therefore be genuinely in the public interest if the institutions were to clarify and codify exactly how they understand the term 'the public interest' in pursuit of the obligations of their charters, and produce (as for ethics) a rigorous, harmonised view of their expectations, both on behalf of themselves and of their members. This would include articulating the issues that arise, engaging with the public, raising the profile of public interest with members (as for ethical issues) and giving them the practical guidance – specifically as to the extent to which their conduct and practice should be modified to acknowledge a duty that extends beyond the immediate one owed to clients.

Current challenges

There are clearly a series of threats facing the construction industry professions that are large and inchoate, but also very present. The Edge Commission report (2015, p10) included a diagram with the following general threats to professional organizations: performance gap, climate change, loss of public trust, globalization, low level of member engagement, growing scepticism of younger professionals, siloed education, proliferations of 'professions', rise of megafirms and limited learned society role. To these could also be added the rise of automation and the 'gig' economy, the social divisions highlighted by the vote for the UK to leave the European Union and the reinvigoration and in some place elections of populist politicians across the world.

Some of these issues have already been addressed in previous sections, but together they appear to add up to a 'perfect storm' that it is difficult to see the institutions making preparations for. The challenge to individual practitioners will be examined in the next chapter but this chapter now turns to the challenges facing the institutions themselves and how they might respond.

Climate change

The issue of sustainability and the challenge of reducing energy use have been at the forefront of professional concern since the oil crisis of the 1970s. Climate change and the need to cut the level of carbon emissions entering the atmosphere joined them following the 'Earth Summit' held in Rio de Janeiro in 1992.[3] Since then the subject has dominated both public and scientific discussion as it has become clear that the danger of global warming and destabilized climate conditions are real. In December 2015, in Paris, governments and other international bodies from around the world agreed (United Nations 2015, Article 2a) to hold:

> the increase in the global average temperature to well below 2°C above pre-industrial levels and pursu[e] efforts to limit the temperature increase to 1.5°C above pre-industrial levels, recognizing that this would significantly reduce the risks and impacts of climate change.

This agreement is now ratified by 142 of the parties to the UN Convention on Climate Change, including the UK.

The construction industry institutions have taken an active part in the discussions and actions around dealing with climate change. Out of the many responses made by them over the years the RICS, RIBA and RTPI grouped together under the umbrella of the Global Alliance for Building Sustainability at the World Summit on Sustainable Development in Johannesburg and issued a Charter for Action (2002). The RIBA launched its 'Combating Climate Change', comprising a series of publications from October 2007 on, and the UK Green Building Council was launched in 2007 with the support of CIBSE and CIOB as well as the professional umbrella group the Construction Industry Council. There was even a proposal from a group co-ordinated by the RICS in 2010, following their Global Climate Change Policy adopted in 2009, for the institutions to collectively apply to the Privy Council to modify their Charters to include clauses that required members to 'meet the challenges of climate change, population growth and the depletion of natural resources' (All Profession Charter Reform Group 2010). The suggestion was not taken up.

Following the burst of activity on the part of the professional institutions on climate change at the time of the UK's 2008 Climate Change Act and the 2009 UN Climate Change Conference (COP 15) in Copenhagen, the change of UK government in 2010 pushed the issue down their lists of priorities. They haven't entirely forgotten its significance and continue to issue briefing documents and respond to consultations, especially on resilience issues such as flooding, but the degree of urgency has fallen away. The institutions appeared unwilling to continue looking for answers to the climate challenge or how, for example, to achieve the UK's commitment to reduce carbon emissions under either the five yearly Carbon Budgets produced by the Committee on Climate Change or the Paris Agreement of 2016.

Certainly the institutions have not taken the opportunity to make themselves essential repositories of knowledge or independent and unbiased sources of advice and expertise on matters relating to climate change and the built environment. None of them would argue with the finding that 'the amount of CO_2 emissions that the construction industry can influence is significant, accounting for almost 47% of total CO_2 emissions of the UK' (BIS 2010, p3), yet action has been limited. This response from the institutions seems surprising given their overall ambition to influence government policy makers and be consulted by government on all-important matters in the sectors they cover. Instead the role of advising government since the demise of CABE has fallen to a number of the larger PSFs, including engineering consultancies, but particularly the major management consultancy firms.

Similarly the institutions have made little attempt to require their members to reduce the carbon emissions from their projects. The UK government's Committee on Climate Change, amongst others, made it clear that the built environment sector would need to make an above average contribution to achieving the 80% reduction in emissions by 2050 required by the 2008 Climate Change Act (Committee on Climate Change 2008, p79), but responding to the challenge, beyond compliance with regulations, has continued to be a voluntary activity.

Membership demographics

The institutions are generally reticent about their membership numbers, with relatively few publishing their annual totals broken down by membership class. The ICE and CIBSE provide membership data most years, but the RIBA and RICS only give the headline total for their members and even then not on a regular basis. None of the institutions provide regular information on the gender split in their membership, if they provide it at all. The registration bodies are more reliable and provide comparable information on an annual basis. Data are not only useful to understand the demographics of the industry, but serve as a useful tool to monitor whether specific policies and strategies are meeting their intended goals.

Despite the variability of the data the numbers for chartered members based in the UK appear to have stayed remarkably stable over the last ten years, although it is likely that some institutions have lost market share as the number of non-chartered professionals has increased. The RIBA reported in 2015 'that the proportion of UK domiciled registered architects who are RIBA members declined to just over 70%, compared with 80% historically' (RIBA 2015a, p6). This is borne out by the number of UK-based ARB registered architects, which rose 20% between 2000 and 2015 (ARB 2000 & 2015), while the RIBA membership remained steady.

As noted, the data on gender in the professions has been provided only sporadically, but 2014–15 was a good point for assessing the position with three of the institutions providing information alongside the more regular data

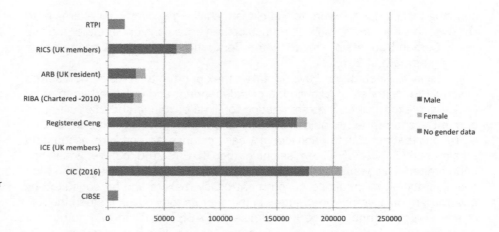

Figure 9.3
Membership and gender
data for institutions and
registration bodies from
2015 annual reports

from the ARB and the Engineering Council (see Figure 9.3). See Chapter 7 for a longer discussion on diversity in the professions.

In 2016 the Construction Industry Council (CIC) accessed institutional data to publish a study, *A Blueprint for Change*, which provided more detailed insight into the make-up of the professional construction workforce, even though their information sources were not disclosed. A sample population of 207 564 professionals produced a 14.1 % female to 85.9 % male gender split.

The CIC's study also included a rare breakdown by age of professional membership drawn from a much smaller sample of 36 756 individuals (see Figure 9.4). The report noted 'the potential for significant progress' implied by the 78 %–22 % male–female ratio in the under-25 age group and the importance of retaining this cohort in the industry if it was to make a long-term difference, but also remarked that 'more people are due to leave the industry due to age retirement, than are joining at a young age' (2016, pp5–6).

The age–gender profiles show the (slightly) higher proportion of women in the younger age groups, especially amongst those aged 25 and under. While this holds out some hope for greater equality in the future (and as shown in Chapter 7 the proportion of female professionals has been very slowly rising) it is also known that there is a significant attrition rate between the stages of: a) studying a professional discipline in the industry, b) qualification and c) five years after qualification. The published numbers on this are minimal, but there have been a number of significant studies, including one commissioned by the RIBA, *Why Do Women Leave Architecture?* (Manley *et al.* 2003) and another, *An Investigation into Why the UK Has the Lowest Proportion of Female Engineers in the EU*, by Engineering UK (Kiwana *et al.* 2011). Neither study provides definitive answers to their headline questions. The RIBA study suggests 16 'identifiable problems',[4] whereas Engineering UK relies on the findings of a study from Enterprising Women, *SET Women* (2002):

> 76 % of women in the UK who finish their education and are qualified in SET are not employed in the SET sector. In addition to this, two thirds of women do not return after maternity leave, therefore, at best, only 24 % of SET qualified women make a long term career in the sector, and at worst only 8 %.

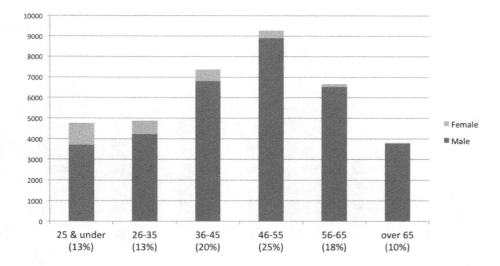

Figure 9.4
Age and gender profiles
of 36 756 construction
professionals, 2016

Source: CIC (2016, p5).

Both studies, along with many others, make extensive recommendations and urge action. It would help considerably if the institutions started to respond by regularly publishing meaningful data.

Such data as exists presents strong evidence for an aging institutional membership that will eventually retire faster than it can be replaced. The increase in the average age (of 2.8 years) for the chartered engineers (see Figure 9.5) may partly be due to postponed retirements for many engineers and some professionals may be joining their institutions later in their careers than previous generations, but there is still likely to be a rapid fall off in membership in the years ahead. The welcome increase in female professionals in recent years, rising amongst registered architects from 12% in 1999 to 25% in 2015, and for chartered engineers from 3.9% in 2008 to 5.0% in 2015, has helped considerably to rebalance the age profile in professions overall, but only if they are not later lost to the industry.

The Edge Commission on the Future of Professionalism addressed the perceptible generational change in the industry professions directly in one of its sessions, having noted that 'some professions are projecting a loss of 20% of their membership over the next decade' (Morrell 2015, p31), hearing from a number of younger professionals, chartered and unchartered, who had themselves gathered opinions from an even wider group of their peers. One unchartered structural engineer, having spoken to 'a wide net' of professional contacts (Malik 2014, p11) noted that:

> There was also an overwhelming consensus that the process to be professionally qualified is unnecessarily confusing, unclear and largely dependent on the company that you're working for. As our sector evolves and requires specialists in new fields, young professionals are finding that they are falling between institutions and join smaller organisations to fit their niche role. I don't find it surprising that with these varying introductions young professionals don't feel connected to their institutions let alone other institutions.

Paul Morrell in his Commission report, *Collaboration for Change* (2015, p31), noted the:

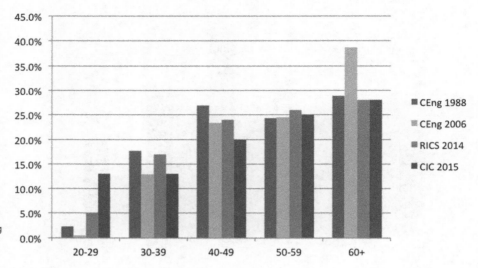

Figure 9.5
Age profiles of
professionals from the
Engineering Council (CEng
1988, 2006), RICS (2014)
and CIC (2015)

call from the upcoming generations for the institutions to look forward and outwards, rather than backwards and inwards: to replace a world view that is too often negative with a positive one; to 'get ahead of the curve' on great social issues; and above all to demonstrate leadership [as well as observing that a] continuing failure to attract the young is potentially terminal.

The demographic challenge to the institutions – from the changing demands and choices of young professionals, new career paths, alternative sources of professional and peer support, to the fundamentals of an aging population – sets out one of the most important tasks for their future, one that has yet to produce a coherent narrative as to the way forward for them.

Silos

In her book, *The Silo Effect*, *Financial Times* journalist Gillian Tett noted 'our world does not function effectively if it is always rigidly streamlined' (2015, p254). The division of labour described by Adam Smith in the *Wealth of Nations* does not necessarily benefit either the work or the worker in a world of complex multi-disciplinary projects, especially where creative input and lateral thinking is required. The construction industry has generally addressed this by almost always assembling teams of professionals from different disciplines to work together to deliver projects. More recently staff in professional service firms have had the opportunity to work and move across disciplines as the need and their developing skills allowed. Yet, in the almost certain face of a multi-disciplinary career, the training for professionals and their professional organizations is determinedly mono-disciplinary. Many have asked whether this approach can last much longer, especially in the era of building information modelling. As Macdonald and Mills (2011, pp1–7) write:

[BIM] requires a cultural shift within the industry from that of fragmentation and litigation to collaborative working and information sharing. It also requires some significant change to the way in which future architects, engineers and construction professionals involved in the building industry are educated.

Efforts are being made by a number of UK universities – including University College London, Bath University and the University of the West of England, which are all running combined courses for engineering and architecture in various configurations, and Cambridge University, which organizes the Interdisciplinary Design for the Built Environment (IDBE) master's course – to address the need and the demand for cross-disciplinary skills, but the dismantling of barriers has yet to develop at a professional level.

As another witness to the Edge Commission (Froud 2014, p3) noted:

interdisciplinary collaboration between professionals should be nurtured by their institutes for the many benefits this may bring to both their interests and that of the public i.e. in doing a job well. They should aim to promote a culture where individual professions, rather than expanding their own territory, move easily within those ease of others.

Young professionals, along with many of their peers in the millennial and post-millennial generations, expect to have portfolio careers, as first described by Charles Handy in 1994 (p175), built up from many different strands and opportunities, paid and unpaid:

Going portfolio means exchanging full-time employment for independence. The portfolio is a collection of different bits and pieces of work for different clients. The word 'job' now means a client.

This offers the potential for far greater professional freedom and autonomy but also the potential for exploitation. It may mean a hand-to-mouth existence with a series of zero hours contracts in the 'gig economy'. Either way professionals without the opportunity or possibly the desire for long-term job security need, even more than others, the support of professional bodies even when their affiliations may not match the work they are currently doing.

In many ways the world of the institutions with their focus on the closely defined specialist skills of individuals and firms no longer provides a good match to the world of work with its more fluid requirements and practices. This is true of both PSFs and individual portfolio workers but also of small companies operating in fairly traditional ways. Everyone is being encouraged to branch out, to develop new skills and to go where the opportunities, or their hearts, lead. The institutions need to develop new ways of supporting and helping them, encouraging the development of new curricula and accreditation of new (interdisciplinary and specialist) courses, even if that means facilitating a move to another professional body.

Automation

The rise of automation and the potential for it to replace, or at least comprehensively change, many professional jobs also represents a serious challenge to the institutions. Automation is currently threatening to cut a swathe through the traditional professions as artificial intelligence-based chatbots have been trained to successfully appeal against parking tickets (Gibbs 2016) and the IBM question answering system, Watson, fresh from winning the US TV game show, Jeopardy (Leske 2013), is being increasingly used to deliver medical diagnoses.

Scepticism has been expressed as to whether construction professionals will suffer the same fate, for as Handy wrote in 2001 (p105):

> Many types of work will continue as they always have done, albeit computer-enhanced, but some will go, never to return, and many new ones will be invented. Town planners, architects and designers may use computers to turn their ideas into working models that they can walk through and around on screen, but there will still be town planners, architects and designers in twenty years' time even if we call them by more grandiose names by then.

But other writers have pointed at products such as Autodesk's Project Dreamcatcher (2017), which allows users:

> to input specific design objectives, including functional requirements, material type, manufacturing method, performance criteria, and cost restrictions. Loaded with design requirements, the system then searches a procedurally synthesized design space to evaluate a vast number of generated designs for satisfying the design requirements.

In their March 2017 UK Economic Outlook report, the consultancy PwC calculated 'that around 30% of jobs in the UK are at potential risk of automation' by the early 2030s, and breaking down that 30%, sector by sector, suggest that 780 000 jobs (25.6%) are at risk in the professional, scientific and technical sector and 520 000 (23.7%) in construction. They anticipate that legal and regulatory barriers to such losses will be gradually overcome and that 'there is therefore a case for some form of government intervention to ensure that the potential gains from automation are shared more widely across society' (PwC 2017, pp3 35, 30). There is also a strong case for preparing a shared institutional strategy and even intervention, well in advance of government being forced to address the issue by a growing crisis.

Conclusion

The short life of CABE shows that it is possible to provide professional and industry support in a very different way from the evolving model run by the institutions for the best part of two centuries. It was not necessarily more successful, as their comparative longevities might indicate, but it has provided an insight into an alternative approach, as have other single-issue support organizations. As the number of challenges to the institutions increase, forced change is becoming more likely and needs to be considered in depth and well prepared for. The next two chapters examine the possibilities in the future for individual professions and the organizations that qualify, support and fight their corner for them – while simultaneously always acting in the public interest.

Notes

1   For example: the RICS information and archive service, the ICE's Ask Brunel, the RIBA's information line and CIBSE's Knowledge Portal.
2   Both the RICS and the RIBA are featured in the UK 'Business Superbrand 2017' list – see www.superbrands.uk.com/b2b-results
3   The United Nations Conference on Environment and Development (UNCED) held in Rio de Janeiro in June 1992.
4   *Why Do Women Leave Architecture?* (Manley *et al.* 2003, p3) reports:

> that there was no definitive answer to the central question. A number of identifiable problems did however come to light. The reasons why women left tended to be a combination of a number of factors and or a 'final straw' moment. Some of the key issues are as follows: low pay; unequal pay; long working hours; inflexible/unfamily friendly working hours; sidelining; limited areas of work; glass ceiling; stressful working conditions; protective paternalism preventing development of experience; macho culture; sexism; redundancy and or dismissal; high litigation risk and high insurance costs; lack of returner training; more job satisfaction elsewhere.

10

Expectations

A new industry professional has passed her exams, breezed through the interview, the subscription to the institution has been paid and she has a certificate, a plastic card with her membership number and a new set of letters to use after her name. There may even have been a ceremony at her Institution's HQ with cakes and sparkling wine, speeches and photos. But what should she expect after all the effort, time and resources it has taken to get to this point? They will have been considerable, with contributions from family and friends, teachers and lecturers, employers and colleagues – she will also, more than likely, be in serious long-term debt. Nonetheless one of life's significant hurdles has been successfully overcome and even if it doesn't make any immediate material difference or change today's work routine, it is a moment of great achievement. Professional accreditation has been gained, but what is anticipated of her as she commences her professional career?

Our new professional has many different idealized versions of her possible professional development to negotiate:

Her own

Her view of professionalism will have been shaped by her experience of training and work, by the people she has met who have supported, encouraged and mentored her, become role models and set examples. She may have a number of professionals she has studied, read about or seen expounding the philosophy that underpins their work. She may have passionately held beliefs that she can do something useful, serve a cause and make a difference. The chances are that most attention during training will have been given to techniques, buildings, history and some of the outsized personalities in the sector. The principles of professionalism and the challenges of a working life will have been peripheral concerns and have only entered the picture relatively late on in the course.

Her immediate focus will be on the quality of her prospective career. She wants to do interesting and creative work that will fulfil her ambitions and earn her a reasonable starting salary. Getting worthwhile experience at this stage may be more important than long-term prospects in a company and job security. Her belief system has driven her to seek out a particular area of work that she is keen to specialize in or a firm whose ethos matches her own. She is keen to 'do the right thing', to do work that contributes to society and the protection of the environment, but she is also keen to impress her new employer in order to get ahead, as well as her peer group, family and friends.

She has an outline career plan that necessarily still needs more strategic and detailed thought and she knows that making a success of it will largely be down to her own efforts, but serendipity, the people she knows and meets and how much the world wants what she can offer it will also have parts to play. A number of red lines already exist in her head, ruling out projects or clients she will not accept, work practices that she wants to have nothing to do with and sectors she will work hard to avoid. In short: she is keen to establish useful and relevant skills and develop capabilities that will further her, somewhat sketchy, plan.

The approach she has to the concept of professionalism is still fairly fluid, but she is confident that she will work to the highest of standards, keep up to date with developments in the industry, comply with all necessary laws and regulations and definitely not contravene any parts of her institution's code of conduct. At the same time she will have high expectations of the professionalism of others who she'll work with. She will expect her employers to treat her with respect, take her opinions and judgement seriously, provide her with good work opportunities, recognize and reward her contribution to projects, provide a good working environment and not show any unfair bias against her or demand unreasonable working hours. Other colleagues will be expected to behave within acceptable social and professional norms and not take advantage of her lack of experience, vulnerabilities and gender, or their seniority and power. They should pull their weight and act as team players.

Outside the office she will hope for due respect from other consultants, contractors and officials based more on the principle that she is there to do a job than her job title, but even then anticipating that this will add to the value with which she is credited. She expects that others will do their work as promised and will not try to cheat the system, lie, steal or take credit where it isn't due. She will learn the importance of professional reciprocity in this respect and find that without it the currently available system won't work.

She will also need support for her professionalism from many different people and sources. First and foremost this will be at work from colleagues, line managers and senior members of the firm. They need to provide advice and guidance, training in techniques and practice and access to their working methods. She will need to reciprocate by sharing knowledge and skills she has learnt during her education and training. Even though fully qualified she needs to keep learning and developing. Part of this will be delivered through formal training, including, amongst many other options, in-service training, conferences, short CPD sessions, online courses and directed reading, other elements through observation and being given the space to try things out by herself and to practice, practice, practice.

Part of her training should be designed to help her undertake her current workload but also to acquire new skills and cultivate old ones to support her career. She will need to develop new technical capabilities, but also soft skills in communication, interpersonal abilities and team leadership. Part of this training should be aimed at reinforcing and helping support her professional proficiency and confidence in upholding principles and maintaining, for example, sustainability criteria or professional standards of behaviour.

Her institution should be helping her with training but also with networking. This is to ensure she stays in contact and shares experiences with peers and colleagues and is exposed to broader conversations, opportunities and knowledge than are available at work. Other membership groups, organizations and specialist media may offer opportunities with invaluable exposure to detailed information, debate and new ideas. They may also provide an invaluable way of building up a contact list and a network of support that can be turned to for help, information and a way to open doors throughout a career.

Her firm and all the other support organizations will help to provide the background against which she can define the standards she will work to. Her firm will set and police its own standards, possibly in codified form, but also as a means of informal control over the work she does that is presented and issued to clients and others. The institutions attempt, sometimes successfully, to set clear standards, but it is mainly through discussions with her peers and by making comparisons with, often misleading, material published in professional and trade journals and online that her standards will develop and she will know how they fit with society's expectations of her role.

Professionalism for her means both doing her job well and using her own set of moral standards, as well as the system of ethics set out by her profession. But professionalism is also a support system that she relies on to perform to the best of her ability and that will enable her to progress her career.

There will, however, be many others' expectations that she will need to deal with.

Tutors

Both school and university will have made clear their hopes for her career and attempted to shape her ambition and guide her in her choices. They may have helped her to be pragmatic, but nonetheless many students emerge from training with career plans that focus on emulating celebrated luminaries in their fields and then find themselves rapidly and unexpectedly hitting a hard wall of reality when they first experience current employment practice.

Her teachers will have illustrated their lectures with images from the rich and exciting history of built environment design and taken her on visits to thought-provoking places. She may have attended inspirational lectures by some of the individuals behind such projects and admired their work in books, journals and on social media. The cultural, political and technical aspects of these projects will have been discussed at length and the lifestyles of their creators admired, but relatively little will have been said about the team of workers who produced them or the professional challenges of doing so. As a result her expectations for her professional career may have been set unrealistically high, or for others the intensely aspirational level of the goal may have occasioned disillusionment and a strong fear of failure. Either way they may be misleading and make it more difficult to establish useful foundations for her professional career.

The best tutors will get the mix of ambition and practicality about right and will have helped her to develop the ability to think for herself, to know how to access useful aids and resources for effective practice and to have the beginnings of good teamworking skills; and these, together with a knowledgeable and empathic employer, will help make the transition from academia to practice work smoothly. For many others it may be a bumpier ride.

Family and friends

The expectations of family and friends may overlap with her own but they probably know little of the inside detail. They may only see and comment on the headline issues and attempt to respond to the symptoms of a professional life without a deep or adequate level of understanding, but the expectations are real enough. Friends and family can supply plentiful and invaluable support, but also huge pressure, hopes and fears alongside a demand for attention, time, space and energy for themselves. Friends and family, having been with her through a long process can invest emotionally in a professionals' career and are keen to see successful results, to receive a string of highlights for their vicarious pride and pleasure. They will want to hear news of all the good things: promotions, completed projects, press articles, awards and so on. Others will want to hear about all the bad things as well. The pressure will be on her to deliver.

The closest family members will also depend on her in more straightforward ways: physically, materially and emotionally. They and she will expect their urgent needs to overrule her professionalism when necessary, but at times they will suffer relative neglect as professional requirements are put first. The personal and the professional will both place tremendous demands on her and need to be balanced and juggled. Occasionally one of them will get dropped with all the damage and the need to staunch it that may result.

The family and friends' view of professionalism can be an additional demand for career performance and possibly lifestyle and material rewards, no-one else outside the work team is going to feel such intense pride in her achievements, but it is a major source of stability and support when strength is needed and may provide the ethical grounding and real world outlook that is necessary to make difficult decisions.

Employer

After friends and family her employer is going to have the greatest interest in her career development and she, in turn, may spend more of her waking hours around her employer than anyone else. As with the family and friends their interests will be closely linked, but in a different way. The employer will have a clear interest in her being happy and fulfilled in her work and in her emotional wellbeing but on the professional side will be interested in her capacity to deliver and to do so consistently and without damaging mistakes and errors.

A conscientious (and professional) employer will allow for the possibility of a long-term relationship, intended to be beneficial to both parties, and will guide her career development from a junior position to one that incrementally takes on more responsibilities, gradually becoming more senior and in turn delivers ever greater value back to the firm. It may not turn out that way, but it does not stop both of them behaving as if it should. The employer will nonetheless remain wary that she will move on to another, better position elsewhere. She, similarly, may feel that the professional relationship is no longer working and that she is under pressure to go.

The employer's interest in her professionalism works on many different levels:

- There is a straightforward demand that she does her work well and in accordance with the standards and working practices set by the firm. It should be done with minimum errors, to time and without unnecessary disruption to others.
- There will be standards of dress, behaviour, respect towards others and a need for a positive attitude to the firm itself that will need to be understood and complied with, whether they are made explicit or are only implicit in the actions of others.
- The firm will have agreed working hours policies, including provisions for holidays and sick leave, which she will need to comply with; but there may also be pressure or calls for additional hours or days to be worked, for homeworking or for being on-call when officially off. They may be framed as calls upon her professionalism and a willingness to go the extra mile, and be difficult to resist.
- She will be required to be capable of making judgements, both for herself and on behalf of the company, even though these may be guided at the outset and only gradually increase in significance.
- She may be expected to bring a special area of expertise or knowledge to the firm, and she will be required to share it with others. She may have been hired for that very purpose. This may cause conflicts of interest.
- The firm may offer a formal mentoring programme, working both in and outside the company, for her to participate in both as mentee and mentor. In the latter role she will have a responsibility to share her skill and knowledge, to give time to and have a duty of care for the both the progress and the emotional wellbeing of others.
- It will be anticipated that she contributes her expectations of professional behaviour to the firm, to raise the standards that it both espouses and delivers and to use her conscience and sense of 'rightness' to help keep the firm honest, directed and strong.
- Her skills, experience, contribution and qualifications are all valuable assets to the employer in promoting themselves to clients and decision-makers and may need to be enhanced as necessary.
- It will be envisaged that she keeps up to date with current knowledge and technical skills and will develop new areas of expertise and capacity as necessary to work in or pursue new areas of work for her and/or the firm.
- As a professional it will be anticipated that she is self-motivated and, in part, self-directed and that although she will have line management, regular reviews and there may be an HR department to offer help and guidance, personal initiative is an intrinsic part of her responsibilities.
- The firm's relationship with its clients and others who may be important in bringing it work will be of critical concern to it and she will be expected to maintain good relationships with them, in the quality of work, in meetings and communications, and socially. Getting it wrong may be very damaging to her prospects.
- She, and especially as she becomes more senior, will be expected to maintain the reputation of the firm and possibly become a successful ambassador

for its interests. This will require both technical knowledge and good presentation skills. It is likely to affect the way she presents herself and behaves, both at and away from work.

- She will be expected to work well with other members of the firm and with external project teams for the sake of each project, regardless of personal chemistry, taking on management roles if necessary.
- Teambuilding, social and co-operative activities may be part of the firm's usual programme of activities and it will be anticipated that she will be an active participant and possibly initiator and organizer.
- The firm may have a charitable or outreach programme, which she will be expected to participate in, representing the firm and possibly giving a good deal of herself to it.
- It will be anticipated that she contributes to the health and future direction of the firm, possibly bringing in new work and helping it to develop in fresh directions. It may even be expected, possibly unreasonably, that she puts the interests of the firm first and, if necessary, sacrifices her future prospects for that of the company.
- In the longer term it may be hoped that she will become part of the ownership structure of the company, either as a partner, director or shareholder, and take on direct legal responsibility for its actions, to other members of staff, to clients and the public and to her fellow partners or directors.
- None of these things will be expected immediately, it will be anticipated that she will need to grow into the role, learn a large number of new skills as well as the ways of the office and, initially at least, adjust to the world of full-time work after years of study and student life.
- Having left the firm, for whatever reason, it will be expected, by them at least, that she keeps confidences, gives credit where it belongs and does not deliberately damage the reputation of her former company. They may also hope and try to control, possibly in vain, whether she takes clients and other members of staff away with her.

Over and above an employer's expectations of her to put their interests to the fore they should also expect her to look after her own, her colleagues', her profession's and the public's welfare. Much of this ground is covered elsewhere in this chapter, but there should be no surprise from her employers if she demands her legal rights as an employee, refuses to act in any way she may consider unsafe or unprofessional or exercises her rights to join a trade union. They should expect to provide a written contract of employment and, ideally, maintain and comply with the terms of a grievance policy.

Owners and shareholders

The owner's interest in and expectations for the young professional's career is a specific subset of the employers' and has only come to the fore in recent times as many larger firms have gained remote ownership structures, including by corporate shareholding and investment houses. As a result a gap has opened between the motivations and requirements of the ultimate owners and those who run companies.

Table 10.1 Duties of a Company Director (UK Companies Act 2006, Clause 172–1)

A director of a company must act in the way he considers, in good faith, would be most likely to promote the success of the company for the benefit of its members as a whole, and in doing so have regard (amongst other matters) to –

(a) the likely consequences of any decision in the long term,

(b) the interests of the company's employees,

(c) the need to foster the company's business relationships with suppliers, customers and others,

(d) the impact of the company's operations on the community and the environment,

(e) the desirability of the company maintaining a reputation for high standards of business conduct, and

(f) the need to act fairly as between members of the company.

Under the 2006 Companies Act the directors have a 'Duty to promote the success of the company'. This is often interpreted as being 'maximizing the return to the shareholders' but it is both a good deal wider and more detailed than that and includes six specific duties – see Table 10.1.

A professional working for a UK registered company will need to have regard for all these points, for which they will inherit vicarious responsibility from the directors. The duties are not dissimilar to clauses in several professional codes of conduct, but in several instances they not only overlap with, but they also go well beyond them, requiring, for example, concern for the community and the environment that not all the codes include. Professionals may also have a responsibility to bring the six duties to individual directors' attention if they believe that any of them is being given insufficient attention.

The owners and shareholders of a business should reasonably expect to be kept well informed on the running and corporate health of the company by the management team they've put in place for the purpose, but there may come times when a professional will be aware of activities that might damage the company, for example illegal, unethical or unsafe practices, that she considers might be being hidden or withheld from them. In such circumstances an employee should usually report their concerns to the company management and they may have a whistleblowing policy that provides advice on how to go about this. But there may be a need, especially if concerns do not appear to be being appropriately dealt with, to report directly to the owners or possibly, if it is the public interest, to disclose wrongdoing, in confidence and with legal protection, to a 'prescribed person', legal adviser or MP. The owners of the company should expect and encourage all their employees to report their concerns to an appropriate level of management or externally as necessary and professionals should feel it is part of their professional obligations and should be able to rely on the support of their institution. It is clear, however, that many companies would prefer their employees not to cause them any unnecessary trouble and may not want to hear bad news. Advice on whistleblowing is available from the UK government website www.gov.uk/whistleblowing, which states that:

You're protected by law if you report any of the following:

- a criminal offence, e.g. fraud
- someone's health and safety is in danger

- risk or actual damage to the environment
- a miscarriage of justice
- the company is breaking the law, e.g. doesn't have the right insurance
- you believe someone is covering up wrongdoing.

It should be anticipated that remote owners are looking for economic benefits above anything else, albeit possibly tempered by a desire for risk avoidance and, if that is unavoidable, mitigation. Predictability and reliability will probably be seen as virtues, but the most attractive characteristic will be the ability to contribute to growing profits, keep the business out of trouble and extract maximum value at minimum cost. The challenge for a professional employee is when there is a conflict between the aims and company methods set out by shareholders and her professional duties. This may mean being expected, for example, to deliver projects with insufficient time or resources; lower specifications than she believes are appropriate or warranted; or that damage the environment (see the discussion in Chapter 8 on the research carried out by the Association of Consulting Engineers on the willingness of engineering consultants to reduce the quality of their services). The company's instructions to their employees may be entirely legal, but not meet a professional standard. In such circumstances a professional employee may need to take action that could, potentially, lead to her leaving her job.

Owners and shareholders might conceive professionalism in a very different way: with a focus on good management skills performed with their benefit in mind, combined with the ability to achieve business growth and the provision of a greater amount of fee generating service at less cost and lower risk.

Staff

The obverse to the employer who hires professionals is the professional who hires staff. There is an expectation that she will treat those who work for her well and there is a higher standard that applies because of her professional status.

The RIBA spells out its stance in its *Employment Policy* (2004, p1):

The RIBA believes that good employment practice, by and for its members, will contribute positively to the effectiveness and influence of the architectural profession. It will also improve business opportunities, employment diversity and personal development, and is vital to the profession's role in raising the quality of our built environment and benefiting society.

The policy is general and encouraging rather than prescriptive, with circular references to the RIBA Code of Conduct and back, but the Institute's criteria for membership of its Chartered Practice scheme are more specific, including requirements for an employment policy, quality management system, CPD framework, making annual returns to the RIBA's benchmarking survey (RIBA 2012a) and an obligation to pay at least the minimum wage to students of architecture working within the practice (RIBA 2012b). The RIBA's Code of Conduct is the only one of the institutional codes surveyed, specifically to address the need to comply with good employment practice.

The RICS runs a voluntary system, the Inclusive Employer Quality Mark (IEQM) (2016c):

The IEQM asks employers to pledge their commitment to adopting and continually improving against six globally-applicable principles.

1. Leadership & vision: Demonstrable commitment at the highest level to increasing the diversity of the workforce.
2. Recruitment: Engage and attract new people to the industry from under-represented groups; best practice recruitment methods.
3. Staff development: Training and promotion policies that offer equal access to career progression for all members of the workforce.
4. Staff retention: Flexible working arrangements and adaptive working practices that provide opportunities for all to perform at their highest levels.
5. Staff engagement: An inclusive culture where all staff engage with developing, delivering, monitoring and assessing the diversity and inclusivity policies.
6. Continuous improvement: Continually refreshing and renewing the firm's commitment to being the best employer; sharing and learning from best practice across the industry.

The CIOB (2008, p9) 'recommends and supports':

- Promoting best practice in management processes.
- Greater training and educational focus on business skills, including human resource management, financial management and legal studies.
- Further onus on information technology skills.
- The use and promotion of effective time management tools.
- The development of a code of practice for the management and control of time on construction projects.
- Continuing Professional Development (CPD), training and education to raise the standards of planning, programming and scheduling.

The importance of retaining staff should lead to good employment practice, but there is still a hire and fire approach in many parts of the sector that manages the ebb and flow of a professional firm using abrupt changes to staffing levels. Staff expect professional firms to be professional about their employment practices and that they ensure fairness, team development and individual opportunities.

Insurers

Predictability is everything to insurers, an attitude that leads to a strong interest in information and applied research. Risk is something that they are willing to engage with so long as it is quantified, mitigation measures are taken and there is access to good, unbiased data. As a result insurers expect construction professionals to respect the importance of information and data; to methodically collect it and to learn and apply lessons from project to project. These tasks help to avoid making the same mistake over again. If things have gone wrong, they hope that a well-logged record of the decision-making process will show that any error was made in good faith following a reasonable assessment of the evidence, that the right choice was made in the circumstances and that it was a decision any equivalent and conscientious professional would have made.

Given this two-pronged interest in the effective collection and use of information, for improving and making more consistent both product and process, insurers encourage professionals in the deliberate and careful use of reliable evidence to inform practice and to follow up by being rigorous in their measuring and record keeping. This approach underpins the quality assurance (QA) process aimed at 'providing confidence that quality requirements will be fulfilled' (ISO 2005, Clause 3.2.11). Information is also to be shared with them whenever it is reasonable and appropriate, in particular the early disclosure of any possibility of a claim.

For insurers professionalism means consistently applying an appropriate quality assurance process, with good record keeping, an evidence-based approach and a high level of risk awareness.

Work colleagues

On the basis that we learn most from our work colleagues this group has possibly the greatest influence on the development of professionalism. The reliance on each other, the demand that everyone keeps up to the mark, the examples that can be set, absorbed and understood, all encourage a collective professionalism to emerge. Joint endeavour rewards and hones useful expertise and knowledge, keeping it sharp and effective through testing and practice. It encourages calmness under fire and a willingness to persevere until problems are satisfactorily solved. It forgives imperfections, but probably not perfection. It expects dependability, openness, recall and generosity.

The professionalism learnt at work is gained through sharing and doing. It combines methods, routines and skills brought from other offices and workplaces, learnt on training courses, born from new ideas and established by long-standing custom. Practices are adopted because they are effective and enjoyable, but they are not necessarily the most efficient and they frequently accrete as office practice until they eventually collapse or are swept away by an office reform. The universalizing effect of management and modelling software, including the several varieties of BIM, may cut through some of this but may also spread habits across a far wider professional community as it develops its own odd routines and sub-routines.

As there is no standard version of professionalism, most habits are learnt through imitation and repetition. With luck they are learnt from true adepts who show and teach by force of personality how the job can be done with verve, confidence and skill. Astute students seek out such adepts and will cluster around them in their office to learn from them during their professional formation. For others it will be more random and haphazard.

The professionalism learnt from colleagues can feature inter-reliance, exchange and inter-group competitiveness. It can also inculcate bad habits and out-dated practice.

Clients

The experienced client possibly knows more about what constitutes effective professional practice than anyone else and will be in the habit of demanding and coaching best practice from any newcomers to their pool of consultants. Repeat clients may use multiple professionals, working across many disciplines, and will see a huge variety of different approaches, styles and success rates as a result. They will know what works, for them at least, who delivers the most effective user experience and how.

The knowing client is looking for and demanding delivery that matches both their brief and a knowingly discounted version of the promise made by their consultant team during the early stages of the project, either at feasibility or bid stage. The discounting comes from the knowledge that very few projects deliver on their promise, but the extent of the discount is a measure of the belief they have in that consultant to turn their earliest assessment into reality. That trust may be only given out in small packets: to one firm or team for design and to get through the planning process; to another for cost cutting and developing a technical solution; and to yet another to facilitate and administer the construction; or it may be given, in its entirety, to a single team that will last the course from project inception to occupation and use. The client will make the decision based on their knowledge as to which professional approach works for them and the project best.

Clients are looking for their professional teams to have technical skills but also a raft of professional attributes including:

- A client-first approach. This may be difficult to reconcile with a professional duty to place the public interest first but for many services there is no problem, and client monies in particular should be treated with scrupulous care. Client confidentiality may need to be negotiated in order to agree what aspects of the project can be shared or publicized and how.
- Good user experience. This may have little relationship to the quality of the product, but if clients are treated well, believe they have been listened to, communicated effectively with and enjoy the process of working with the professionals in the team then they are far more likely to repeat the process in future.
- A safe pair of hands. The ability to deliver without dramatics can be far more compelling than design quality for many clients.
- Continuity. Clients often come to professional firms in order to work with selected individuals. They may find that they are then passed on to others, but this cannot happen too often or too wilfully.
- Confidence and assuredness at providing solutions to the problem and a reasonably rapid delivery. No one likes to be kept waiting. This may be combined with a knowledge of technical limitations and clarity about when issues have to be referred elsewhere.
- Up to date expertise at dealing with their particular type of project. Experienced clients do not expect to be used as guinea pigs, but may occasionally be happy to develop new ideas through a joint process of exploration, trialling and correcting. Every so often a client confident in their own ideas appoints an *ingénue* consultant to a project in pursuit of fresh thinking,

in order to ensure their own control over it or even as an act of considered patronage. Such projects are very rare exceptions to the rule.

- Specialist knowledge. The client may have a particular objective that can only be delivered by one of very few individuals or companies being included in the design team.
- Problem solving skills. Finding effective and possibly innovative resolutions to difficult challenges whether financial, technical, legal or to enable project delivery may unlock a project and make it worth a client's interest.
- Reputation. This will be important to clients for all sorts of reasons and may at times have only a relatively slender link with reality. Professional reputation is one of a professional's most important possessions but also one of the most fragile and difficult to rebuild should it get damaged.
- Creativity and flair. This may be an optional extra, but can be essential on projects that are challenging technically or require the persuasiveness of an exceptional design and its cheerleaders to make it through a difficult planning process or to convince funders to invest. For some clients creativity, as opposed to quality design, is what gets them going and why they do their job, for others it would be a distraction.
- Ability to teamwork and come together around a project's needs for the duration of its delivery. This includes the capacity to resource a project with adequate numbers of qualified people, equipment and finance, possibly at short notice.
- A talent for convincing other stakeholders of a project's worth and viability. This may include investors, planners and other regulators and possibly the local community and a range of other interested parties including the users, parish councils, design panels, heritage groups and environmental campaigners.
- The ability to communicate with and undertake meaningful and effective consultations with stakeholder groups and organizations, local residents and businesses as well, in all likelihood, as pressure groups and protestors.
- Resilience. The determination to see a project through against the odds and in support of the client may be what makes it happen.
- Risk management. Clients or others may express an interest in innovative or novel approaches, but expect to be informed about potential levels of risk as well as how that risk is managed and apportioned.

Not every professional on every project needs to have all these attributes but the expectation will be that they will have, or gain, as many of them as possible, especially as they become more senior in their organization.

The client expectations of professionalism hinge on their user experience as well as the quality of the product provided. Both are necessary. They anticipate commitment, experience and skill as well as the safe pair of hands to be able to reliably sub-contract many of the intricacies of delivery. They are keen that it should be done to budget, programme and quality, but they also want it carried through with a positive attitude, confidence and a smile.

Co-professionals

A team of professionals is essential for the delivery of almost every project and team members will have strong expectations of each other's professionalism.

In part this will be about ensuring team effort and being able to move together trusting that others have got their contribution right, allowing the work to progress at a reasonable pace. It also is a hope that any failures have been picked up and gaps filled; that everyone is supportive of others' work and reputation.

The ability to communicate and work across disciplines while providing in-depth expertise in a particular area, the so-called 'T-shaped' professionalism, is essential (Guest 1991). A willingness to resist groupthink and fight one's corner may be admired, so long as it is accompanied by enough self-knowledge to know if and when to concede and move on. Teams clearly need to work together: sharing ideas, crossing discipline boundaries as necessary and maintaining good working relationships, but individuals must also play their own, individual part strongly and be aware of their boundaries and limitations. Teams work on the basis of mutual trust and rely on everyone supporting collective ethical standards, possibly keeping each other up to the mark and providing support if they find other team members under pressure to make unwarranted compromises.

The team worker version of professionalism emphasizes sharing knowledge and expertise, commitment to realizing a project, reliability, common ethical standards and excellent communication skills.

Contractors

Construction contracts bring together larger numbers of workers, suppliers and equipment in complex, changing and often-dangerous circumstances with almost everyone working principally for their own advantage, hoping to make their particular contribution to the whole, to get paid and to move seamlessly on to the next project. It requires good management, logistical and financial skills to integrate it successfully and to deliver a profit at the end. In recent times it has also demanded a focus on maintaining safe working conditions and ensuring that the many routines and practices that provide these are used effectively.

Professionals work in many different roles on each project, frequently representing parties in, or seeking to be in, contract with each other for construction works and supplies or acting as an impartial contract administrator. Despite the best efforts of Latham, Egan and many others to persuade the industry to adopt more integrated methods, contracting remains a process that uses an adversarial approach to resolve disagreements. Professionals are expected to have the skills to cope with differences of opinion or outlook and to help make an often-fraught process a success. These include the capacity to remain calm under fire, to project assurance and to have the necessary facts and information always available for deployment. Fairness and good diplomatic skills are important, as is the ability to remain on friendly terms, even while fervently disagreeing.

This a version of professionalism that brings together many skills, particularly organization and time management. It requires an overall command of project knowledge, health and safety issues, contract law, building regulations, costs, programming and the ability to predict the implications of any changes. It demands accuracy and rapid thinking with a willingness to take decisions, including the ability, when required, to say 'No' or to hold-off, to take further advice, before reaching a decision.

Table 10.2 CDM (2015), Regulation 9, Duties of designers

(1) A designer must not commence work in relation to a project unless satisfied that the client is aware of the duties owed by the client under these Regulations.

(2) When preparing or modifying a design the designer must take into account the general principles of prevention and any pre-construction information to eliminate, so far as is reasonably practicable, foreseeable risks to the health or safety of any person –

    (a) carrying out or liable to be affected by construction work;

    (b) maintaining or cleaning a structure; or

    (c) using a structure designed as a workplace.

(3) If it is not possible to eliminate these risks, the designer must, so far as is reasonably practicable –

    (a) take steps to reduce or, if that is not possible, control the risks through the subsequent design process;

    (b) provide information about those risks to the principal designer; and

    (c) ensure appropriate information is included in the health and safety file.

(4) A designer must take all reasonable steps to provide, with the design, sufficient information about the design, construction or maintenance of the structure, to adequately assist the client, other designers and contractors to comply with their duties under these Regulations.

Workers

The issue of health and safety on site has become a major concern for professionals in the industry both as designers and through their presence on site, whether they have direct supervision responsibilities or not. In the UK, under the Construction (Design and Management), or CDM, Regulations, now in their third, 2015, iteration, there is an expectation that everyone involved with construction projects should:

- sensibly plan the work so the risks involved are managed from start to finish
- have the right people for the right job at the right time
- cooperate and coordinate your work with others
- have the right information about the risks and how they are being managed
- communicate this information effectively to those who need to know
- consult and engage with workers about the risks and how they are being managed.

The CDM Regulations (HSE 2015a, p26) themselves turn these expectations into duties (see Table 10.2).

The duties of principal designers and principal contractors extend further, but in practice everyone working on a project has responsibilities, including: 'the need for cooperation between dutyholders, reporting anything likely to endanger health and safety and ensuring information and instruction is understandable', and the onus on professionals is that much greater (HSE 2015a, p23):

A designer (including a principal designer) or contractor (including a principal contractor) appointed to work on a project must have the skills, knowledge and experience and, if they are an organisation, the organisational capability, necessary to fulfil the role that they are appointed to undertake, in a manner that secures the health and safety of any person affected by the project.

The professional world took a long time to acknowledge its responsibility for health and safety on both building sites and in completed projects, but following the passing of the first version of the CDM Regulations by the British government in 1994, there is no longer any doubt that health and safety is an essential part of a construction professional's duties. It needs to be part of education and training and should be a regular component of CPD courses. But if this has become clear in a UK context, some UK accredited professionals working abroad still seem conflicted by the issue.

There has been a great deal of discussion in recent years over the responsibilities of professionals who take on and become involved with overseas projects where the conditions for individual workers are unsafe or where their freedoms are restricted, even though they may have no direct involvement. This discussion has become especially heated in the light of various media *exposés* of working and living conditions for immigrant construction workers in the United Arab Emirates (UAE), including high numbers of deaths, and subsequently arrangements for the construction of facilities for the 2022 FIFA World Cup in Qatar (Business & Human Rights Resource Centre 2016).

The response of UK companies was varied. In 2014 Zaha Hadid, the architect at that point only working on the design stages of the Al-Wakrah stadium in Qatar said 'I have nothing to do with the workers. I think that's an issue the government – if there's a problem – should pick up. Hopefully, these things will be resolved' (Riach 2014). Atkins responded to a 2016 survey saying 'the health, safety and welfare of all people associated with Atkins projects, regardless of whether they are our employees, is always our number one priority and this can be evidenced across our global activities'. At the same time *Building* magazine quoted one anonymous 'professional working in Qatar' saying: 'there are pockets of horrendous practices. I have seen things here that I've never seen in my life ... I have seen accommodation and welfare standards that should not be permitted, shocking conditions' (Withers 2016).

As indicated by the Atkins response, construction workers should be able to expect, not only in the Middle East, but also in the UK and everywhere else, that the involvement of an industry professional should give them some protection against dangerous working conditions and that all professionals carry a duty of care for all those involved in delivering their projects. How they can exercise that duty remains a more difficult issue. Effective whistleblowing provisions and firms that take their staff's concerns very seriously may be one response, but many professionals will be naturally cautious about with whom and where they work.

Users and occupants

The one group who might be entitled to expect the most from construction and property professionals is those who inhabit and use the buildings and environments they design, create and manage. They will expect that their needs, welfare and activities are anticipated, understood and given a high degree of priority as design and construction decisions are made. Against this they may fear, based on previous experience that, unless they have some economic leverage, they will find themselves in less than perfect surroundings: cheaply built, poorly performing, expensive to look after and with potential threats to their health and safety. The presence of professionals in positions of influence should, but may not, offer them some reassurance of quality and performance.

The central expectation is that buildings and environments should provide safe, healthy, comfortable and well-serviced environments that do not cost the earth to run and maintain, and at the very least they should perform in accordance with regulations and standards. This requires a degree of skill and experience but it also needs to learn lessons from what works and where failures have occurred in the past. Post-occupancy evaluation work, including the PROBE (CIBSE 2017) and Building Use (BUS) (Arup 2017) studies, has shown that the performance of buildings in use is frequently well below the original promise.

While an awareness of the gap between predicted and built performance has developed in recent years there has been an accompanying interest in the idea of occupant and user wellbeing. The Office for National Statistics (ONS) started measuring national wellbeing in the UK in 2012 and 2013 saw the launch of the WELL building standard, covering seven categories (air, water, nourishment, light, fitness, comfort and mind). There is also a range of standards and technical memoranda from organizations such as CIBSE in the UK and ASHRAE in the US providing detailed guidance on the environmental performance and health impacts of buildings. The Soft Landings programme (BSRIA 2017) is also intended to 'ensure the design meets the needs of occupiers'.

In addition to environmental performance, facilities need to function effectively in many other ways: to provide effective working or living conditions, to optimize the outcomes of activities that take place there, to be robust and safe in use and easy to keep clean. They should even be enjoyable to be in.

It is reasonable for users to expect a professional design team to include people with expertise in assessing how a building or facility will function in use and for all team members to have an up to date understanding and knowledge about building performance with the ability to grasp how a building works from users' perspectives. They would reasonably expect that the whole team would also receive feedback following commissioning and subsequent monitoring covering a wide range of both technical and social issues during its use, so that they can learn relevant lessons to apply on future projects.

A building users' list of professional virtues is likely to start with a good degree of empathy, including the ability to listen and process sometimes imprecisely expressed needs into workable solutions, and a commitment to a duty of care and to defending users' interests. With this would go the professional knowledge and competence to deliver environments that achieve the goals prescribed and described. It will be expected that professionals will provide follow up care and evaluate outcomes, learn lessons and continuously improve their skill and abilities.

Regulators

There are numerous different regulators who work closely with the construction industry, including planners, building inspectors and health and safety inspectors. Although a significant professional group in their own right they place particular demands on the consultants and contractors they engage with on a daily basis.

Regulators tend to deal first with the imagined, virtual version of projects before dealing with their real life equivalents and the process of production, assessing them against laws and regulations, permissions and licences, guidance and best practice. It is meant to be well defined, but frequently isn't and relies heavily on their individual experience and judgement. They are looking for accurate and sufficient information that provides proof of compliance and require from the professionals that they deal with adequate knowledge of and respect for the process and regulation they implement, and expect clear explanations of how policy and legal requirements will be met.

In practice regulators are reliant on professionals to carry out their work honestly, accurately and with diligence. There is only so much that they can inspect or track and the system doesn't work if there are attempts to cheat its provisions or to circumvent inspection regimes. This only exposes professionals to additional risk and to potential problems later on.

Regulators expect a level of professionalism from the construction team that delivers care, diligence and accurate information. It is about ticking boxes and demonstrating compliance, but it needs to be done positively and with the conviction that it is being done for a proper purpose.

The public

A relationship of trust between public and professionals is a key component in the effective running of a modern market economy (see Chapters 3 and 8). But as philosopher Onora O'Neill said in her Reith Lecture (2002a, p17):

> In judging whether to place our trust in others' words or undertakings, or to refuse that trust, we need information and we need the means to judge that information. To place trust reasonably we need to discover not only which claims or undertakings we are invited to trust, but what we might reasonably think about them.

This requires professionals not only to make good decisions, taken with the public interest in mind and based on sound evidence, but critically also to communicate effectively about how and why their recommendations were derived. It is very likely that that process itself will have involved the public, or their representatives, at various stages and it will have been reasonable for them to have expected accountability, transparency, clarity and integrity during this. In 2005 O'Neill (pp85–6) expanded on the types of 'intelligent accountability' that should be provided by professionals:

> The benchmarks for intelligent accountability are informed and independent judgement of performance, complemented by intelligible communication of those judgements to relevant audiences. Each of these requirements matters.

Her benchmarks are onerous and require professionals to communicate: 'what ought to have been done', 'what was actually done' and the 'adequacy with which performance matched primary obligations'. Independence means that professionals should not only exercise it in their work, but also they should be seen to do so and take measures to avoid suspicions of 'corruption, producer capture and professional cosiness'. O'Neill states that:

> such communication is intelligible only if it is both *accessible to* and *assess-able by* relevant audiences, so has to differ for different audiences. Mere transparency will not make complex matters intelligible to all relevant audiences.

Following on from the professional obligation to put the public interest to the fore, the public has the right to expect that construction industry professionals address wider questions of social and environmental benefit when they make judgements and deploy their expertise. This expectation has multiple aspects. The public should reasonably expect professionals to:

1. Balance the needs of their immediate clients with those of society, the environment and the local area;
2. Balance the long-term view over short-term requirements;
3. Be honest and independent brokers when called upon to reconcile or find fair solutions to conflicting demands; and
4. Show impartiality, truthfulness and comprehensiveness in their advice and guidance, even when they are acting as advocates for one side or position.

In addition to these 'occupational' requirements there are further generic expectations of professionals to be upstanding and contributing members of society. These may be tinged with cynicism in the expectation that many, if not all, professionals will follow the bidding of whoever is paying their fee. Nonetheless, the struggle for respectability, hard won by professionals in the 19th century, has left a substantial legacy to today's professional class, which they are expected to maintain.

Society has invested trust in its professional class and expects the integrity that features so frequently in the various codes of conduct. It requires that advice is expert and up to date and that high standards are maintained, even if the definition of these can be vague. The individual professional is expected by society to behave and do their job in good faith and in line with contemporary mores. Professional standards adjust themselves over time and may at present appear fairly relaxed, but they still have currency and bite.

Community

Communities are a special category of 'the public' with more particular values and characteristics with places, activities or beliefs in common, but also with specific strengths and vulnerabilities. The social value embodied in a community can be considerable and is often impossible to recreate once it has been eroded. It is for this reason that communities feature so significantly in political discourse and are an essential professional concern. Both the ICE and RIBA codes of conduct specifically mention the interests of communities and that they should expect to receive consideration as a specific extension of the professional duty to the public interest. This is not to say that all communities are wholly positive – they may often be inflexible, exclusive, elitist or xeno-phobic – or that they are always viewed in a favourable light by other parts of society, businesses or government. Equally they may simply be an inconvenient group, awkwardly at odds with other, grander plans that may be developing around them.

The interests of society demand that communities are consulted and effect-ively communicated with; their rights championed as the essential, usually locally based, components of social and civic cohesion. Communities should have the right and opportunity to have their voice and opinion heard and their interests properly balanced against those with, possibly, greater power and resources. Professionals are expected to have the ability to connect with all parts of communities, and not just the loudest and most prominent or their, possibly self-appointed, spokespersons. As with the public above, communi-ties will expect that professionals will impartially consider their interests and act as honest brokers in advising on a way forward that deals fairly with all parties, even when they are being paid by one side.

Professionals in their own person or as companies may also belong to or wish to support communities and decide to contribute their expertise and time accordingly. Community participation takes many forms: from being involved in local societies and clubs, to volunteering, to standing for election to a public body. Some participation has little or nothing to do with any specific profes-sional expertise or skill, in other cases it draws directly on a particular capability that can be deployed usefully in a community context and is provided on a *pro bono* or at-cost basis. Volunteering can be a private affair or may be organized on a group basis by a firm or a professional organization. Many companies give their staff a quota of volunteering hours per year or sponsor a particular charity that gives their employees the opportunity to work for a period on very different types of social purpose project.

The status awarded by professional qualification and peer recognition carries responsibility, but also opportunity, which can afford those individuals who want, or feel duty bound, to participate in the public arena a platform to do so. As professionals they will often find they are speaking on behalf of their profes-sion and that will give their words substantially more weight than if they were acting as an individual. It places an obligation on them to use their status wisely and to be sure of their facts as well as an expectation that they truly have the expertise that their professional title suggests.

The planet

The planet cannot articulate expectations; it will deal with whatever is thrown at it, however damaging it might be. But there are many who speak on its behalf, knowingly and unknowingly. They have made it clear that built environ-ment professionals need to improve their practices, now and as a matter of urgency, if the world is going to still be a comfortable place to live by the end of the 21st century.

Sustainability has been on the curricula of accredited courses for many years now and a duty to, at least, recognize the impact of projects on the environ-ment is a feature of most of the institutions' codes of conduct. There can be no lack of awareness but action on the ground has been limited. Business, regulators and society have demanded better energy performance from the built environment than has been achieved in the recent past. Governments, both national and supranational, have made laws, laid down regulations and signed treaties to reduce carbon emissions, protect biodiversity and minimize pollution levels, amongst many other goals. Construction professionals are essential to achieve greater sustainability; to meet the targets set in the 2016 Paris Agreement, the binding elements of the programme laid out by the UK Committee on Climate Change and the legal requirements of the 2008 Climate Change Act. The expectation will be for professionals to be able to deal with the complexities of achieving a built environment that requires low or no carbon emissions for its construction or operation. This will require skills not only for designing new zero impact buildings but also in the conversion and retrofitting of the existing building stock. There is likely to be an increased emphasis on the maintenance of buildings during their lifetime; part of a greater shift that is likely to demand that construction and property professionals act as long-term stewards for the built environment as a whole.

As society makes increasing demands on the built environment, there is no doubt that professionals will be expected to have the skills to deal with issues of climate change and environmental degradation as increasingly society demands more and more of them. The scale of the challenge will be enormous and require new approaches and tools, and will affect all projects, large and small. Professionals of all ages need to take steps to meet the challenge and gain the knowledge and capability to act effectively.

The media

Architecture has long received the majority of built environment media coverage. It is featured in the cultural pages of newspapers, in lifestyle magazines and on television programmes such as Grand Designs. More recently it has developed a large online presence heightened by an image conscious public on platforms such as Pinterest and Instagram and any number of blogs and e-zines. This has given architecture and design an aspirational quality, something reinforced by the notional architectural careers given to creative and sensitive types in films and television series. These often come with a stereotyped all-black uniform and an intense, if occasionally otherworldly and distracted, mien. This image and reputation may stretch to include some of the other industry professionals, but probably not to surveyors, who will forever be suited, or contractors in their helmets and high-viz jackets, who have their own positive avatar in the shape of the animated children's TV character, Bob the Builder.

Simultaneously with these relatively benign, if impossible-to-live-up-to, role models, industry professionals are also held up as uniquely destructive and insensitive to people's needs. They are the carbuncle-creating characters of Prince Charles's bad dreams or the modernists responsible for 'the cult of ugliness' denounced by the *Daily Mail* (Scruton 2009) and others.

If the media has created its own pervasive construction professional 'types', which it encourages individuals to play up to, it is also very dependent on professionals to provide relevant stories but, more significantly, explanatory commentary. Neither is obligatory, but there is an urgent need for individual or organized groups of industry professionals to provide clear and cogent insights into the complexities of built environment decision-making as well as the development and effectiveness of policy changes. Chris Twinn (2013, p128) in the New Professionalism special issue of *Building Research & Information* (*BRI*) has proposed that:

> If building professionals are to make a meaningful contribution in society and to sustainability, they must address two fundamental issues: first to fill the large communication gap in society's appreciation of sustainability; and second to develop reasoned and rounded views about the complex world in which we work.

The institutions and other representative groups have a central role in communicating broader messages from the industry and being trusted and independent sources of information and analysis. The great majority of individual professionals must be prepared to contribute their own knowledge and interpretation of events, not just on behalf of themselves and their employers but also as expert members of and spokespersons for their profession. Individual professionals need the skills to communicate complex issues in clear ways while avoiding the temptation to oversimplify or promise easy solutions.

The media promulgates versions of the professional that are highly creative, if not necessarily socially responsible. It is as if professionalism is good at making landmarks and icons, at providing the very wealthy with highly desirable, if unattainable, levels of luxury, but poor at understanding ordinary wants and needs. The expectation is one of up-market commercial production – yet this is barely professional at all. Against this professionals must engage with the media to get their own messages out – responsibly and effectively.

Peers

The importance to the individual professional of recognition and respect by her peers should not be underestimated. Consideration of their judgement will have a significant role in shaping the version of professionalism she adopts. The law takes a similar approach, the standard for acceptable professional conduct being guided by that expected of a reasonably competent equivalent person. Only by comparison with others is it possible to develop a position that seems both ethical and reasonable and only if others stand alongside her will she have the support to adopt and maintain it.

If a co-operative and collaborative approach is essential to the idea of professionalism it also has a serious downside. The tendency to become inward looking and exclusive creates a disconnect with those outside the group and a loss of empathy with them. The resultant groupthink may help to accelerate work throughput, but does little for its quality or ability to solve real world problems. On a personal level peer opinion can equally be enabling and career enhancing or unthinking, intolerant – even prejudiced – and powerfully destructive. Being out of sympathy, for any number of reasons, with her peers may deeply affect her future prospects.

Professionals look for the affirmation of their peers in many different ways. They judge their performance, their progress and their success by seeing how others have done. Magazines are published and site visits organized in order for professionals to see what their peer group is doing. Review panels, and competitions and awards juries are given validity if they include other professionals who know what it takes to produce good work. The quest for peer esteem drives her forward to do one better, to be more exciting and daring, to impress and become mildly famous, at least in her own professional community. Equally, negative input may hold her back or impair her ability to properly exercise her intellectual and creative faculties. Peer pressure can lead to wrongheaded decision-making, just as peer support can radically enhance it and multiply a professional individual's knowledge and capability. Professional skill and judgement is needed to ensure that she learns how to use peer input to achieve the best outcome.

Peer pressure encourages higher professional standards as everyone competes to achieve their best. But it can also encourage bravado and narcissism and predicate short-term flash over careful long-term development for the same reason. Its potency is such that it can create or destroy careers – it is both essential and needs to be treated with extreme caution.

The institutions

As has been seen, only some of the institutions pass their charter obligations directly onto their members, but they all oblige them to comply with a set of rules and regulations intended to encourage or compel professional behaviour. It is important to institutions that their members operate with a shared set of standards and from an early stage they attempt to shape students as model professionals through obligatory course and examination content. Once they have passed those tests and become members, professionals are encouraged to follow best practice guidance, use standard forms, methods and protocols and obey regulations and codes of practice under threat of disciplinary action. Institutional culture naturally has a normalizing effect on professional activity.

The institutions' expectations that their members should perform their work in a precisely prescribed pattern have relaxed considerably since the imposed liberalization of the 1980s. Strict demarcation lines have been abandoned; advertising, competitive fee bidding and commercial relationships with construction companies became permissible; and project protocols (e.g. the RIBA Plan of Work) have changed from being *de rigueur* to becoming voluntary and informative. Nevertheless a great deal of custom and practice still remains, particularly around the general definition of an individual professional's role and the scope of work that she will undertake, and this effectively defines, barring a few renegades, most professionals' working lives.

Through the output of appointment and contract documents, aids, guidance and benchmarks for practice and standardized certificates and forms the institutions, together with aligned bodies such as the Joint Contracts Tribunal (JCT), exercise an effective and largely beneficial control over the way almost all jobs in the industry are run and managed. They leave relatively little possibility for manoeuvre outside the enclosures they have erected. The established routines have in turn developed standard sub- and sub-sub-routines and with them the potential for micro-management of professional work into a series of delivery procedures, some of which are well suited for automation.

As well as encouraging standardization, the institutions also gently attempt to challenge their members to differentiate themselves, to think independently, to develop new skills or to excel in their field. The institutions even give them awards when they do so. The effort encourages a few, but the great majority prefer to succumb to the comforting pull of standardization and homogenization.

The professionalism encouraged by the institutions is one of compliance and consistency. Work can rely upon the use of established methods and a series of well-defined rituals that can be adapted 'to almost any occasion'.

The law

Surprisingly several of the professional codes of conduct find it necessary to state that their members must obey the law. The RIBA's 2015 code is most explicit when it says 'Members must comply with all relevant legal obligations', but engineers appear to have slightly more latitude, however illusory, when they are also told to 'give due weight to all relevant law', although they also told to 'ensure that all work is lawful and justified' (Engineering Council/Royal Academy of Engineering 2014, Principle 3). Other codes, if they mention the law at all, only refer to international law.

In practice there are few legal obligations in the UK that are different for construction professionals than any other member of society, although under both contract and tort law professionals will be expected to offer a greater level of skill and duty of care and be held to a higher standard. Legislation aimed specifically at built environment professionals is largely limited to the Architects Act 1997 and the Construction (Design and Management) Regulations 2015. Even legislation such as the Housing Grants, Construction and Regeneration Act 1996 has no special provisions for professionals.

Nonetheless construction professionals are expected to have some specialist expertise in the law and be able to administer the law relevant to their discipline in areas including:

- Property
- Planning
- Building regulations
- The environment
- Liability
- Procurement
- Contract administration
- Dispute resolution
- Intellectual property
- Safety
- Standards and certification
- Employment

Government

Professionalism is an important delivery tool for governments trying to ensure that the buildings and infrastructure in their jurisdiction are maintained, rejuvenated and renewed. Almost all government policies and actions are dependent on the built environment in one form or other and the built environment professional service sector is an important part of the economy. In 2005/6, the last year with any accurate figures, the professional sector of the UK construction industry earned £13.9 billion – roughly 1% of economic output – employed approximately 270 000 people (0.9% of the working age population) and earned £2.5 billion from work overseas, 4.6% of the annual trade in services (excluding travel, transport and banking) (CIC 2006; ONS 2015, 2016; Rhodes 2015). Government has a clear interest in keeping the sector in good health.

Since the free-market interventions of the 1980s successive governments have preferred to leave the industry's professions well alone, making interventions only when necessary. In the past quarter century the only significant change has been to the registration system for architects, made in the wake of the 1993 Warne Report. The report's recommendation that the profession lose its protected title was rejected after a noisy campaign, with the result that a much slimmer Architects Registration Board (ARB) replaced the old Architects' Registration Committee of the United Kingdom (ARCUK) (The Architects Act 1997). This led to a number of additional obligations being placed on those practising using the title of 'architect', including the need to maintain professional indemnity insurance and the introduction of a new, more rules-based, code of conduct. The changes were essentially technocratic and even if they did little to resolve the continued friction between the ARB and RIBA, there was no attempt to impose political direction on the professions or how they operate. The UK government's approach remains determinedly hands-off.

Construction professionals have been frustrated that they and their advice has been ignored by successive governments, despite the best efforts of the institutions to propose policy ideas, lobby decision-makers and get close to ministers. With some exceptions, particularly the role that the ICE plays in relation to infrastructure planning, the government perceives the professions as presenting them with 'problems' rather than 'solutions' to their policy quandaries. The large professional service companies are far better at making themselves 'useful' to government interests. Nor have the institutions been able to present themselves as non-partisan stewards of the built environment – their degree of self-interest being seen to outweigh any chance of neutrality.

The existence of CABE (see Chapter 9), from 1999 to 2011, provided a conduit between the construction professions and government. This was closed when the organization was merged into the Design Council. Similarly the creation of the role of Government Chief Construction Adviser in 2008 and its occupation by two senior industry professionals, Paul Morrell and Peter Hansford, provided a direct line into government before it was also abolished in 2015. The adviser's role was replaced by a strengthened Construction Leadership Council (2013–), which reduced the voice of professions to a relatively minor position. Possibly the nadir of the government–professional relationship came with the declaration by the then Minister for Justice, Michael Gove that 'people in this country have had enough of experts' (Mance 2016).

Despite this invective governments rely on individual professionals for good advice and there are a range of construction professionals working as civil servants, sitting on committees and providing services either through their companies or in person. There is a high expectation that professionals can provide good, evidence-backed guidance, accompanied by credible, planned and costed means of delivery. Government helps to fund construction industry research through the various research councils and its 'innovation agency', Innovate UK, to ensure that there is a useful pool of knowledge for it and many others parts of the economy to draw on.

The issue for the individual professional is the perceived lack of value placed upon them and their services by government, making the job of convincing others of their worth that bit harder. A relatively disinterested government may be far better than one imposing contract terms, as it has done with doctors, or closely defining day to day work and closely monitoring the results as it does for teachers. The freedom of most of the construction industry to make its own way is vitally important for its continued, if somewhat frayed, health. The individual built environment professional has kept a degree of freedom, although it may have been in exchange for a riskier future.

Future generations

The group that built environment professionals probably owe most to are the generations who are only just or who are still to be born. The built environment that will be used to serve their needs is largely already built and will be added to in the years to come. They are generations that today's professionals want to do the right thing for, but don't always know how or what to provide.

Future generations require good quality infrastructure and buildings that they can use and that will last, with repairs, for them to pass on to others; buildings that are adaptable and re-usable and infrastructure that is robust and large enough to satisfy future need. We know that this is possible because past generations have done it for us. They need cities that work for the many that live in them, that provide adequate space and do not overheat in summer or require social segregation as a result of their planning and design. Above all they need a world that is not many degrees hotter with a polluted atmosphere and no green space for miles around.

The brief is well enough understood and it is clear that today's construction industry professionals will be needed to provide it. Yet aside from a few committed individuals the call for concerted action is not being heard. Future generations will ask why action wasn't taken given the strength of the evidence and the knowledge available. Their future expectations will, doubtless, be that a forward-looking professionalism should have been put in place and acted on decades earlier; a professionalism that understood and provided for their generation's requirements and gave them the chances that their predecessors had had. A professionalism of responsibility, one that works in collaboration with as many other professionals as possible, to ensure that it protects the shared world rather than causing an extensively predicted, and widespread, tragedy of the commons.

Professional expectations

The expectations placed on the shoulders of professionals in the industry are as vast as they are varied; everything from complying with elaborate sets of rules, to being cool and creative, to saving the planet. Hoping that any individual will manage to meet them is clearly an enormous ask and it seems even less likely that hundreds of thousands of professionals can be persuaded to act in concert. The system of professions exists because it helps this to happen, not for everyone all the time, but for most people most of the time.

Expectations of professionals are malleable and subject to change. Society constantly requires more, whether it is better performance, lower risk, higher efficiency, greater consistency or a higher quality environment. The final chapter looks at how construction industry professionals might respond to such demands and how a new professionalism might help.

11
Potential

The idea of the modern professions was born out of the first industrial revolution and the need to organize expertise in ways to suit an increasingly mechanized and mobile society with new divisions of labour, sources of power and cheaply available print communication. The second industrial revolution, founded on electricity, oil and telecommunications, took off at the start of the 20th century and remoulded the professional idea to suit a mass society run by a centralized and bureaucratic state, but largely predicated on high growth and individual opportunity. It was ideally suited for the new, adapted version of professionalism to thrive. There was a substantial increase in the proportion of the workforce that described itself as professional and many new occupations took on the organizational and institutional trappings of the professional system.

As professionalism thrived and adapted to the new society of the 20th century there was little concern or questioning as some of the original and fundamental aspects of professionalism slipped away and functions were outsourced to others. The system that looked so robust and such an essential part of the modern world was in fact becoming weaker as it failed to maintain high standards or make its value to society sufficiently plain. As a result government policies in the 1980s, inspired by the ideological imperatives of emerging neo-liberal economists, required many professional disciplines, including those in the construction sector, to abandon their distinguishing tradition of maintaining a perceptible distance between themselves and commercial business practice. By the millennium there was little that marked out many professionals from other service providers.

At the start of the 21st century the effective enfeeblement of both the industrial and professional societies coincided with the advent of what has been described as the third industrial revolution (Rifkin 2011), centred on the power of internet-based technologies and the provision of increasingly decentralized and renewably sourced energy. This forecast revolution is anticipated to result in automation of many routine and knowledge-based jobs and in particular to cause serious disruption to and transformation of professional employment.

This leads to the conclusion that if transformation is not led from inside the professions it will be imposed externally and possibly brutally. While professionalism is likely to survive in some form simply because of the value it is capable of delivering, the future size and shape of the system is open to question; but also to influence.

Value

There are a number of potential categories under which professionalism and professional services can continue to provide value to society and business. Many of these are also prone to automation and professional displacement:

Place

Professional work in the construction sector has the capacity to transform low value or underperforming places into productive and socially beneficial environments able to deliver over the long term, although such benefits must be balanced against the potential damage to the natural and historic environments. Attempts have been made in recent years to assess the value of place and other physical assets to check whether change will deliver lasting benefit and not just transient gain to one party (see Mulgan 2005).

Making improvements to buildings and places, especially once there is a broadly agreed means of measuring and predicting asset values, however inaccurately, will almost certainly become subject to algorithmic assessment and automated propositions for betterment. It is already possible to automatically appraise all potential sites, using numerous data sets and geographical information, for development potential as economic conditions, transport, planning and legal constraints change, including the production of fully costed feasibility options, schematic planning applications and regulation-compliant project proposals – all without significant human intervention. Such systems can only increase their proficiency and expertise and become more autonomous in the years ahead.

Construction professionals need to analyse where the value lies in what presently constitutes the majority of their effort and input as well as the core of their chargeable outputs. It is likely that only when other values – primarily cultural, political and social – and skills – including imagination and ingenuity – are required that they will be able to improve on and outdo the physical value provided by emerging expert systems.

Technical

The ability to undertake technical and intellectual tasks with accuracy and efficiency has long been the province of professionals, albeit, often that of subordinate groups within the system; whether it has been calculating strengths and capacities, designing to exacting tolerances or compiling documentation with a high degree of precision (see Abbott 1988, p72). Much of this workload has already become semi-automated with the availability of expert systems and online guidance. Such systems already generate immediately applicable engineering calculations and legal documents.[1] Alternatively such work is also frequently economically outsourced to overseas companies. But if work is capable of being adequately described and allocated to a remote third party then it can also be fully integrated into BIM programmes requiring relatively little further human input to produce complete designs and possibly the full construction of buildings and infrastructure.

The detailed understanding and oversight of standard technical problems will continue to be a valuable professional skill. Truly technical skill will become associated only with problems that require creativity and imagination to run alongside technical ability. By definition these projects will be relatively unusual and rare.

Economic

Professional work has the long-proven ability to add economic value. It generates employment and wealth, whether for single individuals or entire cities and communities. For example, a regeneration project or a new cultural attraction boosts footfall and trade in the surrounding region. Although much added value is contributed by repetitive work (e.g. constructing a series of new supermarkets in accordance with a tried and tested formula), the exceptional value is provided through developing creative solutions that match particular circumstances, finding new ways of achieving a competitive edge or improved performance or, even more rarely, inventing something that everybody will want though they just don't know it yet.

Value can also be effectively created by saving it from diminution. Long-lasting and resilient facilities can be constructed at the outset and effective maintenance and repair regimes established. Buildings can be creatively re-used and repurposed – effectively saving existing value and reducing the requirement for new expenditure and resources.

Construction professionals will almost always intend to provide economic value to their commercial clients (and probably most of their other ones) as a core service function, even if it they describe it in other terms. If they don't succeed they will not survive in business long themselves. The highly prized skills of creativity and ingenuity could be redefined, in this context, as the ability to add value. There is little threat to those able to deliver this, but in practice there are few professionals or firms who can make a career from their purely creative moments, and they have to leaven them with other sorts of more predictable, if duller, work. It is the leavening portion that is in danger of being made uneconomic.

Legal and actuarial

Much professional work is about identifying opportunities (including regulatory loopholes), ascertaining what is permissible and possible and then balancing the risks and potential gains. With onerous policy requirements and complex regulations covering an array of different legal issues, the ability to 'get it right' can deliver substantial value and is much in demand.

Computers and artificial intelligence excel at this kind of work. As extensive and real-time data sets become available covering all aspects of development from planning rules, site ownership, sale and rental values, build costs and availability of finance, they will be linked and analysed by computers to produce answers. Professional expertise in this area will be replaced by the ability to provide light touch oversight and little more. Already programmes are scouring 3-D models of cities, aerial photos and mapping data to assess opportunities. The potential for automation of legal and actuarial services provided by construction professionals is enormous although there is likely to always be a need for some additional human judgement in the process.

Competence and risk

One of the major advantages to a client hiring a professional firm is a guaranteed offer of competence backed by professional indemnity insurance (PII). This assurance has the effect of encouraging clients to outsource work to external professional firms that they might otherwise more logically keep in-house. This, with its option of terminating the arrangement if it appears economically advantageous or for any of a multitude of other possible reasons, offers a significant value proposition. It is presented, often at low cost, by professional firms and should continue to stand them in good stead so long as they and their insurers can continue to carry the risk.

Risk and professionalism are closely, if inversely, related. Many of the traditional components of professionalism – competence, standards and conscientiousness – are intended to significantly reduce risk. More recent innovations, including quality assurance systems and internal and external monitoring, have added additional protections to the package. Professionals generally offer a safe pair of hands, but their willingness to accept risk, if not always deliberately or happily, means professionalism also attracts higher levels than it is strictly advisable to take on. This may follow from the belief that risk should be held by the party best able to manage it and not necessarily the one best able to bear it, but it also means that risk frequently ends up with the party least able to refuse it.

Identifying, apportioning, bearing, managing and containing risk, especially on large and complex projects with innovative financing arrangements, represents a major part of the business and economics of the construction industry. Professionals have a key role in advising clients and others about decisions that incur risk and how it can be managed. Traditionally professional practice meant actively avoiding taking on direct project risk, if not responsibility, but such boundaries and scruples have now long been abandoned. If professional companies are willing and able to take on, with appropriate insurance and controls, considered aspects of risk, it is an offer of significant value to their clients and employers and may yet keep them in business.

Technological

While technology will be the proximate cause of many jobs disappearing in professional services in the near future, it is also likely to generate and demand new skills as smart technology becomes integrated into almost all parts of the built environment.

The transition to Building Information Modelling (BIM) and the creation of information-rich models for projects will inevitably open new opportunities for professional engagement. The way the information is used and exploited will expand and allow new ways of designing, constructing and managing buildings. The scope for this is extensive but will only achieve success if information from completed projects is collected, analysed and shared.

Technology will change the resources available to construction professionals, giving them new tools to design, manage and construct with as well as new materials, components and equipment. The way those resources are used and the value they provide will be down to the ingenuity and creativity of individual professionals, their ability to employ them to optimize their performance and to develop best practice techniques.

As technology develops it will present a constant flow of new opportunities for generating professional value. There will be a constant challenge to discover the most effective ways to provide and improve applications, adapt them to unfamiliar situations, systematize the learning that emerges and capture new knowledge. Technology offers openings even as it destroys old jobs and routines. The task facing the professions is how to stay with it and locate new sources of value.

Co-ordination, collaboration and catalysing

The professions have long been recognized as part of the social glue that allows a liberal economy to function. As sociologist Julia Evetts (2003, p143) has written, 'professions might be one aspect of a state founded on liberal principles, one way of regulating certain spheres of economic life without developing an oppressive central bureaucracy'. The job of 'holding the ring' without need for coercion, is core to many professional functions including the roles of team leader, contract administrator and project manager, but it is also often an informal task taken by professionals who simply want to get the best out of their delivery team and to ensure that their projects run purposefully and with minimum friction. Overall the leadership/co-ordination role represents a considerable part of the value provided by professionals on projects.

As an extension to this leadership role many construction professionals are also project catalysts, pushing forward ideas and getting things to happen that wouldn't otherwise. Although this is far from being an exclusively professional trait, for some in the industry it is an aspect of their approach to generating projects and may involve personal commitment that extends well beyond standard professional service. Both the leadership-coordination and catalyst functions are currently threatened by the tendency towards atomization of professional services into discrete bundles. Whole project oversight is being replaced by short-term specialist contributions, each only focusing on their particular challenges. Generalists who see projects through from beginning to end are increasingly rare and the extinction of such non-specialist roles on many projects may not be too far off. Professionals need to assert the value that they bring to projects that are over and above the requirements of their terms of appointment.

Environmental

Designing and delivering solutions to environmental problems initially appears a sure winner for providing future professional value given the number of urgent environmental challenges affecting the post-industrial world, whether climate change, pollution or the need to protect biodiversity. In practice, government U-turns and policy reversals have plagued the sector as regulations and parallel incentive packages introduced by one administration have been radically altered or cancelled by its successors. For example the UK's Code for Sustainable Homes introduced in 2006 was withdrawn again in 2015; with the requirement for Zero-Carbon Homes, due to come into effect in 2016, disappearing alongside it (Oldfield 2015). Environmental consultants, having invested in extensive training and skill development, have been constantly and consistently left stranded by shifts in policy.

Despite such setbacks, an environmental crisis or even a catastrophe is highly likely to eventually precipitate a need for action, causing standards and regulations to suddenly become highly demanding and rigorous. The principal unknown is when. In this scenario the professional skills that can help deliver projects with minimal environmental harm and with maximum health benefits will become an essential component of design and delivery teams. Expertise will also be urgently needed to advise on effective policy formulation and assessment, particularly on the transformation of the existing building stock at scale. Future demand for proven environmental skills appears potentially limitless, but with multiple examples of start-stop policy in the area it currently appears unwise for large numbers of professionals to rely solely on their environmental expertise to give them employment. Its value is still unstable.

If government lacks consistency on the issue, professionals cannot afford to ignore it and need both individually and collectively to develop the skills and expertise to deal with environmental challenges. Not doing so risks considerable loss of value. The professional institutions need to consider whether to require their members to acquire and exercise higher standards of environmental expertise than presently legally required in reasonable anticipation of problems ahead. They also need to find ways of developing and making available to their membership the necessary knowledge for mitigating and adapting to environmental change before it becomes indispensable.

Ethical

An ethical approach may or may not always be welcome to those paying for professional services, but it clearly has value once success is achieved and a reputation established for fair dealing and high standards. The difference to other forms of value is that its benefits can be far more widely distributed; with the public, the local community, project users, occupants and future generations all potentially benefitting. The clients' and investors' values may be in alignment with those of the professional team, but an ethical approach can have value to them that extends well beyond such considerations. A scheme that delivers on a fairness agenda may find it easier to win planning consent or a highly sustainable design may help to obscure other, less palatable, matters. Buying in an ethical reputation and set of behaviours may be an attractive proposition whatever the underlying motivations.

An ethical reputation can rarely be gained from a single project. Normally it is accumulated over a lengthy period. In contrast it can be lost quickly, sometimes in a moment. Establishing ethical value can be targeted deliberately, if not cynically, but may involve losing work to others who are less scrupulous. There are no guarantees that an ethical reputation can be developed, but once achieved, and provided it is properly maintained, it can be a very potent aspect of professional value that would be difficult to replace either outside the system or through automation.

Knowledge

Specialist knowledge, for centuries vital to professionalism, has already been substantially replaced and extensively augmented by online resources, currently available through any smart phone or voice-activated service. The existence of and access to an up to date and applicable knowledge base adds significant professional value. As available information continues its geometric expansion, having the ability to know what is important and how to find and interpret it has become a specialist skill that commands its own premium. In a world awash with information, the ability to find and deploy the correct and most up to date data has become a highly valued commodity.

Knowledge, following the work of the Hungarian polymath, Michael Polanyi, first published in the 1960s, is conventionally divided into the codified and tacit – sometimes also labelled 'knowing-that' and 'knowing-how' (Ryle 1949). The codified is that knowledge, in the form of either data or techniques, which can be recorded and communicated. Codified knowledge forms the bulk of information found in libraries or delivered through conferences, academic papers and the Internet. It is the formal knowledge of professional institutions made available to members and researchers and the often-shielded knowledge of private firms, used to give them a competitive edge. Tacit knowledge is the knowledge that professionals thrive on: the 'things that we know but cannot tell' (Polanyi 1962) that includes skills, understandings, beliefs and unprocessed feedback from experience. There is a constant, and apparently inexhaustible, flow of knowledge running from the tacit to the codified as lessons are learnt and understood and as experience is converted into useful and reliable knowledge.

Professional value derives from combining different forms of knowledge to create new propositions. Evidence-based (codified) knowledge obtained from analysing feedback from existing projects is essential, but only by including tacit knowledge in the mix are professionals able to stay ahead of automation and machine learning. As tacit knowledge is continuously codified it requires the constant acquisition of new tacit knowledge to maintain its edge.

Political and advisory

Good advice and steady nerves born of experience can help decision-makers navigate through uncertain and complex conditions. Professional skills have considerable value in this respect.

Useful advice needs to be based on well-founded evidence, knowledge and experience, but also on judgement and the ability to make the right call at the right time. Superficially this seems to have much in common with resolving design challenges and designers can be attracted to making policy recommendations on this basis, whether in the form of the strategic plans of planners and architects or the 'system thinking' of engineering professionals. Such is the apparent crossover between policy development and design that 'design thinking' is often proposed as a solution to stalled or ineffective policy generation (e.g. Bason 2014). Yet successful policy implementation generally requires a mindset involving goal setting combined with exceptional navigational skill rather than the design of 'solutions'. Policy makers like to quote the Roman philosopher Seneca (Letter LXX, line 4): 'There are no fair winds for those who don't know where they are heading.'[2] Construction professionals who can translate their knowledge into policy objectives and contingent delivery mechanisms can provide great value to the political system (see Foxell and Cooper 2015). As the former Downing Street policy chief, Geoff Mulgan (2009, p23), has noted:

> Consultants and advisers can become adept at applying generic methods for breaking issues down into their constituent parts, and systematically piecing together implementation plans. These methods have long been the bread and butter of competent administration, but public organizations all too often let their skills atrophy. But such generic methods are less useful in

fields where knowledge is all important (such as medicine) and they provide few insights for the more central tasks of government including legitimation, public value, and, for politicians, how to win re-election.

The challenge of atrophying skills is far from being unique to administrators. Professionals need to retain their own ability to think through unforeseen problems and offer the right guidance. Machines are becoming better at giving 'optimum' advice most of the time, but, as has been seen with automated share dealing, such systems can fail on encountering unexpected conditions, leading to financial crashes and worse. Intelligent learning systems will hit such unforeseen 'black swans' (Taleb 2007) ever more rarely, but when they do, the consequences will be increasingly serious and problematic. At those times professional knowledge becomes indispensable, but the difficulty is how to build up and maintain the ability to cope with infrequent crises. It is a problem familiar to doctors and surgeons who now see certain conditions only very occasionally but who, nonetheless, need to preserve their ability to deal with them. When so much day-to-day work is relatively ordinary it is necessary to make extraordinary effort to be ready to cope with the unexpected.

Social

There is a close relationship between the design and realization of buildings and spaces and the quality of the lives of the people who live, work and play in them. Spaces help or hinder communities to gather, socialize and operate. Common social objectives such as personal privacy, psychological comfort or emotional wellbeing are all aided (or prevented) by the combination and con-figuration of the various parts of buildings and man-made environments. Social value is, in small part, very dependent on the knowledge and expertise of con-struction professionals and is intrinsic to many institutions' founding principles, as expressed in Rule 3 of the ICE's Code of Professional Conduct:

> Members must take account of the broader public interest – the interests of all stakeholders in any project must be taken properly into account, including the impact on future generations. This must include regard for the impact upon the society and quality of life of affected individuals, groups or com-munities, and upon their cultural, archaeological and ethnic heritage, and the broader interests of humanity as a whole.

The provision of social value as either a public or, increasingly, a private issue, is likely to remain at the core of the professional function long into the foreseeable future. For both individual professionals and professional firms it will continue to demand their skills, ingenuity, empathy, time and commitment.

There is an additional urgent need for a collective and collaborative approach from professionals and their organizations for delivering broader, systemic social improvements that help society as a whole and protect and preserve those parts that already provide long-term value. This might be through the development of higher and more effective standards; new approaches to delivery; support for forward planning; and the sharing of knowledge on and research into what works for people using the built environment. It is clearly a task for professional organizations and, as will be argued below, it could be social value that provides the greatest opportunity for the professions to thrive when much other expert work is outsourced or automated.

Cultural

Over the centuries many students, professionals and institutions have become fixated on the cultural value intrinsic to the built environment – sometimes to the exclusion of other forms of value. So far this book has eschewed much discussion of high culture, but there is no doubt that both architecture and engineering make an important cultural contribution to places and through it enrich and enliven peoples' lives. It is one aspect of value that it will be very difficult to replace with algorithms and automation, but it is also arguably not a professional 'value', as the creation of aesthetic worth easily defies professional ownership or jurisdiction.

A deficiency of cultural value will not result in charges of professional misconduct and not every building or project needs to have high cultural aspirations. The importance of cultural value is commonly inculcated during professional education and its provision is later aided and abetted by professional practice, but the linkage between the two is not intrinsic and there are many buildings and structures of recognized cultural value that have managed without professional involvement in their creation just as there are many projects delivered by professionals with minimal cultural value. Nonetheless professional expertise provides a skill set and a confidence with the elements and processes of construction that enables originality and creativity to flourish and, in consequence, professionals are socially valued for their creativity and ability at delivering ideas, projects, buildings and places that enhance others' existences, both real and virtual.

Inquiry

The above suggests that it is the basic skills employed by construction professionals that are most at risk of being downgraded or automated in the future. These include fundamental and hard-earned capabilities such as sound building design and technical mastery of industry processes and methods. In comparison it is the more intangible human attributes – the ones that generate collaborative, ethical, social and cultural value – that present the greatest opportunities for a long lasting career. Some qualities show great promise but are not yet systematized: risk carrying, environmental harm reduction and technology development and use. Making the transition from the current core basic professional activities to more highly valued services will not be popular with many professionals. They may feel it is not where their talents lead or even where they want to go. They may prefer to keep a hands-on role with projects, but may have to accept that they will struggle to be properly rewarded for years of dedication and training. Their choice may be comparable to that of a traditional joiner of wood preferring to maintain the enjoyable and fulfilling skills of their exacting craft in the age of the CNC (computer numerical control) machine.

Although innovation in the construction industry will often be led by individuals and firms the professional organizations are in a position to act to prepare the ground and aid the transition in the world of work. If this latest industrial revolution is going to force major change it raises the question of what plans the professional institutions have made to ensure a healthy future for their members.

To address this question the Edge think tank (comprising a multi-disciplinary group of experienced construction professionals) assembled a Commission of Inquiry in 2014 on the Future of Professionalism, chaired by the former Government Chief Construction Adviser, Paul Morrell. The Commission (previously discussed in Chapter 9), was given the remit:

> to examine how the professional institutions and other industry bodies are planning to address pressing issues facing the design and construction sector and how they can act together to reinforce the value of professionalism in the decades ahead.

The Inquiry took evidence in person from the leaders of the major professional institutions (their presidents and chief executives) as well as a range of other interested parties and thought leaders. Participants were asked the following questions (The Edge 2014b):

- **The Environment**: Should it be a professional requirement to address environmental issues, including responsibility for long-term performance and reporting?
- **The Economy**: How can professionals continue to do what they regard as the right thing, when this is not a priority for their client?
- **Society**: How can professionals working across the built environment and their institutions maintain relevance and deliver value to society?
- **Future value**: How can institutes share and co-operate to improve the quality, standing and value of professionals?

In the evidence sessions the professional institutions showed that they fully recognized the challenge of the issues raised in these pages and in front of the Commission they frankly discussed the challenges involved in addressing them. On institutional change / fitness for purpose:

> "The world is moving on so much faster now that I really do question the capacity of professional bodies to be able to keep up with the changes that society demands" (Chris Blythe, Chief Executive CIOB – Session 1, 5th March 2014).

> "Our world is in a continuous state of change – market forces, culture, nature of regulation, competition and growth of the corporation have all impacted the roles each of us and our members play today. Client behaviour is a consequence of the market conditions at play" (Sean Tompkins, Chief Executive RICS – Session 2, 23rd April 2014).

> "They [the Institutions] are undergoing change as democratic bodies but the pace of change may not be sufficient to meet the rising demands on the built environment" (Barry Clarke, President ICE, 2012–13 – Session 3, 7th May 2014).

> "Our professional engineering institutions, all 36 of them, are a bit like old Routemasters [buses]. We're all very fond of them. They may have done a fantastic job and we don't really want to see them go, but actually they're not really very fit for purpose" (Stephen Matthews, Chief Executive CIBSE – Session 1, 5th March 2014).

"[In] my institution 20 years ago 20% of what came out of it was done by staff and 80% by the members. It's slowly changing. We're now seeing a far more professional team in charge of the Institute and actually achieving far more than the amateurs. Now you would say that the members have lost control, but they haven't, they control in different ways" (Chris Blythe, Chief Executive CIOB – Session 1, 5th March 2014).

On value:

"While professional institutions are doing what they can to enhance value, UK plc itself doesn't appear to place value or even know the value of what professions are. … as a professional body we're seeing huge demand for professionals in the built environment from the future super economic superpowers where there is great middle class growth, a population explosion and urbanisation, but yet I don't hear that same value necessarily being placed within the context of the UK" (Sean Tompkins, Chief Executive RICS – Session 2, 23rd April 2014).

"We can only deliver public benefit, and be relevant to society if people value what we deliver, and to value what we do, they have to understand what it is that we do, and therein lies part of the conundrum" (Sue Illman, President Landscape Institute 2012–14 – Session 3, 7th May 2014).

"Delivering value [is] applying expertise and judgement in ways which are really useful to society beyond the immediate needs of our clients, in ways which society appreciates and values and, essentially, in ways which in one way or another they are willing to pay for, because none of this comes for free" (Colin Haylock, President RTPI 2012–13 – Session 3, 7th May 2014).

On the public interest:

"In the modern world, it becomes a very difficult proposition for professional bodies to keep that focus on the public good, because people are looking for more of a transactional relationship with their professional body than one that is built on the ethos of when they were originally set up" (Chris Blythe, Chief Executive CIOB – Session 1, 5th March 2014).

"The RIBA's mission is to advance architecture in the public interest. We set out to create the conditions in which architects can contribute to economic, social and environmental sustainability" (Stephen Hodder, President RIBA 2012–14 – Session 2, 23rd April 2014).

"Delivering public benefit, and maintaining professional status amongst society is not about dumbing down our role, but through example and explanation, making our contribution recognized and valued as being important" (Sue Illman, President Landscape Institute 2012–14 – Session 3, 7th May 2014).

On ethics:

"There is no point in having a professional body unless you are prepared to make a stand for what you believe is morally right. I think that occupying the moral high ground as best as you can with the evidence available is incredibly important" (Stephen Matthews, Chief Executive CIBSE – Session 1, 5th March 2014).

"Since the 1980s and 90s the profession has been challenged to strike the right balance between embracing business values and competition whilst maintaining ethics and social responsibility" (Stephen Hodder, President RIBA 2012–14 – Session 2, 23rd April 2014).

"Professionals have a very wide remit. It is not so much doing what they feel is the right thing, but explaining to their clients why they are recommending solutions and a course of actions. There is no one correct answer any more" (Nick Russell, President IStructE 2014–15 – Session 2, 23rd April 2014).

"The ethical issue comes down to the dilemma of choosing between public and client interest when there is a conflict between the two – the other issues such as honesty are uncontroversial. The idea that professionals are uniquely ethical is a remnant of class superiority, on which the formation of the professions was based in the first place" (Sunand Prasad, President RIBA 2007–9 – Session 4, 22nd May 2014).

On the environment:

"What society needs [is] that we address environmental matters. Whether professional bodies can do it, and do it to the level required, I doubt" (Chris Blythe, Chief Executive CIOB – Session 1, 5th March 2014).

"Everyone is focused on the short term here-and-now and no one is looking at the long term consequences and impacts and that's probably the gap that professions, along with industry and academia, could [deal with and] pose some of the questions to people who only want short-term answers" (Sean Tompkins, Chief Executive RICS – Session 2, 23rd April 2014).

"Adapting the built environment to cope with a low carbon economy, population growth and the changes in society's expectations and technology are placing great demands on the built environment professional, which requires greater engagement with society in order to help society appreciate the world they live in and the difficult decisions they face in the future" (Barry Clarke, President ICE 2012–13 – Session 3, 7th May 2014).

On evidence:

"We need more evidence about what works in order to do the right thing and demonstrate whole life value. Knowledge is more powerful than opinion" (Stephen Hodder, President RIBA 2012–14 – Session 2, 23rd April 2014).

"Within CIBSE we have over 1500 energy assessors ... [and] we've got a fantastic resource for a feedback loop about how we can create better buildings. In my opinion we don't use it very effectively" (Stephen Matthews, Chief Executive CIBSE – Session 1, 5th March 2014).

"There is a move to evidence-based political decisions; the emergence of big data as a design tool as the concept of smart cities develops is creating opportunities to tackle the global challenges" (Barry Clarke, President ICE 2012–13 – Session 3, 7th May 2014).

On collaboration:

> "Professions need to represent now a much more collective force in the eyes of policy makers and opinion formers ... [and] need to come together to drive greater standards and international consistency" (Sean Tompkins, Chief Executive RICS – Session 2, 23rd April 2014).

> "There is a job to be done to see how the institutions can collaborate further and possibly there has been a little bit of institutions working in silos looking after their own interests, but I think on the ground it's a lot better than you suggest, there is a great deal of collaboration" (Nick Russell, President IStructE 2014–15 – Session 2, 23rd April 2014).

> "I don't care whether you're an engineer an accountant or anybody else. But if you're passionate about making a change in the built environment let's go and do it together" (Stephen Matthews, Chief Executive, CIBSE – Session 1, 5th March 2014).

> "Society ... would expect the professionals in the built environment to understand that they work in a complex multi-disciplinary world, and they would expect us to find mature ways of working with each other" (Colin Haylock, President RTPI 2012–13 – Session 3, 7th May 2014).

> "All built environment institutes ... need to collaborate on relationships with central government, local government and key public institutions. The advantages are huge and obvious" (Sunand Prasad, President RIBA 2007–9 – Session 4, 22nd May 2014).

The final Commission session included several representatives of a younger generation of construction professionals who spoke about the future challenges that needed to be faced, including increases in population, globalization, resource capacity and climate change. Two witnesses, Lee Franck (ICE – Session 4, 22nd May 2014) and Ciaran Malik (ICE Graduate member – Session 4, 22nd May 2014), speaking for a group of young professionals noted:

> Setting minimum standards for ethical behaviour in codes of conduct is not good enough – we want to be challenged, inspired and guided to do things in a better, more responsible way. ... To continue doing the same things and expect a different outcome is just insanity. The Institutions need to change, but not only them, we all need to change by taking a more collaborative, vocal and responsible role within society. We should build on the energy, enthusiasm and optimism of the young to take our professions to even greater heights and to contribute to society in a more profound way [Franck].
>
> The sad news is that young professionals in the built environment feel their role and even their whole sector isn't rated highly enough. They feel underpaid for the work they do and people who are in the first two years of their careers feel that there may come a point where they will have to consider changing profession to start a family. ... I'm not surprised most of us want to keep our heads down and ignore issues on the horizon. This is exactly why the institutes need to lead by example, work together to help us feel proud about what we do and ready to face whole new challenges [Malik].

Other contributors spoke of the context of professionalism in the industry and beyond and, in particular, Matthew Taylor, Chief Executive of the Royal Society for the encouragement of the Arts, Manufactures and Commerce (RSA) and a former head of the Number 10 Policy Unit in Downing Street, suggested (at Session 3, 7th May 2014) a framework for examining the role of professional associations:

We live at the moment in a time of weak hierarchy, relatively weak solidarity and very strong individualism ... you need all those three things: you need authority, you need solidarity, you need shared values and individualism ...

- [The] forms of leadership and authority that work in the modern world are more to do with influence and the capacity to convene than to do with raw power and exercise of control.
- The critical question with solidarity is: Is it an exclusive or inclusive form of solidarity? ... Is it a solidarity which invites other people to participate if they share the values of the organization?
- [The] individualism dimension is ... professional creativity versus a notion of individualism positioned in commercial acquisitiveness.

The ground is shifting beneath our feet very rapidly and we need to be equally able to think quite radically.

Taylor wasn't alone in discussing the need for engaged and inclusive professional bodies: Sean Tompkins, Chief Executive of the RICS, made a similar call in his notes for the second session (23rd April 2014):

We have an obligation to continuously look at the relevance of professions and regularly make changes to how we operate. We have seen in the past when we [aren't ex]plicit in our approach the value of professionalism will be diminished.

- There needs to be greater cross-sector collaboration
- We need the professions looking at how they can support standards that link better across sectors internationally
- We need to be explicit about the benefits of professionalism to wider society
- We need to be more inclusive and diverse.[3]

The Edge Commission report, *Collaboration for Change* (2015, p88), authored by Paul Morrell, discussed in detail the themes raised in the evidence sessions and, in particular, stressed the need for professional inclusivity:

Legitimacy would be increased if efforts are made to include the wider public in both plans and debates on issues directly affecting them (such as fracking, for example), sharing expertise objectively, in substitution for presuming to know best based purely on that expertise – replacing, or at least supplementing, exclusivity in favour of inclusivity.

This is a clear challenge for the institutions. Although they express a desire to be open and to communicate widely, they constantly seek to do this on their own terms and tend to revert to secrecy when left unchallenged. As a minor example, they fail to make their membership numbers, including age and gender breakdowns, publicly available, except on an occasional, non-comparable basis. As Morrell noted (p53):

> with the exception of the RICS (which goes further than publish rulings, and also makes public the disciplinary proceedings held for the more serious charges), all institutions engaged in this exercise declined to offer statistics on disciplinary actions, either through non-response or by citing confidentiality.

The Commission report made 26 recommendations in six areas (p6) (see Appendix D)

> where institutions could, by joint action, demonstrate their effectiveness and thereby enhance their relevance and value:

- **Ethics and the public interest**, and a shared code of conduct
- **Education and competence**
- **Research and a body of knowledge**
- Collaboration on major challenges, including **industry reform** in the interests of a better offer to clients, **climate change** and **building performance**.

Propositions

To a varying extent all these issues have been followed up by the Edge since the publication of the Commission report and they form the basis for the following proposals for moving the UK construction professions forward.

The interlinked propositions are arranged under six headings:

Clarity and transparency

If the professions are to reposition themselves as society and politics go through a period of realignment, they need to rebuild their social capital and regain trust as a source of well-informed, independent and impartial judgement. To achieve this, a slow and gradual process is required to overcome a now long-standing reputation for self-interest. This will require the professional institutions (or others) clearly articulating the expectations that society should have from professionalism and for them to be fully transparent about how those expectations will be delivered and monitored. Such a proposition will need to be accompanied, or preceded, by a substantial and difficult shift in the ingrained habits of individual professionals, professional firms as well as their membership organizations.

The first necessary change will be for professionals, through their professional organizations, to recognize and commit themselves to the position that being professional means having a primary obligation to serve the public interest (including protecting the wider environment and those affected directly by their work), even when their employment and income comes from other sources. Agreeing to such an obligation would not make professionals subservient to members of the public, governments or groups, but it could, for example, require that they should be transparent and open about their actions and the resultant outcomes. As public companies have obligations to prepare and publish annual reports, with the level of disclosure depending on their size, professional firms would be expected to do the same, reporting against a small number of standard and verifiable measures, which might include turnover, staff numbers, diversity statistics and some basic project performance data including (predicted and actual) carbon emissions. Much of this data is already collected and published by various institutions and journals, but there is a need is to make it a publicly accessible resource using standardized metrics and to begin to reform the culture of secrecy surrounding professionalism.

The same principles would apply to the institutions themselves, which would be expected to see the public as their most important audience and to commit themselves to being open and outward facing representatives for their disciplines. Basic aggregated information about their professional member-ship (numbers, geography, age profiles, diversity, etc.) should be made avail-able, as should the outcomes of disciplinary hearings, course accreditations and decisions of their ruling councils. Institution officers should have a remit to engage with the wider public as a normal and everyday activity, providing talks, exhibitions, briefings, children's workshops and more, as indeed some do already. Outreach work is already a primary objective of many institutions, in line with the obligations written into their Royal Charters to promote know-ledge of their discipline, and should be fully acted on.

The Royal Charters and other corporate statements should be reviewed to ensure that they make the purpose of each profession absolutely clear to the public, public authorities, those commissioning professional services and professionals themselves. Charter wording should be revised as neces-sary to put public interest duty first and foremost, leaving no room for doubt, followed by commitments to a number of other professional priorities, possibly including: independence, impartiality, developing and disseminating knowledge and the advancement of the discipline.

Ideally the professions should work collectively when revising their Charter objectives and develop a common approach to reforming their professional agenda. They should aim to develop a standard set of words that could be used, with minor adaptations, by numerous different bodies, and which, through standardization and repetition, could potentially become a familiar text across the world. Only by aligning the many different explanations of professionalism in the construction sector will professional standards become well enough known to the public and commissioning clients to give the words meaning and strength. If the institutions can act in concert on this (and avoid a lowest common denominator outcome) they will make up a formidable force for both clarity and eventually the reputation of professionals in the sector.

The idea of a formal compact between construction and property professionals and those commissioning buildings was suggested in Chapter 1. Such a compact could provide much needed definition to the relationship and clarify expectations on both sides, and should work in tandem with appropri-ately revised professional codes of conduct (see Table 11.1).

Table 11.1 Proposed 'Professional Compact' between construction and property professionals and those commissioning their services – developed in conjunction with the Edge

---

Professionals in the construction and property sectors will deploy:

- expertise, skill, knowledge and experience to deliver agreed services in good faith
- competence, diligence, honesty, integrity and care
- evidence-based judgement to achieve high standards of work and conduct.

Subject to the obligation to:

- put the interests of the wider world and society first and to take protective action when necessary, but otherwise to put clients' interests before their own
- take personal and corporate responsibility for the outcomes of their work
- show proper care, consideration and fairness towards others, especially those involved in realizing projects and those who will live with the outcome
- keep their own knowledge relevant and up to date
- train and help develop the abilities of other members of both their profession and society
- measure, feed back and share relevant information and insights gained from their work in order to develop and improve knowledge and skill across the disciplines.

In exchange for:

- the trust of those commissioning services
- recognition of their independence and right to self-direction
- the grant of respect and status
- a degree of exclusivity over the provision of socially important services
- fair payment for their work.

As recommended in *Collaboration for Change*, the various institutions need to work together on a base framework for their codes that again uses the same structure and wording. Each discipline may need to elaborate on particular areas of the code that are especially applicable to their members and develop discipline-specific guidance that stands alongside it, but there is very little at present in any institution's code that is not widely applicable, although there are plenty of gaps in each of them that could usefully be filled by a more universal approach.

A joint code is essential for construction professionals working together in project teams with appointment documents intended to operate back-to-back between the disciplines. Standards of professional conduct must of necessity follow suit. At present the various codes rarely get discussed openly with clients, but a common code would encourage all parties to get to know it better. The RICS reports that maintaining a clear code is essential to their ability to promote themselves and their members across the world. The same would be even truer if the majority of UK based professional organizations maintained and upheld a shared code. The various engineering professions already do this in the form of the code published jointly by the Engineering Council and Royal Academy of Engineering (2014), and there is no reason for the whole industry not to follow their example. As with the institutions the primary audience for any code should be the public.

A professional code of conduct should be made accessible, clear and at one with all the other professional codes that will be encountered on a project. Working together can only make the individual professional bodies stronger.

A draft version of a possible joint code is included in Appendix E. Like the compact it has been developed with input from members of the Edge. The draft code is intended to be an example of what a joint code might look like and draws on the current codes of a range of different organizations.[4] In line with the principles discussed above it is structured to make the obligations and duties to various parties as clear and unambiguous as possible, in the following order:

- The wider world
- Society
- Those commissioning services
- Those in the workplace
- The profession
- Oneself

A code of conduct is only as good as its enforcement. All the institutions need to work with their members to implement a purposeful and rigorous checking regime. Possibly this would be similar to the one the RICS already has in place. The institutions would also need to make public the disciplinary measures they apply when the code is contravened.

Individual professionals and companies will need to follow the lead of the institutions and become more open and transparent in their work, as well as building on all the outreach work that the professions carry out. This will include the need to feed back data and experience, which is considered below. The intention is to eventually restore the reputation of the professions as honest, independent brokers able to mediate between the demands of different parties, if only because society needs reliable intermediaries; intermediaries who are in turn governed by an effective system of accountability and discipline.

Responsibility (for outcomes and performance)

The second proposition is that construction and property professionals should take responsibility for the performance of their projects and stand by the claims they make for it.

There is a well-known problem in the industry concerning the difference between energy-use predictions made for buildings and the reality of energy usage in practice. It well enough known for it to have a familiar name: the performance gap. The same problem almost certainly applies to all other claims made at the design stage about the outcomes of projects, including functional efficacy, usability, environmental standards and social impact. In practice there is very little available evidence for the way projects work once they are in use, as they have rarely been rigorously examined once they have been handed over to their users. Post-occupancy evaluation (POE) studies are still few in number decades after a systematic method was developed at Strathclyde University,[5] and subsequently refined by Building Use Studies (Leaman *et al.* 2010) and the Usable Buildings Trust. The methodology was demonstrated in the series of PROBE studies published in the *CIBSE Journal* from 1995 to 2001 and in a series of *Building Research & Information* special issues,[6] but construction professionals, for want of feedback from completed projects, are still largely operating in the dark.

POE is nonetheless gradually gaining momentum. Several architecture and engineering firms have employed researchers to undertake studies across a range of their completed projects.[7] There is a range of university departments and other researchers involved in better understanding building performance, but progress is slow. The UK's Government Soft Landings programme (GSL) (BSRIA 2017) demands POE and, since 2008, the RIBA and CIBSE have collaborated on the CarbonBuzz initiative, designed to collate anonymized data on the energy and carbon performance of buildings. Even insurers, traditionally hostile to the discovery of bad news relating to building performance, have made cautious announcements about permitting its use as a means of containing risk and promoting continuous improvement.[8]

Information on the performance of completed projects is essential if design standards are to be improved, something that applies equally to work practice, customer satisfaction and problem solving in general. Companies in other sectors constantly monitor their products and their production and are judged on the quality of their aftercare.

Understanding and influencing the behaviours and practices of occupants is a vital part of a low-carbon strategy. ICT companies regularly perform carefully monitored micro-experiments on their service models to see if they can make improvements. This approach is now used by government programmes such as the UK's Behavioural Insights Team (also known as the 'Nudge Unit') that attempt to use and test ways of presenting public services to enable people to make 'better choices for themselves'.

The relatively small size of many companies in the construction sector means they will only directly generate a limited amount of information that can be fed back and analysed. This will inhibit the potential to make comparisons with industry best practice or the work of their competitors. Data collection from a much larger pool of projects is required in order to enable all sizes of company to learn from across the industry and implement best practice. Such a task is best suited to one or more of the professional bodies, industry organizations or an independent organization outside the industry. Certainly it is the responsibility of the institutions to ensure that their members have means to acquire the appropriate skills and tools to fulfil their roles as 'experts' in the industry.

A great deal of industry data is already collected, including the client and user feedback collected on projects that underpin various benchmarking initiatives.[9] Nonetheless intelligent collection and use of the performance data that is already available on projects is still woefully underdeveloped and there is huge scope for its expansion and the development of an essential knowledge resource. The key question is who will, or should, do the collection and analysis, and will it be done independently and in the broader public interest?

The collection, analysis and sharing of data through an arms length organization,[10] jointly established by a wide range of professional institutions, with a remit to stay impartial and independent,[11] would be an immense boost to the usefulness and standing of professional institutions. It would provide essential information for both practice and policy makers to make better decisions and to improve overall built environment strategy. Such a body would need to partner with other research organizations and universities and develop good links with private companies, government departments and other equivalent bodies elsewhere in the world; developing and promoting UK research and expertise as it grows in effectiveness.

The sources of data that such a body might draw on are extensive and would include: existing and developing statistical sets relating to buildings and infrastructure projects, feeds from the many monitoring devices permanently installed in facilities and information from logged BIM models; but the core source would be professionals working in the built environment reporting on their projects as part of their normal professional obligations. It would be a big data project in every sense, with all the concerns that such projects raise about privacy and confidentiality, but is essential to validate and improve built environment performance and outcomes.

The impact of the availability of good data from built projects, further explored in *Retropioneers: Architecture redefined* (Foxell 2017, pp36–8), could potentially lead to several significant changes in professional practice, including:

- Using evidence culled from a wide range of earlier projects practitioners would be able to more accurately predict the performance of their designs in use and, as result begin to use realistic outcomes as a positive means of marketing and distinguishing themselves. This would eventually develop to the point where they would be able to offer performance guarantees.
- With a better grasp of project outcomes, professionals and their insurers would become happier to accept and encourage user feedback and reporting. The many rating websites that will inevitably develop along the lines of TripAdvisor and Uber, offering anecdotal information and opinion, would be superceded by a more accurate and reliable system based on evidence, potentially run by or on behalf of the professional institutions.
- Designers prepared to stand by their performance claims and accept user feedback as standard will generate a capability to involve themselves fully in the whole life cycle of buildings and other facilities. This should lead towards responsible long-term stewardship of buildings and a pattern of much closer engagement with the buildings themselves, users, owners and investors; much as a cathedral architect or surveyor of the fabric does at present. Such engagement would provide further insights and data for feeding back through the system.

Taking responsibility, given the many very different forms of procurement with their varied lines of accountability, is not simple or straightforward. Rather it will be a gradual process of re-establishing control over the process, which if beneficial to project owners will develop its own momentum. For the moment professionals need to accept accountability at least for those aspects of projects that are in their control.

Knowledge and career

An expert knowledge base, as discussed extensively in the preceding chapters, is at the core of the professional system, even though the knowledge function itself was long ago outsourced to academia. While knowledge at the beginning of the professional period was growing at an invigorating pace, today it is expanding far faster than any individual can cope with and much faster than any print-based system is capable of keeping up with. By practical necessity as well as convenience professionals have become almost entirely reliant on online resources and with them the inherent filtering and categorization systems of the Internet, over which they have only limited control.

On the surface the professional organizations have little chance of competing against this global system of knowledge production, prioritization and distribution. Their libraries have become repositories of historical documents rather than current information and their resources are slim. But in practice they have at least two essential tasks to fulfil in order to maintain the usefulness of the professional knowledge base and help both society and their membership keep up with the continual input of data, research findings and semi-random information.

Firstly the institutions need to ensure that new knowledge relevant to their disciplines is being continually identified and curated; enabling them to notify members and others about important issues, index and provide *précis* of the most significant findings and help with access to the original source material whenever and wherever possible. There are major research findings being generated annually by universities, both UK and overseas, government bodies and many other organizations, but the great majority of that work will remain unknown and inaccessible to practitioners unless there are means of reaching and exploring it. Professionals are presently failing to keep up to date with new research and this seriously challenges their claims of expertise and is an issue the institutions urgently need to address.

The extent and multi-disciplinary nature of the task suggests it needs to be carried out by multiple institutions, universities, research organizations and funding bodies working in partnership. Since 2015 a number of UK institutions, led by the ICE and CIBSE, have been collaborating on a research digest publication along these lines, with an online journal, *Prospective*, due to be launched in 2018. Such work needs to become a mainstream activity of professional institutions and would form a significant service to both their members and the wider public.

Secondly, as discussed in the previous section, the institutions need to ensure that data fed back from their members is collected and fully analysed. This may best be delegated to a separate and independent body, but the institutions must play their part in ensuring that findings and insights are communicated back to their contributing source in an accessible and useful form. They must also maintain effective debate amongst their membership about how to interpret and improve the quality of both the information provided and the resulting feedback.

As the institutions return to the business of possessing and providing a live knowledge base, their own ability to use that base will also grow; generating opportunities for providing authoritative information and presenting a more confident public profile. The institutions either individually or collectively need to fully develop their grasp of the information relevant to each discipline and become the prime source of considered opinion drawn from that data for the media, government and others. Several of the institutions already fulfil this function in a number of areas, for example, the RICS's *Market Surveys*, the RIBA's *Future Trends Workload Index* and CIBSE's *Technical Memoranda*; but there is scope for going much further.

The institutions should also develop their role as accessible knowledge hubs dealing with specialized subjects. They should develop research literacy amongst both students and members to enable them to understand and use research. Practitioners should be encouraged and helped to undertake their own research and make their findings widely available through the hub. The institutions should offer membership to researchers from a range of different backgrounds and with a variety of qualifications (or none) who are adding to the useful knowledge of the profession. For some researchers the built environment will be the subject of a lifetime's research, but even if it is a subject that will concern them only for a limited period in their career, they should be welcomed in. The institutions have long welcomed non-members onto their advisory groups, committees and particularly into their special interest groups and societies but it may be time to expand the practice to become more inclusive and open.

Potentially, and possibly advisably, the institutions should decide to embrace the increasingly transient nature of professional careers and learn to welcome those who are intending to spend only a short time within a discipline or have a career straddling several different topic areas. This would mean focusing more on defined areas of professional knowledge, practice and experience than on the comings and goings of a membership body marked out by their first choice of tertiary education. This would have the advantage of making institutions more inclusive organizations, able to contain everyone, from school children to government ministers, interested in their disciplines or working in their topic area.

Such an approach might result in a less coherent set of members and reliable income stream, but the institutions are already moving in this direction with their ever expanding membership categories and desire to find new sources of members. Historically the institutions recognized the importance of having significant and well-known non-professional affiliates, with their inclusion of honorary members amongst their earliest supporters and even non-professional presidents. To date none of the institutions has made the decision to disconnect professional qualification from full institutional membership, but it may come as professional qualifications themselves destabilize.

Qualifications in the modern world need to become more flexible and better at allowing professionals to cross sectoral boundaries and to find other routes of entry into the professions, if only to match current international practice. A particular qualification should no longer be expected to constrain a career over a full lifetime; sideways movement should be encouraged and enabled. Adapting to a more fluid system of qualification will be greatly eased once the tight lock between a professional discipline and its cadre of practitioners is broken, but such a move will depend on more responsive knowledge distribution and an accreditation system based on a mixture of computer checks and peer recognition, rather than formal exams.

Collaboration

The need for the professions to work together more consistently has already been raised several times and is the central premise of the Edge Commission report *Collaboration for Change*. In practice and behind the scenes the institutions continue to work together at a variety of levels as well as through their umbrella body, the Construction Industry Council (CIC), with its membership of 32 professional bodies. There have also been a few higher profile collaborations between them in recent years, notably on the development of BIM, but in general they choose not to be seen working together beyond the occasional shared endorsement of a report or position paper. On the big issues, whether education, procurement, climate change, industry transform-ation or even awards, they prefer to work alone.

The need to show independence may be a response to member pressure. This reinforces the false notion of the professional exceptionalism of each dis-cipline and the myth that their president or chief executive has a close rela-tionship with senior politicians and decision-makers. Projects of any size are, without exception, run by multi-disciplinary teams, usually with individual team members working across sector boundaries, in an industry that is used to working together. If there is a divide it is between the large and small firms rather than between disciplines. The challenges that face the industry cut right across the different professions and can only be tackled by joint programmes of work. The institutions need to overcome their reluctance to be seen collab-orating or to allow others to lead, to solve them.

Valuing independence

The professions have always thrived on their ability to provide independent and objective advice – an approach enshrined in article 1(b) of the RTPI's *Code of Professional Conduct*:

> Members … shall fearlessly and impartially exercise their independent pro-fessional judgement to the best of their skill and understanding.

Yet the independence of professional work has been progressively weakened in the transition from occupational to organizational values over the last two decades, as documented over the period by Julia Evetts, and by the culture of control, regulation, monitoring and enforcement over the professions, noted by Onora O'Neill in 2002. The continued development in ICT looks set to expand the surveillance and checking of work even further as detailed data is permanently captured precisely detailing location, health, tasks, work rate and accuracy as well as any overall evaluations. Such systems may not be able to reveal some of the more critical human aspects of professional work, but will record enough to deny professionals their long established autonomy and ability to self-direct.

As with the RTPI the professions must develop a robust culture of self-determination in professionalism and be prepared to defend individual professionals who insist on following where the evidence and best practice guidance leads them. Impartiality and independence have been carelessly jettisoned in recent years and need to be reclaimed and championed as defining features of professional work by institutions, educators, employers and individuals alike.

Institutional change

To paraphrase Giuseppe Tomasi, Prince of Lampedusa, in *The Leopard* (1958): 'If the institutions want things to stay as they are, things will have to change', and to a great extent change has been continuous since the 1820s when the first modern professions were established. The pace of change will only increase over the coming period and while the institutions are unlikely to go out of business, they need to work differently if they are to remain relevant and work as a force for the public good. As Paul Morrell notes in *Collaboration for Change*, 'the threats and pressures for change that the professions face, if not yet existential, are real and profound, and demand change' (2015, p5).

The challenge is that in the process of modernizing and professionalizing themselves the institutions have become estranged from their public interest role and remit, and they need to heed the warning from Matthew Taylor of the RSA at the third session of the Edge Commission hearings:

> Legitimacy is derived from the fact that you believe you're in the business of balancing professional interest and public interest, and if you abandon that, or even the pursuit of that, you abandon your legitimacy, and you become simply a trade association.

The legitimacy of organizations has become a major issue in recent times with the perception across the western world that a widening gap exists between the governing and the governed, a gap that has the potential to turn into serious social division. The UK institutions representing the construction professions are not immune to this concern and need to ensure that they, along with many other parts of civic society, are in robust democratic shape.

Conclusion

These propositions form an outline for a six-point plan. Together they will help, if not ensure that, the professions remain an effective and essential part of both society and the economy; a part that allows the system to operate smoothly and easily, but also fairly and to the benefit of all.

In the 1680s, as Samuel Pepys began to reorganize the naval profession in the wake of the Anglo-Dutch wars, he proposed reforms based on three observations (Frankopan 2015, p268):

1. Specialist, heavy vessels were more effective than light cruisers;
2. Experience could teach better lessons than theory – sharing knowledge and learning was crucial in making the navy the best in the world; and
3. The Navy Office needed to support sailors and officers effectively as an institution.

In the 21st century, albeit with a very different set of institutions and conditions, these lessons are still relevant, but the challenge of reforming the professions seems that much greater.

Notes

1   See, for example www.onlinestructuraldesign.com, one of many structural design calculators, or www.donotpay.co.uk, which generates specific legal letters to cope with a variety of common situations.
2   '*ignoranti quem portum petat nullus suus ventus est*'.
3   All the above quotations have been taken from the transcripts of and prepared notes for the four sessions held in spring 2014 by the Edge Commission of Inquiry on the Future of Professionalism. Summary versions are available on the Edge website www.edgedebate.com. They have, in places, been lightly edited for sense.

4   Sources for the draft code include those of the Society of Construction Law, RIBA, ARB, EC-RAEng, ICE, RICS, CIAT, CIBSE, CIOB, IStructE, LI and RTPI as well as the International Ethics Standard, the Nolan seven principles of public life and the Elements of a New Professionalism (Bordass and Leaman 2013, p6).

5   Principally by Tom Markus at the Building Performance Research Unit (BPRU), University of Strathclyde.

6   See www.tandfonline.com/toc/rbri20/29/2 and www.tandfonline.com/toc/rbri20/33/4

7   Firms include AHMM, Arup, Bennetts Associates, Buro Happold, Feilden Clegg Bradley Studios, Max Fordham and Penoyre & Prasad amongst others.

8   'The RIBA Insurance Agency has confirmed that architects with an RIBA Insurance Agency professional indemnity insurance policy are covered to undertake Post Occupancy Evaluation/Building Performance Evaluation services, but recommend that practices that are going to offer these services inform the RIBA Insurance Agency of their intention to do so' (RIBA 2015b).

9   For example, see Glenigan, CITB *et al.* (2016) or Fees Bureau (2016).

10 An Institute for Building Performance has been argued for for many years by the building physicist Bill Bordass and others – see for example Bordass and Leaman 2013, p7.

11 A model for such a body might be the Institute for Fiscal Studies.

Appendix A
Professionalism – defined and discussed

We trust our health to the physician; our fortune and sometimes our life and our reputation to the lawyer and attorney. Such confidence could not safely be reposed in people of a very low or mean condition. Their reward must be such, therefore, as may give them that rank in the society which so important a trust requires. The long time and great expense which must be laid out in their education, when combined with this circumstance, necessarily enhance still further the price of their labour.

Adam Smith, *An Inquiry into the Nature and Causes of the Wealth of Nations* (1776, Chapter X)

The importance of the professions and the professional classes can hardly be overrated, they form the head of the great English middle class, maintain its tone of independence, keep up to the mark its standard of morality, and direct its intelligence.

H. Byerley Thompson, *The Choice of Profession* (1857, p5)

'You are mixing up two very different things in your head, I think, Brown' said the master putting down the empty saucer, 'and you ought to get clear about them'. You talk of 'working to get your living' and 'doing some real good in the world' in the same breath. Now you may be getting a good living in a profession, and yet not doing any good at all in the world,'

Thomas Hughes, *Tom Brown's School-days at Rugby* (1857)

A profession. The word was understood well enough throughout the known world. It signified a calling by which a gentleman, not born to the inheritance of a gentleman's allowance of good things, might ingeniously obtain the same by some exercise of his abilities.

Anthony Trollope, *The Bertrams* (1858)

At least there is tolerably general agreement about what an University is not. It is not a place of professional education. Universities are not intended to teach the knowledge required to fit men for some special mode of gaining their livelihood. Their object is not to make skilful lawyers, or physicians, or engineers, but capable and cultivated human beings. It is very right that there should be public facilities for the study of professions. It is well that there should be Schools of Law, and of Medicine, and it would be well if there were schools of engineering, and the industrial arts. The countries which have such institutions are greatly the better for them; and there is something to be said for having them in the same localities, and under the same general superintendence, as the establishments devoted to education properly so called. But these things are no part of what every generation owes to the next, as that on which its civilization and worth will principally depend. They are needed only by a comparatively few, who are under the strongest private inducements to acquire them by their own efforts; and even those few do not require them until after their education, in the ordinary sense, has been completed. Whether those whose speciality they are, will learn them as a branch of intelligence or as a mere trade, and whether, having learnt them, they will make a wise and conscientious use of them or the reverse, depends less on the manner in which they are taught their profession, than upon what sort of minds they bring to it, what kind of intelligence, and of conscience, the general system of education has developed in them. Men are men before they are lawyers, or physicians, or merchants, or manufacturers; and if you make them capable and sensible men, they will make themselves capable and sensible lawyers or physicians.

John Stuart Mill, Inaugural address delivered
to the University of St. Andrews (1st February 1867)

A profession is a vocation founded upon specialised educational training, the purpose of which is to supply disinterested counsel and service to others, for a direct and definite compensation, wholly apart from expectation of other business gain.

Sidney Webb and Beatrice Webb, *New Statesman*
(21st April 1917)

The fundamental consideration in the work of an engineer – if he is ever to pull himself out of the his present state of being a hired servant – is that he shall make public interest the master test of his work.

Morris Cook, Letter to A.G. Christie (9th June 1921)

Professions begin with the establishment of professional associations that have explicit membership rules to exclude the unqualified. Second they change their names, in order to lose their past, to assert their monopoly, and, most importantly to give themselves a label capable of legislative restriction. Third, they set up a code of ethics to assert their social utility, to further regulate the incompetent, and to reduce internal competition. Fourth they agitate politically to obtain legal recognition, aiming at first to limit the professional title and later to criminalize unlicensed work in their jurisdiction.

Theodore Caplow, *The Sociology of Work* (1954, pp139–40)

The professional has been defined as one whose decisions are concerned with the total performance of the building as seen by the community and by his client.

Royal Institute of British Architects,
The Architect and his Office (1962, p80)

Specialization prevents mutual understanding ... The man of today is no longer able to understand his neighbour because his profession is his whole life, and the technical specialization of his life has forced him to live in a closed universe.

Jacques Ellul, The Technological Society (1964, p164)

[while] men in other professions may blunder or play false with more or less impunity ... the mistakes of the engineer are quick to find him out and proclaim his incompetence. He is the one professional man who is obliged to be right.

Monte Calvert, The Mechanical Engineer in America (1967, p265)

Instead of engaging in a power contest between the haves and the have-nots, the [professional] association undertakes to protect and expand the knowledge base, enforce standards of learning, entry and performance, and engage in similar activities designed to enhance the position of the practitioner while simultaneously purporting to protect the welfare of the public in the person of the client. Indeed, professional claims concerning the primacy of the public good over the practitioner's own private benefits might be viewed as a critical difference between the professionalizing and the unionizing modes of mobility, were it not for the considerable evidence that the claims are watered down with rhetoric.

M. Haug and M. Sussman, 'Professionalization and unionism' (1971)

Architecture, whether private or official, was considered by Cicero and Vitruvius as 'one of the learned professions for which men of good birth and good education are best suited'.

Martin S. Briggs, The Architect in History (1974, p35)

Profession is more often defined as *an occupation which tends to be colleague-orientated*, rather than client orientated.

Magali Sarfatti Larson, Introduction, The Rise of Professionalism (1977, p226)

Anyone coming to this country from the Continent – perhaps also coming from America – is puzzled beyond words by the careful and rigid division of professional and commercial activities in Britain. This, he feels, must be a nation riddled with demarcation lines, and with professional taboos and rituals.

Otto Kahn-Freund, Thank offering to Britain lecture (1979)

Salaried mental workers who do not own the means of production and whose main function in the social division of labor may be described broadly as the reproduction of capitalist culture and capitalist class relations.

Barbara and John Ehrenreich, The Professional Management Class (1979)

A 'profession' was defined as an occupation which possessed given traits: a collegial and hierarchical organisation; group control of recruitment, training and entry qualifications; a self-imposed code of personal behaviour and group practice; a service ethos; a claim to monopoly in the chosen area of practical service on the basis of a defined and standardised body of intellectual knowledge and expertise; considerable individual autonomy; altruism, *ésprit de corps*.

Rosemary Day, 'The professions in early modern England' (1986)

For some professionalism was a means to control a difficult social relation; for others, a species of corporate extortion. For still others its importance lay in building individual achievement channels, while a fourth group emphasised how it helped or hindered general social functions like health and justice.

Andrew Abbott, *The System of Professions* (1988, p7)

The replacement of gentlemanliness by scientism, efficiency and accountability has drastically reshaped the contents and results of interprofessional competition.

Andrew Abbott, *The System of Professions* (1988, p144)

Knowledge replacement becomes a serious career problem when a total turnover of effective knowledge takes less than the thirty years that span a typical career.

Andrew Abbott, *The System of Professions* (1988, p180)

Professions evolve together. Each shapes the others. By understanding where work comes from, who does it, and how they keep it to themselves, we can understand why professions evolve as they do.

Andrew Abbott, *The System of Professions* (1988, p297)

Professionalism has been the main way of institutionalizing expertise in industrialised countries. There are, as we sometimes forget, many alternatives: the generalised expertise of the imperial civil services, the lay practitioners of certain religious groups, the popular diffusion of expertise characteristic of microcomputing. The contrasting examples show the essence of professionalism: professionalism's expertise is abstract, but not too abstract; it is not generally diffused; its practitioners work full time in particular areas. But professionalism shares with these alternatives the quality of institutionalizing expertise in people.

Andrew Abbott, *The System of Professions* (1988, p323)

An occupation which so effectively controlled its labour market that it never had to behave like a trade union.

Harold Perkin, *The Rise of Professional Society* (1989, p23)

The professional is offering a service that is ... esoteric, evanescent and fiduciary – beyond a layman's knowledge or judgement, impossible to pin down or fault even when it fails, and which must therefore be taken on trust – he is dependant on persuading the client to accept his valuation of the service rather than allowing it to find its own value in the marketplace. His interest therefore is to persuade society to set aside a secure income, or a monopolistic level of fees, to enable him to perform the service rather than jeopardize it by subjecting it to the rigours of capitalist competition in the conventional free market. It is the success of such persuasion which raises him (when he succeeds) above the economic battle, and gives him a stake in creating a society which plays down class conflict (in the long if not in the short term) and plays up mutual service and responsibility and the efficient use of human resources.

Harold Perkin, *The Rise of Professional Society* (1989, p117)

An intellectual service occupation having a body of specialist knowledge and skill, long and difficult training, a prestigious social position, and some legal mechanism to enforce its monopoly over the provision of its service and use of its title.

Vincent Clark, *A Struggle for Existence* (1990)

Practitioners are forever torn between their desire for material self-aggrandisement and an ethos of altruistic service. How experts resolve that fundamental dilemma between narrow selfishness and broader public interest will determine whether the professions are a scourge or a benefit to humanity.

Konrad Jarausch, *German Professions 1800–1950* (1990, p8)

A profession:

- Is a full-time liberal (non-manual) occupation;
- Establishes a monopoly in the labor market for expert services;
- Attains self-governance or autonomy, that is, freedom from control by outsiders, whether the state, clients, laymen or others;
- Training is specialized and yet also systematic and scholarly;
- Examinations, diplomas and titles control entry to occupational practice and also sanction the monopoly;
- Member rewards, both material and symbolic, are tied not only to occupational competence and workplace ethics but also to contemporaries' belief that their expert services are 'of special importance for society and the common weal'.

Michael Burrage, Konrad Jarausch and Hannes Siegrist,
An Actor-Based Framework for the Study of the Professions (1990)

1. The profession must be controlled by a governing body, which in professional matters directs the behaviour of its members. For their part the members have a responsibility to subordinate their selfish private interests in favour of support for the governing body.
2. The governing body must set adequate standards of education as a condition of entry and thereafter ensure that students obtain an acceptable standard of professional competence. Training and education do not stop at qualification. They must continue throughout the member's professional life.
3. The governing body must set the ethical rules and professional standards that are to be observed by the members. They should be higher than those established by the general law.

4. The rules and standards enforced by the governing body should be designed for the benefit of the public and not for the private advantage of the members.
5. The governing body must take disciplinary action, if necessary expulsion from membership, should the rules and standards it lays down not be observed, or should a member be guilty of bad professional work.
6. Work is often reserved to a profession by statute – not because it was for the advantage of the member, but because of the protection of the public, it should be carried out only by persons with the requisite training, standards and disciplines.
7. The governing body must satisfy itself that there is fair and open competition in the practice of the profession so that the public are not at risk of being exploited. It follows that members in practice must give information to the public about their experience, competence, capacity to do the work and the fees payable.
8. The members of the profession, whether in practice or in employment, must be independent in thought and outlook. They must not allow themselves to be put under the control or dominance of any persons or organization that could impair that independence.
9. In its specific field of learning, a profession must give leadership to the public it serves.

> Lord Benson, Criteria for a group to be considered a profession,
> Hansard (Lords), 8th July 1992, (pp1206–7)

Only in England did the professional associations proliferate, from 1850 on, outside the university, because only in England were efforts to increase group status attempted without university training. Elsewhere enhancement of professional guild power has always involved the university.

> Elliott Krause, *Death of the Guilds* (1996, p13)

In the pre-Thatcher years, regardless of whether Laborites or Conservatives were in power, the professions were respected by government, which sought their input on the shape and nature of reforms before implementing them. Because the professions tended to defend their own interests in this process, change was slow, but when it occurred, it was with the (sometimes grudging) support of the groups involved.

> Elliott Krause, *Death of the Guilds* (1996, p265)

Professions remain primarily national creations, international conferences notwithstanding. But capitalism, international to some extent by 1960 and to a much greater extent by 1990, began to view professionals as obstacles to economic progress.

> Elliott Krause, *Death of the Guilds* (1996, p273)

The concept of 'professionalism' can no longer be sustained given education as a dynamic matrix beyond traditional levels of control. Consequently the ability of any institute to effectively guard professional standards will be progressively eroded. However, such challenges are the easy bit – the cultural context in the next century is likely to pose far more radical challenges.

Allan Cunningham, 'Getting other, not better' (1999)

Professionalism may be said to exist when an organised occupation gains the power to determine who is qualified to perform a defined set of tasks, to prevent all others from performing that work, and to control the criteria by which to evaluate performance.

Eliot Freidson, *Professionalism: The third logic* (2001, p12)

The ideology of professionalism asserts knowledge that is not merely the narrow depth of a technician, or the shallow breadth of a generalist, but rather a wedding of the two in a unique marriage.

Eliot Freidson, *Professionalism: The third logic* (2001, p121)

The interdependent elements of the ideal type, professionalism, are:

1 specialized work in the officially recognized economy that is believed to be grounded in a body of theoretically based discretionary knowledge and skill and that is accordingly given special status in the labor force;
2 exclusive jurisdiction in a particular division of labor created and controlled by occupational negotiation;
3 a sheltered position in both external and internal labor markets that is based on qualifying credential, which is controlled by the occupation and associated with higher education; and
4 an ideology that asserts greater commitment to doing good work than to economic gain and to the quality rather than the economic efficiency of work.

Eliot Freidson, *Professionalism: The third logic* (2001, p127)

What is likely to be most at risk for the professions is their freedom to set their own agenda for the development of their discipline and to assume responsibility for its use. Thus the most important problem for the future of professionalism is neither economic nor structural but cultural and ideological. The most important is its soul.

Eliot Freidson, *Professionalism: The third logic* (2001, p213)

If professionalism is to be reasserted and again regain some of its influence, it must not only elaborate and refine its codes of ethics but also strengthen its methods of adjudicating and correcting their violation.

Eliot Freidson, *Professionalism: The third logic* (2001, p216)

Because professional work is sheltered from ordinary market processes, maximising gain is clearly a violation of the terms of legitimising that shelter. There can be no ethical justification for professionals who put personal gain above the obligation to do good work for all who need it, even at the expense of some potential income.

Eliot Freidson, *Professionalism: The third logic* (2001, p218)

Professionalization is the project of constituting a profession as well as that of controlling a labor market.

Elizabeth Popp Berman, 'Before the professional project:
Success and failure at creating an organizational representative
for English doctors' (2006)

Professionalism as defined by the Royal College of Physicians is essentially: 'a set of values, behaviours, and relationships that underpins the trust the public has in doctors'.

Hazel Thornton, 'Understanding the role of the doctor',
BMJ (2008, 337:a3035)

As professionals learn to work in more heterogeneous teams [they] learn to see other professional communities and nonprofessionals as sources of learning and support rather than as interference.

Paul Adler, Seok-Woo Kwon and Charles Heckscher,
'Professional Work: The Emergence of Collaborative
Community' (2008, p371)

The image of the ideology of professionalism as an occupational value that is so appealing involves a number of different aspects. Some might never have been operational; some might have been operational for short periods in a limited number of occupational groups. Aspects include:

- Control of the work systems, processes, procedures, priorities to be determined primarily by the practitioner/s;
- Professional institutions/associations as the main providers of codes of ethics, constructors of the discourse of professionalism, providers of licensing and admission procedures, controllers of competences and their acquisition and maintenance, overseeing discipline, due investigation of complaints and appropriate sanctions in cases of professional incompetence;
- Collegial authority, legitimacy, mutual support and cooperation;
- Common and lengthy (perhaps expensive) periods of shared education, training, apprenticeship;
- Development of strong occupational identities and work cultures;
- Strong sense of purpose and of the importance, function, contribution and significance of the work;
- Discretionary judgement, assessment evaluation and decision making, often in highly complex cases, and of confidential advice-giving, treatment, and means of taking forward;
- Trust and confidence characterize the relations between practitioner/ client, practitioner/employer and fellow practitioners.

Julia Evetts, 'Professionalism in turbulent times' (2012, pp11–12)

I see professionalization as the process by which producers of special services sought to constitute and control a market for their expertise. Because marketable expertise is a crucial element in the structure of modern inequality, professionalization appears also as a collective assertion of special social status and as a collective process of upward social mobility.

Magali Sarfatti Larson, *The Rise of Professionalism* (2013, p.xvi)

The model of profession emerged during 'the great transformation' and was originally shaped by the historical matrix of competitive capitalism. Since then the conditions of professional work have changed, so that the predominant pattern is no longer that of the free practitioners but that of the salaried specialist in a large organization.

Magali Sarfatti Larson, *The Rise of Professionalism* (2013, p.xviii)

A sanctioned profession's superior competence is what ensures that the greater good is served better than lesser (and unsanctioned) rivals would serve it. Knowledge, in other words, comes before morality.

Magali Sarfatti Larson, *The Rise of Professionalism* (2013, p.xxv)

If the public does not know what most experts are good at, what they do, or who controls them, broad-based problems of accountability become insurmountable.

Magali Sarfatti Larson, *The Rise of Professionalism* (2013, p.xxxi)

Anyone who does an intervention to somebody else has a professional and moral responsibility to be able to describe what they do and defend how well they do it. That is the essence of professionalism. They should be happy to share that with their patients. In a sense what this endeavour does is demonstrate a new level of professionalism.

Sir Bruce Keogh (19th November 2014, www.theguardian.com/society/2014/nov/19/nhs-chief-surgeons-moral-responsibility-publish-death-rates)

People with power based on their expertise, neither knights nor peasants but able to front the middle to tell both what to do.

Michael Pye, *The Edge of the World* (2014, p154)

Professionals play such a central role in our lives that we can barely imagine different ways of tackling the problems that they sort out for us. But the professions are not immutable. They are an artefact that we have built to meet a particular set of needs in a print-based industrial society. As we progress into a technology-based Internet society, however we claim that the professions in their current form will no longer be the best answer to those needs.

Richard Susskind and Daniel Susskind,
The Future of the Professions (2015, p3)

Rendition of the grand bargain:
In acknowledgement of and in return for their expertise, experience and judgement, which they are expected to apply in delivering affordable, accessible, up-to-date, reassuring and reliable services, and on the understanding that they will curate and update their own knowledge and methods, train

their members, set and enforce standards for the quality of their work, and that they will only admit appropriately qualified individuals into their ranks, and that they will always act honestly, in good faith, putting the interests of clients ahead of their own, we (society) place our trust in the professions in granting them exclusivity over a wide range of socially significant services and activities, by paying them a fair wage, by conferring upon them independence, autonomy, rights of self-determination and by according them respect and status.

Richard Susskind and Daniel Susskind,
The Future of the Professions (2015, p22)

Engineering has this curious position at a crossroads where politics and commercial interests all come together.

Mike McCormack, *The Guardian* (24th June 2017)

Appendix B

Codes of professional conduct – ICE, RIBA, RICS

Institution of Civil Engineers (ICE)

1823 – Regulations of the Institution of Civil Engineers

If at any time there shall appear cause for the expulsion of any Member or Associate, it shall be competent for the Council to call a general meeting of the Institution for that purpose, and if one half of the members then present, agree that such Member or Associate be expelled, the President, or other Member in the Chair, shall declare the same according, and the Secretary shall forthwith communicate the same to the such Member by the Letter (D) in the Appendix.

Revisions to the Regulations were published in 1824, 1827, 1830, 1831, 1837, 1839, 1846, 1857, 1877, 1878, 1886, 1887 and 1896.

1896 – Bye-laws and Regulations of the Institution of Civil Engineers, as amended 15 April 1896

The Council shall have the right, by a majority of two thirds of those present at a meeting of Council, to expel from the Institution any Member or Associate who may be convicted by a competent tribunal, of felony, embezzlement, larceny or misdemeanour, or other offence which, in the opinion of the Council, renders him unfit to be a member; and in case the expulsion of any Member or Associate shall be judged expedient on any ground whatever by twenty Corporate Members, and they shall think fit to draw up and sign a proposal requiring such expulsion, the same being delivered to the Secretary, shall be by him laid before Council for consideration.

In 1908 a Special Committee established to 'consider the role of the Institution in promoting professional conduct' reported, proposing rules of conduct to be incorporated in the Institution's By-Laws. This was accomplished in 1910 (By-Laws: section IV, Professional Conduct). The By-Laws were updated in a series of Supplemental Charters in 1915, 1922, 1935 and 1945 (Watson 1988).

1915 – Bye-laws and Regulations of The Institution of Civil Engineers
As amended 27 April 1915

Section IV. PROFESSIONAL CONDUCT

Every Corporate Member of the Institution shall observe and be bound by
the following Regulations:

1. He shall act in all professional matters strictly in a fiduciary manner with
   regard to any Clients whom he may advise, and his charges to such
   Clients shall constitute his only remuneration in connection with such
   work, except as provided by Clause 4.
2. He shall not accept any trade commissions, discounts, allowances, or any
   indirect profit in connection with any work he is designed or to superintend, or with any professional business which may be entrusted to him.
3. He shall not, while acting in a professional capacity, be at the same time,
   without disclosing the fact in writing to his Clients, a Director or member of,
   or a Shareholder in, or act as an Agent for any contracting or manufacturing
   company or firm or business with which he may have occasion to deal on
   behalf of his Clients, or have any financial interest in such a business.
4. He shall not receive, directly or indirectly, any Royalty, Gratuity or
   Commission, on any patented or protected article or process used on
   work which he is carrying out for his Clients, unless and until such Royalty,
   Gratuity or Commission has been authorized in writing by those Clients.
5. He shall not improperly solicit professional work, either directly or by any
   agent, nor shall he pay, by commission or otherwise, any person who may
   introduce Clients to him.
6. He shall not be the medium of payments made on his Clients' behalf to any
   Contractor, or business firm (unless specially so requested by his Clients), but
   shall only issue certificates or recommendations for payment by his Clients.

Any alleged breach of these regulations or any alleged professional
misconduct by a Corporate Member which may be brought before the
Council, properly vouched for and supported by sufficient evidence, shall
be investigated, and if proved, shall be dealt with by the Council, either by
expulsion of the offender from The Institution or in such other manner as the
Council may think fit.

1936 – A new preamble and five additional clauses (for consultants)
were added:

All Corporate members of the Institution are required to order their conduct so as to uphold the reputation of the Institution and the dignity of the
profession of Civil Engineer.
Every Corporate Member of the Institution shall observe and be bound by
the following regulations:
A Corporate Member may make his professional knowledge and experience available to others either as a consultant in return for fees, or as a

salaried employee, or as a teacher of engineering science, or as a person engaged in the design and construction of engineering works, or a person engaged in the design and manufacture of articles of commerce. ...

f) He shall not offer his professional services (other than for a salaried appointment) by advertisement or circular; nor shall he allow his name to appear in any advertisement or circular offering his services.
g) He shall not compete with another Corporate Member for employment on the basis of professional charges.
h) He shall not attempt, directly or indirectly, to supplant another Corporate Member.
i) He shall not review or take over work of another Corporate Member acting as a consultant for the same client, until he has either obtained the consent of such Corporate Member or satisfied himself so far as he can reasonably do so that the connection of such Corporate Member with the work has been terminated.
j) He shall not take part in a competition involving the submission of proposals and designs for engineering work unless an assessor who shall be an engineer of acknowledged standing has been appointed to whom all such proposals and designs are to be submitted for adjudication.

A Corporate Member shall not act in a dual capacity as a consultant and as a contractor.

1946 – A further clause was added:

a) He shall not practise in Great Britain or Northern Ireland in the name of a limited company or otherwise under the protection of limited liability.

1963 – RULES FOR PROFESSIONAL CONDUCT.

Rules made by Council of the Institution on the 19th March, 1963 in accordance with By-law 29.

1. A member, in his responsibility to his Employer and to the profession, shall have full regard to the public interest.
2. A member shall order his conduct so as to uphold the dignity, standing and reputation of the profession.
3. A member shall discharge his duties to his Employer with complete fidelity. He shall not accept remuneration for services rendered other than from his Employer or with his Employer's permission.
4. A member shall not maliciously or recklessly injure or attempt to injure, whether directly or indirectly, the professional reputation, prospects or business of another Engineer.
5. A member shall not improperly canvass or solicit professional employment nor offer to make by way of commission or otherwise payment for the introduction of such employment.

6. A member shall not, in self-laudatory language or in any manner derogatory to the dignity of the profession, advertise or write articles for publication, nor shall he authorize such advertisements to be written or published by any other person.

7. A member, without disclosing the fact to his Employer in writing, shall not be a director of nor have a substantial financial interest in, nor be agent for any company, firm or person carrying on any contracting, consulting or manufacturing business which is or may be involved in the work to which his employment relates; nor shall he receive directly or indirectly any royalty, gratuity or commission on any article or process used in or for the purposes of the work in respect of which he is employed unless or until such royalty, gratuity, or commission has been authorized in writing by his Employer.

8. A member shall not use the advantages of a salaried position to compete unfairly with other engineers.

9. A member in connexion with work in a country other his own shall order his conduct according to these Rules, so far as they are applicable; but where there are recognized standards of professional conduct, he shall adhere to them.

10. A member who shall be convicted by a competent tribunal of a criminal offence which in the opinion of the disciplinary body renders him unfit to be a member shall be deemed to have been guilty of improper conduct.

11. A member shall not, directly or indirectly, attempt to supplant another Engineer; nor shall he intervene or attempt to intervene in or in connexion with engineering work of any kind which to his knowledge has already been entrusted to another Engineer.

12. A member shall not be the medium of payments made on his Employer's behalf unless so requested by his Employer; nor shall he in connexion with work on which he is employed place contracts or orders except with the authority of and on behalf of his Employer.

13. A member shall not knowingly compete on the basis of professional charges with another Engineer.

The Rules for Professional Conduct were modified in 1971, 1973, 1992 and 1999:

1971 – Rule 13 was deleted.
1973 – Rule 1 had the phrase 'particularly in matters of health and safety' added.

Rule 2 was rewritten to read:

A member shall discharge his professional responsibilities with integrity.

A new Rule 13 was added:

A member shall afford such assistance as he may reasonably be able to give to further the Education and Training of candidates for the Profession.

1992 – A new Rule 14 was added:

A member, either himself or through his organisation as an employer, shall afford such assistance as may be necessary, to further the formation and professional development of himself and of other members and prospective members of the profession in accordance with the recommendations made by the Council from time to time.

1999 – The language of the Rules was modernised and they were re-numbered from 1 to 16.

2004 – Code of Professional Conduct

Made by the Council on 21 July 2004.
The Code was separated into 1) Introduction, 2) The Rules of Professional Conduct and 3) Guidance notes on the interpretation and application of the Rules of Professional Conduct.

2004 – Code of Professional Conduct
The current version reads:

INTRODUCTION

…
The duty to behave ethically

The duty upon members of the ICE to behave ethically is, in effect, the duty to behave honourably; in modern words, 'to do the right thing'. At its most basic, it means that members should be truthful and honest in dealings with clients, colleagues, other professionals, and anyone else they come into contact with in the course of their duties. Being a member of the ICE is a badge of probity and good faith, and members should do nothing that in any way could diminish the high standing of the profession. This includes any aspect of a member's personal conduct which could have a negative impact upon the profession.

Members of the ICE should always be aware of their overriding responsibility to the public good. A member's obligations to the client can never override this, and members of the ICE should not enter undertakings which compromise this responsibility. The 'public good' includes care and respect for humanity's cultural, historical and archaeological heritage, in addition to the duties specified in the Rules of Professional Conduct to protect the health and well being of present and future generations and to show due regard for the environment and for the sustainable management of natural resources.

THE RULES OF PROFESSIONAL CONDUCT

1. All members shall discharge their professional duties with integrity and shall behave with integrity in relation to all conduct bearing upon the standing, reputation and dignity of the Institution and of the profession of civil engineering.
2. All members shall only undertake work that they are competent to do.

3. All members shall have full regard for the public interest, particularly in relation to matters of health and safety, and in relation to the well-being of future generations.

4. All members shall show due regard for the environment and for the sustainable management of natural resources.

5. All members shall develop their professional knowledge, skills and competence on a continuing basis and shall give all reasonable assistance to further the education, training and continuing professional development of others.

6. All members shall:

   a. notify the Institution if convicted of a criminal offence;

   b. notify the Institution upon becoming bankrupt or disqualified as a Company Director;

   c. notify the Institution of any significant breach of the Rules of Professional Conduct by another member.

Royal Institute of British Architects (RIBA)

1834 – The original by-laws of the RIBA (see Mace 1986, p59)

The following shall be deemed grounds for the expulsion of any Fellow or Associate viz. for having engaged since his election in the measurement, valuation or estimation of any works undertaken or proposed to be undertaken by any building artificer, except such as are proposed to be executed or have been executed under the member's own designs or directions; or for the receipt or acceptance of any pecuniary consideration or emolument from any builder or other tradesman whose works he may have been engaged to superintend; or for having any interest or participation in any trade, contract or materials supplied at any works; or for any conduct which, in the opinion of the Council, shall be derogatory to his professional character.

RIBA Codes of Professional Conduct were issued in 1929, 1950, 1976, 1981, 1997 and 2005

1929 – Code of Professional Practice

A member of the R.I.B.A. is governed by the Charter and Bye-laws of the Royal Institute. The following clauses indicate in a general way the standard of conduct which members of the R.I.B.A. must adhere to, failing which the Council may judge a member guilty of unprofessional conduct, and either reprimand, suspend or expel him or her.

Cases of unprofessional conduct not specifically covered by these clauses are dealt with by the Council having regard to the particular circumstances of the case.

1. An Architect is remunerated solely by his professional fees, and is debarred from any other source of remuneration in connection with the

works and duties entrusted to him. It is the duty of an Architect to uphold in every way possible the Scale of Professional Charges adopted by the Royal Institute

2. An Architect must not accept any work which involves the giving or receiving of discounts or commission from contractors or tradesmen, whether employed upon his works or not.

An Architect may be an architectural consultant, adviser, or assistant to building contractors, decorators, furniture designers, estate development firms or companies, or firms or companies trading in materials used in or whose activities are otherwise connected with the building industry, provided that:-

   1. He is paid by salary or fee and not by commission.
   2. He does not either directly or indirectly solicit orders for the firm or company.

He may use his affix in connection with his appointment and his name and affix may appear on stationery of the firm or company as architectural consultant or adviser.

His name and affix may appear in advertisements inserted by the firm or company in the press provided that it is done in an unostentatious manner.

An Architect may be a director of any company including a building society registered under the Building Societies Acts, except a company trading in materials used in or whose activities are otherwise connected with the building industry or engaged in the financing or erection of buildings.+

His name and affix may appear on the notepaper of the company.

An Architect may not carry on or act as principal, partner or manager of any firm carrying on any of the above excepted trades or businesses.

3(a) An Architect must not advertise nor offer his services by means of circulars or otherwise, nor may he make paid announcements in the press. He may, however notify his correspondents by post once of any change of address.

(b) Though there is no objection to an Architect allowing signed illustrations and descriptions of his work to be published in the press, with reference to such illustrations or descriptions, it is contrary to professional ethics –

   (1) To give monetary consideration for such insertions.
   (2) To allow such insertions to be used by the publishers for extorting advertisements from unwilling contributors

(c) An Architect may consent to the publication of a series of illustrations either in circular, brochure or book form, with or without descriptive letterpress, of any building or buildings for which he has been responsible, provided that –

   (a) if advertisements appear, Clause (b) (2) of the Code is complied with, and
   (b) there is no attempt to distribute the publication to potential clients.

(d) An Architect may sign his buildings and may exhibit his name outside his office and on buildings in course of construction, alteration and/or extension provided that it is done in an unostentatious manner and the lettering does not exceed two inches in height.

If the client so desires the Architect's name may remain on the building for a period not exceeding twelve months after its completion providing that the board does not display 'To Let' or 'For Sale' or similar notices. A notice may, however, indicate that the plans can be seen at the Architect's office.

Auctioning and House Agency are inconsistent with and must not form part of the practice of an Architect.

(e) Architects who are appointed Surveyors to recognised estates* may announce land or sites or premises for sale or letting in connection with their appointments.

When Architects are acting as Surveyors or Town Planners in connection with the development of land, announcements may be made in the press and on notice boards in connection with such development, provided that such announcements are made in an unostentatious manner.

4. An Architect must not attempt to supplant another Architect, nor must he compete with another Architect by means of a reduction in fees or by other inducement.

5. An Architect, on receiving instructions to proceed with certain work which was previously entrusted to another Architect, shall, before proceeding with such work, communicate with the Architect previously employed and inquire and ensure the fact that his engagement has been properly terminated.

6. In all cases of dispute between employer and contractor the Architect must act in an impartial manner. He must interpret the conditions of a contract with entire fairness as between the employer and the contractor.

7. An Architect must not permit the insertion of any clause in tenders, bills of quantities, or other contract documents which provide for payment to be made to him by the contractor (except for duplicate copies of drawings or documents) whatever may be the consideration, unless with the full knowledge and approval of his client.

8. An Architect should not take any part in a competition as to which the pre-liminary warning of the Royal Institute has been issued, and must not take part in a competition as to which the Council of the Royal Institute shall have declared by a Resolution published in the JOURNAL of the Royal Institute that Members must not take part because the Conditions are not in accordance with the published Regulations of the Royal Institute for Architectural Competitions, nor must he be associated in any way with the carrying out of a design selected as the result of a competition as to which the Council shall have declared by a Resolution published in the JOURNAL that members must not take part.

Architects asked to take part in a limited competition must at once notify the Secretary of the R.I.B.A., submitting particulars of the competition.

Note:- A formal invitation to two or more architects to prepare designs in competition for the same project is deemed a limited competition.

9. An Architect must not act as Architect or Joint Architect for a work which is or has been the subject of a competition in which he is or has been engaged as Assessor.

An Assessor must not act as Consulting Architect, unless he has been appointed as such before the inception of the competition, nor in any other professional capacity in any matter connected with the work which has been the subject of a competition, provided always that he may act as Arbitrator in any dispute between the Promoters and the selected Architect.

If an Architect is officially approached by the Promoters for advice as to the holding of a competition with a view to his acting as Assessor, and eventually it is decided not to hold a competition but to appoint an Architect to carry out the work, the Architect originally approached in an advisory capacity is precluded from acting as Architect for the work in question.

10. It is desirable that in cases where the Architect takes out the Quantities for his buildings he should be paid directly by the client and not through the Contractor except with the previous consent of the client.

+ Note. – The regulation with regard to directorships will come into force as from the 23rd March 1935, but in cases where Architects are already directors of companies carrying on any of the above excepted trades or businesses the regulation will not become effective until 23rd March 1937.

x Note – This does not refer to advertisements or letters respecting appointments open or wanted, nor to the insertion of one notice of change of address in the professional press.

* The term 'recognised estates' is intended to apply to family and trustee estates such as the Bedford, Grosvenor, Berners and Howard de Walden.

(RIBA, as revised 23rd March 1935)

1976 – Code of Professional Conduct

Principle 1

A member shall faithfully carry out the duties which he undertakes. He shall also have a proper regard for the interests both of those who commission and of those who may be expected to use or enjoy the product of his work.

Rules

1.1 A member in private practice shall inform his client in advance of the Conditions of Engagement and the scale of charges and agree with his client that those Conditions shall be the basis of his appointment and shall not accept any other payment or consideration for the duties entrusted to him.

1.2 A member shall arrange that the work of his office and any branch office insofar as it relates to architecture is under the control of an architect.

1.3 A member shall not sub-commission or sub-let work for which he has been commissioned without the prior agreement of his client to any change in responsibilities nor without defining the responsibilities of those concerned.

1.4 A member shall act impartially in all cases in which he is acting between parties. Where he has responsibilities as architect under a building contract, or is acting for the supervising officer, he shall interpret the conditions of such contract with fairness.

Principle 2

A member shall avoid actions and situations inconsistent with his professional obligations or likely to raise doubts about his integrity.

Rules

2.1 A member shall not have an interest in or connection with any business as would or might breach this principle. In particular he shall not be, nor act as, proprietor, director, principal, partner or (subject to 2.6) manager of any body corporate or unincorporate engaging in any of the following occupations: carrying on the business of trading in land or buildings, property developers, auctioneers, house agents or estate agents; or contractors, subcontractors, manufacturers or suppliers in or to the building industry.

2.2 A member shall not take discounts, commissions or gifts as an inducement to show favours to any person or body; nor shall he recommend or allow his name to be used as recommending any service or product in advertisements.

2.3 A member shall not improperly influence the granting of planning consents or statutory approvals.

2.4 A member shall not carry on his practice in the form of a limited company.*

2.5 A member shall not act in disregard of the professional obligations or qualifications of those from whom he receives or to whom he gives authority, responsibility or employment, nor of those whom he is professionally associated.

2.6 A member who is appointed to superintend or control the architectural business of any body corporate or unincorporated (including a central government department, a local authority, public board or corporation or commercial firm or company) shall arrange with his employer that the business of that body so far as it relates to architecture is conducted in conformity with this Code. A member who is unable to ensure that the business of his employer is so conducted shall furnish the RIBA with a written declaration of the facts.

2.7 A member shall not have or take as partner or co-director in his firm any person who is disqualified for registration by reason of the fact that his name has been removed from the register under Section 7 of the Architects (Registration) Act 1931; any person disqualified for membership of the RIBA by reason of expulsion under Bye law 5.1; any person disqualified for membership of another professional institution by reason of expulsion under the relevant disciplinary regulations (unless the RIBA otherwise allows); any person who engages in any of the occupations proscribed under Rule 2.1 even though that person engages in any such occupation in a firm or company separate from the architectural firm.

2.8 A member who in circumstances not specifically covered in these Rules finds that his interests whether professional or personal conflict so as to risk a breach of this principle shall, as the circumstances may require, either withdraw from the situation or remove the source of conflict or declare it and obtain the agreement of the parties concerned to the continuance of his engagement.

Principle 3

A member shall rely only on ability and achievement as the basis for his advancement.

Rules

3.1 A member shall not give discounts, commission, gifts or other inducements for the introduction of clients or of work.
3.2 A member shall uphold and apply the RIBA Conditions of Engagement.
3.3 A member shall not work without payment or at a reduced fee except as permitted in the RIBA Conditions of Engagement; nor shall he prepare designs in competition with other members for a client without payment or for a reduced fee except in a competition conducted in accordance with the RIBA Regulations for the promotion and conduct of competitions or a completion otherwise approved by the RIBA and the International Union of Architects.
3.4 A member shall not attempt to supplant another architect.
3.5 A member on being approached or instructed to proceed with work upon which he knows, or can ascertain by reasonable inquiry, that another architect is or has been engaged by the same client shall notify the fact to such architect.
3.6 A member shall not solicit either a commission or engagement for himself or business for a client or employer, but he may make his or his practice's availability and experience known by giving information which in substance and in presentation is factual, relevant and neither misleading nor unfair to others nor discreditable to the profession, either in response to a direct request or in accordance with the Notes to this Rule.

* **Overseas Work**: for the purposes only of overseas work, and as a conditional exception to this Rule, a member may carry on his practice in the form of a limited liability company incorporated in the United Kingdom or elsewhere (if permitted by local codes of conduct), **provided** that control of a company is vested in the professional partners of the practice: and further **provided** that such a company shall not accept a commission for part or all of the Normal services in respect of work which is intended to be constructed within the United Kingdom.

2005 – Code of Professional Conduct

Introduction

1. This Code and its accompanying Guidance Notes set out and explain the standards of professional conduct and practice that the Royal Institute requires of its members.
2. This Code comprises:

   - **three principles** of professional conduct
   - **professional values** that support those principles
   - **Guidance Notes** which explain how the principles can be upheld.

The Royal Institute's Values

Honesty, integrity and competency, as well as concern for others and for the environment, are the foundations of the Royal Institute's three principles of professional conduct set out below. All members of the Royal Institute are required to comply.

The Three Principles

Principle 1: Integrity
Members shall act with honesty and integrity at all times.
Principle 2: Competence
In the performance of their work Members shall act competently, conscientiously and responsibly. Members must be able to provide the knowledge, the ability and the financial and technical resources appropriate for their work.
Principle 3: Relationships
Members shall respect the relevant rights and interests of others.

(January 2005)

Royal Institution of Chartered Surveyors (RICS)

1902 – New By-Laws were introduced in 1902 which not only gave the Council a power of expulsion which it had not formerly possessed, but also contained a set of fundamental rules whose violation constituted unprofessional conduct. These rules amounted, in effect, to the first code of conduct to be issued by the Institution.

Members became liable to suspension or expulsion:

1) If they became connected with any occupation inconsistent with a surveyor's profession
2) If they shared commissions with solicitors, accountants or liquidators
3) If they gave any secret or illicit trade discounts or commissions
4) If they were convicted of a misdemeanour, fraud or other offence
5) If they were bankrupt

6) For using any professional designation to which they had no entitlement
7) For describing any partnership which included non-members as an Institution firm
8) For any act or default disgraceful to a surveyor

(Thompson 1968, pp307–8)

1932 – First full code of conduct
1934 – Code of Conduct (In accord with that issued by the Land Agents' Society, the Auctioneers' and Estate Agents' Institute and the Incorporated Society of Auctioneers and Landed Property Agents).

I. Relationship with other Professions and Occupations

a) No member shall engage in or be connected with any occupation or business which in the opinion of the Council is inconsistent with membership of the Institution.
b) No member shall directly or indirectly allow or agree to allow any person, other than a member of his own profession, to participate in his remuneration.
c) No member shall engage in work recognised as being properly that of a solicitor.

II. Members' Responsibilities

a) In order that members shall accept full personal liability for advice given to clients, they will not be permitted to convert their firms into limited liability companies unless special circumstances, approved by the Council, exist.
b) Members who are principals will be held responsible to Council for the acts of their partners and staffs so far as they relate to matters coming within the scope of their practice.

III. Sales and Lettings

a) No member shall, either directly or indirectly, in writing or verbally, seek instructions for business which he knows, or with ordinary care could have ascertained, is in the hands of another agent; nor, in other cases, without a definite intimation that if another agent has already been retained, instructions can only be accepted from, and as sub-agent to, that agent.
A member who seeks instructions to deal with a property on which another agent's board or notice is exhibited will be deemed to have disregarded knowingly this injunction.
Canvassing by personal call or by telephone by a member or one of his staff is prohibited.
b) No member shall offer any financial inducement to secure instructions, nor shall a member charge for effecting a sale, purchase, letting or taking, an amount of commission which in the opinion of the Council would be unfair to other members.

IV. Surveys, Valuations and work other than Sales and Lettings

a) No member shall in any circumstances solicit instructions in any manner whatsoever. This injunction does not apply in the case of a client for whom a member regularly acts.

b) No member shall offer to accept instructions on the basis that no charge will be made unless a successful result is attained, nor shall a member undertake work for charges which in the opinion of the Council would be unfair to other members.

V. Advertisements

Members shall ensure that advertisements and other public announcements are not such as would bring the Institution into disrepute.

VI. Generally

No member shall conduct himself in such a manner as to prejudice his professional status or the reputation of the Institution.

2002 – The Royal Institution of Chartered Surveyors' Core Values

The strategy formulated by the RICS's Ethics Conduct and Consumer Policy Committee (ECCPC) and reported to General Council (RICS 2002a) has an underpinning philosophy which is that all regulatory requirements should be grounded in the 'Nine Core Values'. These are set out as rule 3 (RICS 2002b, p.c5), in Part II of the RICS's Rules of Conduct, together with a brief explanatory statement (RICS 2002c). Within the Guidance to Rules of Conduct (RICS 2002c, p8), the core values are prefaced, with: 'RICS has a set of overarching values it expects members to apply in their work. Adherence to such a set of values is one of the key features that defines the professionalism of all chartered and technical surveyors. Members are expects to determine all their actions and judgements on the basis of them.

(Plimmer and Sayce 2003)

Nine Core Values

1. *Act with integrity*. Never put your own gain above the welfare of your clients or others to whom you have a professional responsibility. Respect their confidentiality at all times and always consider the wider interests of society in your judgements.
2. *Always be honest*. Be trustworthy in all that you do – never deliberately mislead, whether by withholding or distorting information.
3. *Be open, and transparent in your dealings*. Share the full facts with your clients, making things as plain and intelligible as possible.
4. *Be accountable for your actions*. Take full responsibility for your actions and don't blame others if things go wrong.
5. *Know and act within your limitations*. Be aware of the limits of your competence and don't be tempted to work beyond these. Never commit to more than you can deliver.

6. *Be objective at all times.* Give clear and appropriate advice. Never let your own sentiments or your own interests cloud your judgement.
7. *Always treat others with respect.* Never discriminate against others.
8. *Set a good example.* Remember that both your public and private behaviour could affect your own, RICS's and other members' reputations.
9. *Have the courage to make a stand.* Be prepared to act if you suspect a risk to safety or malpractice of any sort.

2007 – Rules of Conduct for Members

Part One: General

Interpretation

1. In these Rules, unless the context otherwise requires, 'Member' means a Fellow, Professional Member, Technical Member or Honorary Member of RICS or a member of the Attached Classes.

Service of documents

2. Any notice or other document required by or for the purposes of these Rules to be given or sent to a Member may be given to them personally or sent by post to their last address notified to RICS

Part Two: Personal and Professional Standards

Ethical behaviour

3. Members shall at all times act with integrity and avoid conflicts of interest and any actions or situations that are inconsistent with their professional obligations.

Competence

4. Members shall carry out their professional work with due skill, care and diligence and with proper regard for the technical standards expected of them.

Service

5. Members shall carry out their professional work in a timely manner and with proper regard for standards of service and customer care expected of them.

Lifelong learning

6. Members shall undertake and record appropriate lifelong learning and, on request, provide RICS with evidence that they have done so.

Solvency

7. Members shall ensure that their personal and professional finances are managed appropriately.

Information to RICS

8. Members shall submit in a timely manner such information, and in such form, as the Regulatory Board may reasonably require.

Co-operation

9. Members shall co-operate fully with RICS staff and any person appointed by the Regulatory Board.

(version 4.01 January 2011)

2010 – The nine core principles were reorganised, amended slightly and augmented by a further three principles in a RICS Guidance Sheet: *Maintaining Ethical Standards* (2010)

The three additional principles require members to:

10. Comply with relevant laws and regulations and avoid any action, illegal or litigious that may bring the profession into disrepute;
11. Avoid conflicts of interest and declare any potential conflicts of interest, personal or professional, to all relevant parties; and
12. Respect confidentiality; maintain the confidentiality of your clients' affairs. Never divulge information to others, unless it is necessary.

2012 – Rules of Conduct for Firms

Part II Conduct of Business

Professional behaviour

3. A Firm shall at all times act with integrity and avoid conflicts of interest and avoid any actions or situations that are inconsistent with its professional obligations.

Competence

4. A Firm shall carry out its professional work with due skill, care and diligence and with proper regard for the technical standards expected of it.

Service

5. Firm shall carry out its professional work with expedition and with proper regard for standards of service and customer care expected of it.

Training and Continuing Professional Development (CPD)

6. A Firm shall have in place the necessary procedures to ensure that all its staff are properly trained and competent to do their work.

Complaints handling

7. A Firm shall operate a complaints handling procedure and maintain a complaints log. The complaints handling procedure must include an

Alternative Dispute Resolution (ADR) mechanism that is approved by the Regulatory Board.

Clients' money

8. A Firm shall preserve the security of clients' money entrusted to its care in the course of its practice or business.

Professional indemnity insurance

9. A firm shall ensure that all previous and current professional work is covered by adequate and appropriate professional indemnity insurance cover which meets standards approved by the Regulatory Board.

(version 5, January 2012)

2012 – The Global Professional and Ethical Standards

There are five standards. All members must be able to demonstrate that they:

- Act with integrity
- Always provide a high standard of service
- Act in a way that promotes trust in the profession
- Treat others with respect
- Take responsibility

Appendix C
Le Code Guadet (1895)

PROFESSIONAL DUTIES OF THE ARCHITECT TOWARDS HIMSELF, COLLEAGUES, CLIENTS AND CONTRACTORS

The *Société centrale des architectes français*:

considering that it is necessary to define the moral obligations, which have always underlain the conduct and honour of architects truly worthy of the name,

and that it is necessary for the public, clients and administrations to be aware of the guarantees that they are entitled to expect from architects honourably exercising their profession;

declares that the principles governing the conduct of architects in their dealings with their colleagues, clients, contractors and building staff are as follows:

I

DUTIES OF THE ARCHITECT TOWARDS HIMSELF AND HIS COLLEAGUES

1. The architect is defined by the *Dictionnaire de l'Académie française* as:

2. 'An artist who designs buildings, determines the proportions, the arrangements and the decorations, executes them under his instructions and regulates the expenses.' [1878 edition]

    In consequence the architect is both an artist and a practitioner. His function is to conceive and study the composition of a building, direct and supervise its execution, verify and settle the accounts of the expenses relating thereto.

3. He practices a liberal and a non-commercial profession, which is incompatible with the role of a contractor, industrialist or the supplier of materials and articles used in construction.

    He shall be remunerated solely by payment of a fee, to the exclusion of other sources of professional income, at an appropriate time or following the completion of his duties.

4. If an architect has taken out a patent for a product relating to the building industry, he must not exploit it personally, but should sell it to an industrialist, assigning all the property rights for the operation.

5. The architect, being neither a businessman nor an agent, is prohibited from undertaking any activity that would give rise to discounts or commissions.

    He shall refrain from placing advertisements or making offers of services in newspapers, on posters, signs and prospectuses or through any other form of advertising.

6. It is forbidden for him to tout for work or clients by means of concessions, commissions, remittances on fees or other advantages offered to intermediaries; including managers, businessmen and owners' agents; and generally through any deals that are intended to be kept secret from current or potential clients.

7. The architect is forbidden to plagiarise the work of his colleagues and must follow his conscience in order to maintain good professional relations with his peers.

    He must not approach clients who have appointed another architect. However, if he is called upon to take over a project, following the death, voluntary retirement or the termination of another architect's appointment, the new architect should consider himself the guardian of the honour and interests of his colleague.

8. He must recognize the status of colleagues and grant the title of architect to anyone working honourably in the profession.

    He should give a high priority to meetings with colleagues, including at conferences, receptions, etc. Where there is a meeting between several architects, they should take place in the office of the oldest.

9. When the architect employs, as draftsmen or clerks, young people on a vocational training course, he should make his experience available to for their benefit and treat them as colleagues.

II

DUTIES OF THE ARCHITECT TOWARDS HIS CLIENTS

10. The architect should provide his client with:

    - The benefit of all his knowledge and experience in relation to the project under commission, the direction and supervision of the work and to any advice given;
    - Full commitment to the interests of the job entrusted to him.

11. However, the architect must not involve himself in actions, even if required to do so by the client, which would be liable to prejudice the rights of third parties.

Nor must he act in any way that could compromise those rights, endanger third parties or cause risk of accidents.

In such cases, he must warn his clients that he cannot fulfil their request.

12. He must warn his client when the client, through changes to the planned work, is exposed to a potential increase in costs.

13. He must be paid by his client, and by his client alone, in the form of professional fees. In consequence he must receive no remuneration, in any form whatsoever, from contractors, suppliers, sellers or buyers of land or property, who are under contract to, or who might wish to contract with his client. In addition when the remuneration for his work is ultimately to be borne by a third party, his client must pay any fees due to him, except when reimbursed under the law.

14. He shall provide his client with copies of the plans, specifications and contracts used for the award of the contract. He should maintain his own records as well as all the preparatory studies and details of execution. He should also provide his client with the accounts of the contractor that he has checked and settled.

15. In the case of works in connection with maintenance, administration, vacations, etc., the architect should ordinarily produce an annual bill for fees. For new works or major repairs, he should receive fees proportional to the sums expended.

16. The architect must refuse an appointment as an expert in a case where one of his clients is involved. The same shall apply if he has already given an opinion on a matter in dispute.

When he is appointed as an expert by a client, for example over a question of insurance, registration, etc., he no longer acts as the agent of his client. He is now solely an expert.

When he acts as arbitrator, the same obligations apply.

III

DUTIES OF THE ARCHITECT TO CONTRACTORS AND BUILDING STAFF

17. The architect should employ his moral authority to make implementing his projects as straightforward as possible and ensure that good relations, sociability and a sense of honour are developed and maintained between everyone involved in executing them.

18. The architect shall refrain from receiving any remission, commission or gift from contractors or suppliers, whether in money or in kind, and whether or not such contractors or suppliers are employed on his projects.

19. The architect must refrain from inserting into the contractors' specifications and contracts any clause obliging them to pay him expenses, such as for reimbursement of costs for travel, vacations, etc., or any other allocation of expenses, with the sole exception of the cost of reproducing drawings and copies of specifications to be charged to the contractor; provided that a clause to that effect is included in the specifications or other documents, and agreed and signed by the client.

20. He must issue to the Contractor notice for advance payments or balances according to market conditions or, in the absence of a contract, after the progress of the work.

    In settling accounts for building works, he must give, without making any alterations, a clear explanation to the contractor for its records. He must check and settle any claims.

    Unless under special instruction from the client, he should not handle any payments directly.

21. If an architect has a contractor or a developer as his client, he must be solely remunerated by a fee. He should never be exposed to the potential for gains or losses, which is commercial activity and runs counter to the liberal profession of the architect.

22. An architect who becomes a contractor, a contractor's clerk, quantity surveyor or measurer loses the status of architect.

    He does not lose it by becoming an architect's clerk.

The Rapporteur – J. Guadet

Witnessed: The President of the *Société centrale des architectes français*, member of the *Institut*, Ch. Garnier

This report to the *Société centrale des architectes français* was adopted unanimously by the Congress of Architects held in Bordeaux in 1895.

Source: Guadet, J., (1895), Les devoirs professionnels de l'architecte, *L'Architecture*, 10 août 1895, reprinted in Guadet, J., (1905) *Éléments et Théorie de l'architecture*, Paris, La Construction Moderne, Vol. IV, pp. 504–7.
Trans. SF June 2017

Appendix D

Collaboration for Change

The Edge Commission Report on the Future of Professionalism (2015)

Summary of recommendations (Morrell 2015, pp92–4)

Ethics and the public interest

A-1)   Develop and standardise a national code of conduct/ethics across the built environment professions, building on shared experience in the UK and internationally.

A-2)   Take a lead in raising the awareness of members to a shared understanding of ethical issues, creating guidance (rather than prescription), and monitoring both individuals and practices.

A-3)   Define and harmonise the commitment to the public interest at institutional, practice and individual level, again raising awareness and creating guidance.

A-4)   Make public and clear the procedures for complaint and the institution's sanctioning process, details of members who have been sanctioned, and the grounds for doing so.

Education, competence, and the development of a body of knowledge

B-1)   For the built environment institutions to commit to a cross-disciplinary review of the silo nature of the education system, to see how they can use their badging to encourage greater integration.

B-2)   Collectively promote the built environment as a career path of choice, demonstrating the relevance of the institutions (and membership thereof) in a way that engages current and future generations.

B-3)   Provide the means of allowing and encouraging greater movement between professions during a career.

B-4)   Provide access to young members by accepting a donation of time in return for a lower subscription.

B-5)   Improve the 'guarantee' of a particular quality of individual – for example by benchmarking the expertise of members.

B-6)   Become agents for disclosure as guardians of quality – e.g., a TripAdvisor-type public feedback system for individual/practice performance.

B-7)   As learned societies, engage with and disseminate research and best practice, including agenda setting; pulling together some of the knowledge coherently for members; reviewing how as institutions they themselves should respond to emerging evidence from research and practice; and adjusting requirements for membership, practice and education accordingly.

B-8)   Establish a joint think tank that could pool the resources of the Institutions to conduct research and develop policy for the industry – a King's Fund for the built environment.

B-9)   Determine how the professions can support standards that link better across sectors internationally – e.g. (taking an example from the RICS) linking matters financial to accounting to valuation to measurement to ethics to environment.

Institutional organisation and the relationship with Government

C-1)   Develop and empower the CIC as a shared outlet for joint initiatives and announcements, lobbying, campaigning etc.

Collaboration on strategic issues

On industry rej199form

D-1)   Establish a shared vision as to structural reform of the industry that would improve the industry's offer to client …

D-2)   … and society.

On climate change

D-3)   Set environmental matters high on the professional portfolio both collectively and individually, with action and measurement.

D-4)   Establish a joint institutional position on the right response to the impact of the built environment on climate change (mitigation) …

D-5)   … and the impact of climate change on the built environment (adaptation).

D-6)   Publish cross-disciplinary recommended behaviours for members on designing for climate change.

On building performance

D-7)   Take responsibility for the whole-life of projects, by getting involved from the start and remaining involved beyond project delivery to monitor performance through post-occupancy evaluation.

D-8)   Develop and maintain joint codes and standards/key performance metrics so that all construction professionals are working from a shared understanding of (building) performance.

D-9)   Agree cross-disciplinary recommended behaviours for members on holistic post occupancy evaluation of projects …

D-10) … and an integrated system for publishing this information …

D-11) … against which members should be obliged to report annually.

D-12) Develop the various awards systems to be a truer reflection of the performance of buildings, with the ultimate accolade being reserved for a building that after 10 years of use out-performs others.

Appendix E

Proposal for a shared institutional Code of Professional Conduct

Introduction

A shared Code of Conduct is one of the recommendations (A1) in the Edge Commission report Collaboration for Change (Morrell 2015) – see Appendix D – with the intended outcome that a common understanding on the duties of a construction professional should be developed. The proposed code is intended to demonstrate that a comprehensive common code, shared by the various professional institutions, is feasible and practical.

The draft was drawn and adapted from various other codes, including those of the: Society of Construction Law, RIBA, ARB, EC-RAE, ICE, RICS, CIAT, CIBSE, CIOB, IStructE, LI, RTPI, International Ethics Standard, Nolan, Deloitte, Atkins and Arup – plus the new professionalism code proposed by Bill Bordass (Bordass and Leaman 2013, p6). Rephrasing has been carried out where necessary to achieve consistency and grammatical sense.

The structure of the proposed code is based on whom duties are owed to; as is, for example, le Code Guadet (see Appendix B) or the AIA's code (see Chapter 4). This format was selected in order to provide maximum clarity to all those using the code, including professionals, those commissioning services from them, their employees and wider society.

The code doesn't include institutional housekeeping matters or disciplinary procedures as it is considered that these are generally better located within the rules and regulations of membership.

The proposed code

Duties towards the wider world

The environment:	to minimize any adverse effect on the natural environment now and into the future
Use of resources:	to take into account the limited availability of natural and human resources
Future proofing:	to allow for the needs of future generations taking into account any reasonably predictable circumstances, including the effects of climate and demographic change

Accounting:	to demonstrate for each project, by an appropriate audit trail, that all reasonable steps have been taken to ensure that the above issues have been adequately addressed
Feedback:	to evaluate and reflect upon the performance in-use of projects and feed back the findings

Duties toward society

Public interest:	to act consistently in the public interest and take the interests of all stakeholders in any project properly into account, including future generations
Integrity:	to act with objectivity, responsibility and truthfulness at all times
Impartiality:	to exercise impartial and independent professional judgement
Use of evidence:	to base professional advice on relevant, valid and objective evidence and the best quality knowledge that can be reasonably accessed
Impact:	to consider the broader impact of projects on society, the industry and government
Health and safety:	to take all reasonable steps to protect the health and safety of occupants, users and members of the public affected by projects over their full lifecycle
Responsibility:	to provide timely information and warning of matters, which may adversely affect others, when they become apparent
Disclosure:	to disclose accurate and truthful information on project intentions and outcomes in accordance with industry-wide methodologies and metrics
Fairness:	to treat all persons fairly and with respect and to embrace equality of opportunity, diversity and the elimination of discrimination
Users:	to have a proper concern and due regard for the impact that projects may have on both users and local communities
Bribery:	to reject bribery and all forms of corrupt behaviour and make positive efforts to ensure others do likewise
Value:	to create lasting value and keep options open for the future

Duties towards those commissioning services

Honesty:	to act for each and every one commissioning services in a reliable and trustworthy manner
Duty of care:	to discharge professional duties with fidelity and probity
Transparency:	to keep identified and relevant individuals informed of the progress of projects and any key decisions made
Conflicts of interest:	to identify and declare any potential bias, conflict of interest or undue influence, whether real or perceived
Competence:	to be competent to carry out the professional work undertaken, and if others are engaged, to ensure that their work is also competent and adequately supervised
Diligence:	to apply high standards of skill, knowledge and care in all work undertaken

Knowledge and skills:	to maintain and develop new knowledge and skills to ensure services are kept up to date and effective
Service improvement:	to foster new ideas and service development to improve the value and performance of services over time
Targets:	to use best endeavours to meet agreed time, cost and quality requests
Accountability:	to take full responsibility for services provided
Appointment:	to ensure that terms of appointment, the scope of work and the essential project requirements are clear and recorded in writing, and to explain to service commissioners the implications of any conditions of engagement and the way their fees are to be calculated and charged
Quality systems:	to have systems in place to ensure that projects are run professionally, and are regularly monitored and reviewed
Communications:	to be open and share (as appropriate and necessary) information with service commissioners and/or others in a way that is readily understood
Confidentiality:	to ensure that appropriate security is in place for all records in accordance with the service commissioner's requirements for confidentiality and to ensure compliance with data protection legislation
Money:	to keep proper records of all money held for service commissioners and other parties and be able to account for it whenever required
Insurance:	to maintain appropriate professional indemnity insurance
Follow through:	to provide project follow-through and aftercare when required
Post-project evaluation:	to carry out post-project evaluation, to learn from the evidence gathered and the project experience and to share understanding and admit mistakes
Complaints:	to have in place (or have access to) effective procedures for dealing promptly and appropriately with disputes and complaints

Duties towards those in the workplace

Respect:	to show consideration for colleagues and for all other persons encountered in the course of professional duties. All persons are to be treated with respect and without bias
Equal opportunities:	to avoid discrimination against anyone for whatever reason and ensure that issues of race, gender, sexual orientation, age, size, religion, country of origin or disability have no place in any dealings with other people or business decisions
Employment practice:	to comply with good employment practice both as employer and employee
Payment:	to pay a fair and commensurate reward to all employees and others in the workplace for work carried out

Working hours:	to ensure hours worked are reasonable and allow for a healthy work/life balance
Training:	to provide the training, advice and information necessary for employees and others to operate effectively, gain new skills and progress
Work environment:	to provide a safe working environment
Health and safety:	to take all reasonable steps to protect the health and safety of all those in the workplace and engaged in delivering projects
Collaboration:	to cooperate and integrate proactively with other professionals and to develop trusting relationships with open and honest collaboration
Competition:	to avoid acting maliciously or recklessly when competing with another person or when taking actions likely to adversely affect the professional, business or other interests of another person
Supply chain:	to avoid acts which, directly or indirectly, are likely to result in the unfair treatment of other people or deprive them of a fair reward for their work
Modern day slavery:	to proactively counteract and report abusive labour practices in connection with any projects undertaken
Challenge:	to challenge assumptions and standards. Be honest about what you don't know
Plagiarism:	to respect the intellectual property rights of others and not collude with any form of plagiarism
Risk:	to identify and evaluate and, where possible, quantify risks and to share any concerns with appropriate parties
Whistleblowing:	to report dangerous situations and suspected wrongdoing as soon as possible, to take seriously and investigate any concerns, respect confidentiality and to protect those fairly raising concerns from any repercussions
Dispute resolution:	to encourage, if appropriate, alternative methods of dispute resolution, including mediation or conciliation

Duties towards the profession

Behaviour:	to act in a way that promotes trust in the profession
	Promoting the highest standards globally
	Understanding that being a professional is more than just about how you behave at work; it is also about how you behave in your private life
	Fulfilling your obligations. Doing what you say you will
	Always trying to meet the spirit of your professional standards and not just the letter of the standards
Notify:	to notify the professional body if convicted of a criminal offence or disqualified as a company director

Disclose:	to report, in confidence, to the professional body and subject to any restrictions imposed by law, issues, problems and 'near-misses' that could aid better and more informed future practice and the avoidance of preventable disasters
Professional knowledge:	to contribute to the knowledge base of the profession through sharing appropriate project information and data with accredited research bodies and fellow professionals
Acting on behalf:	to accurately represent the views of the professional body or other organizations when speaking on their behalf and to refrain from promoting personal, employers' or others' interests
Reporting on others:	to report, in confidence, to the professional body and subject to any restrictions imposed by law, any alleged breach of this Code and assist the professional body in its investigations

Duties to oneself

Integrity:	to fearlessly do the right thing, beyond any obligation to whomsoever is paying you
Truthfulness:	to decline to be a party to any statement you know to be untrue, misleading, unfair to others or contrary to your own professional knowledge
Leadership:	to actively promote and robustly support the principles of professionalism and to challenge poor behaviour wherever it occurs
Accountability:	to be personally accountable for decisions and actions you take and submit to the scrutiny necessary to ensure this
Confidentiality:	to avoid taking personal advantage of confidential information or allowing others connected with you to do so
Openness:	to take decisions in an open and transparent manner and not withhold information from the public unless there are clear reasons for doing so
Keeping up to date:	to actively maintain, and where possible encourage others to maintain, professional competence through systematic improvement and broadening of knowledge and skill
CPD:	to maintain, record and provide evidence of your continuing professional development (CPD)
Evaluate and reflect:	to evaluate and reflect on the quality and impact of work carried out and the possibilities for improvement and potential for wider benefit.

Appendix F

The role of the professional in an age of widely accessible knowledge

(Il ruolo del professionista nell'era dell conoscenza accessibile a tutti)

Simon Foxell, Bologna, 2007

I am very pleased to be presenting this in Bologna, in '*un pianeta porticato*' and following the earlier contributions from the US–UK viewpoint I am happy to bring you a further report from Planet Anglo-America.

In the last two decades we have undoubtedly seen the '*professionalization*' of the middle classes, as described by Gerald Hanlon in his paper, and the society that I can describe, that of the UK, now expects a very different service from its professionals than it did only 20 years ago. For that reason my remarks will be less about the profession of architecture, to which I belong, and more about the professions in the UK in general.

In that context the discussion about the legal protection of the professions in Italy was both interesting and relevant. We had a very similar argument about the status of architects in the UK 25 years ago when our government required the profession to abandon their mandatory minimum fee scale. However, the profession is still around and much healthier, despite the fierce arguments over whether we would be forced to sacrifice quality for lower cost. Instead professionals have been forced to explain to their clients, to those who hire us, what it is that we provide for their money and why we should be charging higher fees than the apparently equivalent professional next door and why we believe our service to be better.

We now have to be upfront about the quality of our work; it is something that we have to argue for. If we believe in quality, because that is what we wish to produce and provide, then we must find ways of convincing the general public and our clients that it is something that they want as well.

Because of the challenges to the way professionals work and operate in the UK there has been a perception that we have been going through a crisis in professionalism and, although I suspect that the nature of the challenge in Italy is different, we are all facing constant change. And we don't like change, we would prefer it if the world would just stop for long enough for us to catch up and sort ourselves out. But we, as professionals, need to be better prepared for inevitable change, to look forward to see what changes are approaching and to do our best to be ready and prepared to deal with them.

This is the context for a study I carried out when I was Vice-chair of the think tank Building Futures – a collaboration between the Royal Institute of British Architects and the UK government advisory body The Commission for Architecture and the Built Environment (CABE) – that resulted in a book published in 2003 entitled *The professionals' choice: the future of the built environment professions*. This was followed by another study looking at the general drivers for change in the built environment, *Riding the Rapids* [2004], commissioned from the writer and urban thinker Charles Landry, as well as reports on the future of education and healthcare buildings amongst others.

The professionals' choice was based on a scenario planning exercise under-taken with a wide range of partners, but critically another UK think tank – The Work Foundation, and produced five separate scenarios for the future of the built environment professions. But initially we looked across the range of professions and we saw that the old professions – architecture, law, medicine etc. – were all struggling to maintain their position in a society in which large numbers of people considered themselves to be professional or to be acting in a professional capacity. The study therefore concluded that the old professions needed to be constantly looking ahead and preparing themselves for a very different world.

To help us we looked at definitions of what it was to be a professional and a member of a profession and proposed the following:

- A profession should develop and maintain a body of knowledge
- It should be a formal association of people who have come together with a joint purpose
- Its members should be trustworthy
- They should behave and act according to agreed standards and be expelled if they fall below them
- And finally, and critically, they should act in the public interest.

In addition we identified a number of changes that were affecting society in general and the built environment in particular:

- Firstly a change in the relationship between professionals and society. This has been characterized as 'a question of trust' by the Cambridge philoso-pher and ethicist Onora O'Neill in her 2002 lecture series of that name. Professionals no longer have trust bestowed by their position – their trust-worthiness has to be tested by a constant monitoring and checking, such that it is barely trust at all.
- Secondly a rapidly changing world to work in and for – one that has become globalized but that is now likely to be dominated by the impact of climate change and increasing international tensions

- Thirdly – a whole new toolset for both professionals and their customers, building users and society to use and abuse. Information and communication technology has changed fundamentally the commodity that was originally at the heart of professional organizations – the control of knowledge.

I would like to address these issues in turn, but end by concentrating mainly on the third – the issue of knowledge for the professional and looking at where the value created by the knowledge worker in 21st century economies is likely to be found.

Trust

There is nothing new in a lack of trust in professionals. We only need to remember George Bernard Shaw's much quoted statement that 'the professions are a conspiracy against the laity' from his play *The Doctor's Dilemma* of 1906. And in the UK now there is certainly a very high level of mistrust. In the professions generally this has resulted in a much greater level of legislation, guidance, form filling and the use of performance indicators. There appears to be a greater willingness amongst the public to trust the opinion of a celebrity than believe an 'expert'. Perhaps that is because celebrities stand out against the anonymity of everyday life, they have a track-record, however dubious, that is harder to fake than professional credentials – which as we know can be bought mail-order and in bulk from the Internet.

In architecture this has resulted in schemes to score design quality using complex batteries of questions and increased calls for 'value for money', but more seriously it has led to the downgrading of the profession in the decision-making chain. Frequently the architect will now work for the builder, working to his values, rather than directly for the client or user. This culture of design and build, and latterly 'Public Private Partnerships' and the Private Finance Initiative, has taken responsibility away from the professional. Those who commission projects have frequently chosen not to trust the professional – preferring instead to believe in the power of the market to produce the best outcome.

The need to regulate the market and prevent abuse is of course one of the reasons why the professions were originally established. Perhaps this is a cyclical relationship and one that will continue to evolve if there is again felt a need to find people to trust, possibly because the alternative of not trusting is just too horrific.

Changing world

A rapid turn for the worse in economic, social or environmental conditions may well trigger just such a response and it is one of the jobs of the professions to prepare now for the future, to look ahead and have developed solutions to offer.

Climate change is undoubtedly one of the most serious threats that is facing us and one that requires a collective consolidated professional response. I recently had a part in persuading my professional institution to accept the principle of Contraction and Convergence – a strategy for survival and one that prescribes a degree of global equity and very significant reduction in the emissions of carbon dioxide into the atmosphere. Agreeing with such a framework provides the Royal Institute of British Architects with the ability to consider what very much more specific actions are necessary. For one we believe that professionals are going to have to be much more rigorous in what they do. They need to be far more responsible – that is to behave more ethically. Behaving ethically is, again, one of the central pillars of being a professional.

Last week I saw the 2004 Nobel peace prize winner, Wangari Maathai, from Kenya, speak, in part about climate change but more about her history as a campaigner for women's role in society, for the protection and enhancement of natural habitats, for political rights and the importance of planting trees – as many and as diverse a range as possible. But for an audience of architects she had one very significant story. Twenty years ago she was part of a campaign to prevent the building of a 40-storey tower on a site in the middle of Nairobi – a site that was one of the last areas of park left in the city. The campaign only turned around and succeeded when the Kenyan Association of Architects took out a full page advertisement in a Nairobi paper condemning the proposals. Architects campaigning against buildings – extraordinary, but professionally responsible.

We are all going to have to change the way we behave to reduce the amount of CO_2 we emit, but professionals are essential to help wider society achieve that goal using their skill, their knowledge and above all their willingness to collectively act in ways that they know to be correct.

Knowledge

Which brings me back to knowledge, for undoubtedly professionals are knowledge workers. But it is also important not to confuse various other definitions that are being used, including 'the creative class' and 'the intellectual'. I am a professional, I work in a professional capacity and to professional standards. It happens that as an architect I also use creativity as part of my business offer and I hope I also have a contribution to make as an intellectual. But although these roles are in the same territory and significantly overlap they are distinct and should not be confused, for when we consider the issue of knowledge for the professions it should be addressed with clarity.

The most detailed and wide-ranging knowledge is now widely available to anyone, and especially those with a computer and a broadband connection. Doctors are very used to their patients coming to consultations with in-depth knowledge about their condition and all the treatments and remedies. The patient may know far more about the subject than the doctor they've come to see. But they still come.

I can sit at my computer and do research that would only a very short time ago have required long and lengthy visits to some of the world's best libraries. I can also carry out complex transactions without any other human agency being involved. Many have seen this as inevitably short-circuiting the need for the professional intermediary. In *The professionals' choice* one of the writers – a professor and head of a leading school of construction management in the UK – predicted the automation of much of the UK's building industry. The customer for a new home would in his mischievous view and within 20 years be able to develop the specification, design, make detailed individual choices, get consents, guarantees and loans and finally just push the Return button to start the delivery lorries moving towards the building plot. No architect, engineer or surveyor would have been necessary on that particular building. The systems, once developed would be able to handle almost all of it without them.

So what in a culture of universalized and instantly accessible knowledge will be role of the professional? It is no longer as the sacred keeper of privileged knowledge. It has been suggested that instead of knowledge, that extensive assemblage of facts, rules and methodologies, the professional provides that much rarer commodity – 'judgement'. The professional is the intermediary who can make decisions. Decisions on quality, on cultural value, on diagnoses of conditions and between rival claimants for advantage. This requires a different attitude to knowledge and of the skills of professionals.

It is worthwhile here to defer to the chemist and social philosopher Michael Polanyi (1891–1976) who made the celebrated distinction between codified and tacit knowledge. Codified knowledge is all the information and methodology that can be, in some form or other, captured, recorded and retrieved: essentially all the information we can find in libraries, in cyberspace, in expert systems, that is transmitted directly by education and can be tested by examinations.

Tacit knowledge, in Polanyi's phrase, is 'that which we cannot tell'. Like codified knowledge it covers a broad area including hunches, intuition and ideas, emotional knowledge and acts of creation and discovery. But tacit knowledge is also always in the process of becoming codified knowledge. Yesterday's intuitive response is becoming tomorrow's received and tested wisdom. There is a constant flow of knowledge across the boundary between the two and mostly in the direction of codification. Yet it is tacit knowledge that tomorrow's professionals need to have – it will be their most important item of stock; the only really valuable thing they have to offer the world.

It is true that professionals have always exercised judgement, have always used soft skills in addition to hard knowledge. The difference now, is that it is the only thing they have left that marks them out as different from those simply using knowledge and skill as tools and commodities in the market-place – often with the assistance of formidable computer power. The challenge for professionals now is to be developing more tacit knowledge as that that they already have is formalized and embodied in extensively managed knowledge banks and systems.

But like advance in science this is not something to be done on one's own. There is a role for the individual genius but it is not the most significant one. The development of useful tacit knowledge is a collective and collaborative endeavour that requires another aspect that Polanyi insisted on – 'dialogue within an open community' – the community of shared purpose, of interest, that is another of the pillars of professionalism; although possibly re-defined to be far more open and accepting than is traditional for professional bodies.

So far as we know there is an infinite supply of tacit knowledge, unlike oil or uranium deposits, but mining it can be hard work to sustain over many decades of a professional's career. We need to go into training now to be in shape for the challenge. The new vocation of the professional is to be constantly imaginative, innovative and creative, not for the sake of the new and shocking, but to discover ways of dealing with some of the most pressing problems that the world has to face. That is if we are able to improve our overall position rather than be overwhelmed by the forces affecting the planet.

The professional needs to respond to this by looking outward, to see how their particular set of analytic and synthetic skills can be developed, to see where the challenges are and how to meet them. The danger is a retreat behind professional boundaries and the defence of the *status quo*. One of the most important things we must do is to expand the idea of professionalism again, and to rediscover the importance of working together, of sharing, for the common good.

There is plenty for professionals to do in our more challenging world, a world where much of the work we used to do will have been outsourced overseas; to India, to China, to Vietnam. Instead we need the develop the ability and skills to do other, possibly more interesting things. They will become the real core of what we sell, of what we do, of who we are. But in order to do that, we need to look out, to be open, to embrace development elsewhere; we must not retreat behind professional boundaries and we must not try to defend too hard the *status quo*. We must find new territory to conquer and work together for the common good as professionals within professional bodies, as forward thinking, creative and engaged citizens.

Presented at *La Vita Intellettuale: Professioni, Arti, Impresa in Italia e nel Pianeta*
Simon Foxell – Bologna, 13–14 February 2007

BIBLIOGRAPHY

Abbott, A., (1988), *The System of Professions: An essay on the division of expert labor*, Chicago, University of Chicago Press.

Abel, R., (1989), Between market and state: The legal profession in turmoil, *The Modern Law Review* 52(3), pp285–325.

ABET, (2017), Criteria for Accrediting Engineering Technology Programs, 2017–2018, ABET, available from www.abet.org/accreditation/accreditation-criteria/criteria-for-accrediting-engineering-technology-programs-2017–2018/#program [Accessed 13th November 2017].

ACE, (2004), *The Architectural Profession in Europe: A sector study*, Brussels, Architects' Council of Europe.

ACSA, (2017), *Constitution of the American Society of Civil Engineers*, Reston, VA, American Society of Civil Engineers.

Addison, J., (1711), The three learned professions, *The Spectator* (24 March) 21, pp59–60.

AECOM, (2013), *Social Responsibility: Beyond what the eyes can see*, Los Angeles, CA, AECOM.

AECOM, (2014), *Code of Conduct*, Los Angeles, CA, AECOM.

Agas, R., (1596), *Preparative to Platting of Landes and Tenements for Surveigh*, London.

AIA, (1909), A circular of advice relative to principles of professional practice and the canon of ethics, *The American Architect* 96(1774), pp273–4.

AIA, (2007), History of the American Institute of Architects, available from http://district.bluegrass.kctcs.edu/kevin.murphy4/shared_files/ACH%20260/about_history.htm [Accessed 12th November 2017].

AIA, (2013), *History of the Fellowship in the American Institute of Architects* (pdf), Washington, DC, American Institute of Architects.

AIA, (2016), *The Architecture Student's Handbook of Professional Practice*, 15th edition, Hoboken, NJ, Wiley.

AIA, (2017), *Code of Ethics and Professional Conduct, Ethics from the Office of the General Counsel*, Washington, DC, American Institute of Architects.

Albu, A., (1980), British attitudes to engineering education: A historical perspective, in Pavitt, K. (Ed.), *Technical Innovation and British Economic Performance*, London, Palgrave Macmillan, pp.67–87.

All Profession Charter Reform Group, (2010), RICS actions on climate change, unpublished paper issued by a small group of RICS members.

Anderson, C.J., (1993), *Inigo Jones's Library and the Language of Architectural Classicism in England, 1580–1640*, Cambridge, MA, MIT Press. Available from http://hdl.handle.net/1721.1/12670 [Accessed 14th April 2016].

Anderson, C., (2017), *Overcoming Challenges to Infusing Ethics into the Development of Engineers*, proceedings of a workshop, Center for Engineering Ethics and Society, Washington DC, National Academies Press.

Ansell, G. *et al.*, (1985), *Engineering in Society, Panel on Engineering Interactions With Society*, Washington, DC, National Academies Press.

ARB, (2000 & 2015) ARB Annual Reports 2000 & 2015, Architects Registration Board, available from www.arb.org.uk/annual-report [Accessed 12th October 2017].

ARB, (2010), *Prescription of Qualifications: ARB Criteria at Parts 1, 2 and 3*, London, Architects Registration Board.

ARB, (2017a), Registration Facts and Figures: The Register 2016, Architects Registration Board, available from http://2016.arb.org.uk/facts-figures/registration/ [Accessed 12th October 2017].

ARB, (2017b), *The Architects Code: Standards of professional conduct and practice*, London, Architects Registration Board.

Arcadis, (2014), Arcadis General Business Principles, Arcadis, available from www.arcadis.com/en/united-kingdom/who-we-are/business-practices/arcadis-general-business-principles/ [Accessed 1st May 2018].

Architectural Society (1835), Laws and Regulations of the Architectural Society, 35 Lincoln's Inn Fields. Pamphlet.

Aristotle, (349 bce), *The Nicomachean Ethics*, trans Thomson, J., revised Tredennick, H., (1976), Harmondsworth, Penguin.

Armstrong, W., (1972), The use of information about occupation, in Wrigley, E., (Ed.), *Nineteenth Century Society: Essays in the use of quantitative methods for the study of social data*, Cambridge, Cambridge University Press, pp191–310.

Artandilian, A. and Campbell, P., (1987), *The Implementation of Foreign Real Estate Development Opportunities: A case study of Canary Wharf, London*, Cambridge, MA, MIT Press.

Arup, (2015), Our ethics standards policy, available from www.arup.com/our-policies [Accessed 3rd February 2017].

Arup, (2017), BUS methodology, Building Use Studies Ltd, available from www.busmethodology.org and www.usablebuildings.co.uk

ASCE (1875), *Transactions 4*, New York, American Society of Civil Engineers, pp123–35 and pp208–22.

ASCE, (2007a), ASCE removes price-bidding restrictions from Code of Ethics, available from www.asce.org/question-of-ethics-articles/feb-2007/ [Accessed 5th November 2017].

ASCE, (2007b), Development of the first ASCE Code of Ethics, available from www.asce.org/question-of-ethics-articles/dec-2007/ [Accessed 5th November 2017].

ASCE (2008), Development of Sustainability in ASCE Code of Ethics, available from www.asce.org/question-of-ethics-articles/nov-2008/ [Accessed 8th November 2017].

ASCE, (2014a), Code of Ethics (updated 29th July 2017), available from www.asce.org/code-of-ethics/ [Accessed 30th October 2017].

ASCE, (2014b), *Guide to Professional Engineering Licensure for the Construction Engineer*, Reston, VA, Construction Institute – American Society of Civil Engineers.

ASCE (2016), *Guidelines for Establishing a New ASCE Student Chapter: Student chapter basics*, revised July 2016, Reston, VA, American Society of Civil Engineers.

ASCE Metropolitan Section, (2005), ASCE Founders' Plaque, available from www.ascemetsection.org/committees/history-and-heritage/landmarks/founders-plaque [accessed 5th November 2017].

Atkins, (2013), Celebrating 75 Years: Our Founder Sir William Atkins, available from www.atkinsglobal.com/en-GB/media-centre/features/our-founder-sir-william-atkins [Accessed 30th December 2016].

Atkins, (2016), Atkins response re working conditions in the UAE, Business & Human Rights Resource Centre, available from https://business-humanrights.org/en/atkins-response-re-working-conditions-in-the-uae-0 [Accessed 27th April 2017].

Atkins, (2017), Behaving the Atkins Way: Our code of conduct, pdf, available from https://careers.atkinsglobal.com/File.ashx?path=Root/Documents/north-america-welcome-new-employee/Behaving_the_Atkins_Way-Code_of_Conduct-2014.pdf [Accessed 2nd May 2018].

Audi, R., (Ed.), (1999), *The Cambridge Dictionary of Philosophy*, 2nd edition, Cambridge, Cambridge University Press.

Autodesk, (2017), Project Dreamcatcher, available from https://autodeskresearch.com/projects/dreamcatcher [Accessed 13th April 2017].

Badger, J., (1693), *The Case between Doctor John Badger and the College of Physicians*, London.

Bagihole, B., (2005), Reflections on Women, Civil Engineering and the UK Construction Industry, Paper to the International Conference 'Creating Cultures of Success for Women Engineers', 6th–8th October 2005, pp73–82.

Balfour, A. (1927), *Committee on Trade and Industry: The Balfour Report*, London, HMSO.

Banfield, A., (2005), *Stapleton's Real Estate Management Practice*, 4th edition, London, Routledge.

Barclay, I., (1976), *People Need Roots: The story of the St. Pancras Housing Association*, London, Bedford Square Press.

Barlow, A. (1946) *Scientific Man-power: Report of a Committee appointed by the Lord President of the Council*. London, HMSO.

Barnes, H., Bonjour, D. & Sahin-Dikmen, M., (2002), *Minority Ethnic Students and Practitioners in Architecture: A scoping study for the Commission of Architecture and the Built Environment (CABE)*, London, Policy Studies Institute. Available from http://webarchive.nationalarchives.gov.uk/20110118192038/http://www.cabe.org.uk/files/architecture-and-race.pdf [Accessed 13th October 2017].

Bason, C., (Ed.), (2014), *Design for Policy*, Farnham, Gower.

Bateson, A., (2015), *Soft Landings and Government Soft Landings: A convergence guide for construction projects*, BG 61, Bracknell, BSRIA.

Bauman, Z., (1994), *Alone Again: Ethics after certainty*, London, Demos.

BDP, (2015), Our Values, BDP, available from www.bdp.com/en/about/about-bdp/ [Accessed 3rd February 2017].

Becher, T., (1999), *Professional Practices: Commitment and capability in a changing environment*, London, Transaction Publishers.

Beck, U., (1992 [1986]), *Risk Society: Towards a new modernity*, Lash, S. and Wynne, B. (Trans.), London, Sage Publications.

Beck, U., (2006), Living in the world risk society, Hobhouse Memorial Public Lecture, *Economy and Society* (August) 35(3), pp329–45.

Behavioural Insights Team (2014–17), The Behavioural Insights Team, available from www.behaviouralinsights.co.uk

Bennion, F., (1979), The Sex Disqualification (Removal) Act – 60 inglorious years, *New Law Journal* (8th November) 129, pp1088–9.

Bentham, J., (1789), *An Introduction to the Principles of Morals and Legislation*, Oxford, Clarendon Press.

Biancavilla, D., (2007), A Sketch History of the Central New York Chapter of the AIA (American Institute of Architects) on the occasion of the 120th Anniversary of its founding on October 29th 1887, pdf, available from www.aiacny.org/assets/Uploads/documents/AIACNY-ChapterHistory.pdf [Accessed 2nd May 2018].

BIS, (2010), Estimating the amount of CO_2 emissions that the construction industry can influence: Supporting material for the Low Carbon Construction IGT Report, pdf, available from https://assets.publishing.service.gov.uk/government/uploads/system/uploads/attachment_data/file/31737/10-1316-estimating-co2-emissions-supporting-low-carbon-igt-report.pdf [Accessed 2nd May 2018].

Black, J., (1992), *The British Abroad: The Grand Tour in the eighteenth century*, Sutton Publishing, reprinted (2003), Stroud, The History Press.

Blakemore, M. and Longhorn, R., (2004), Ethics and GIS: The practitioner's dilemma, paper for the AGI 2004 Conference Workshop on 'GIS Ethics'.

Blond, P., Antonacopoulou, E. and Pabst, A., (2015), *In Professions We Trust: Fostering virtuous practitioners in teaching, law and medicine*, London, ResPublica.

BMJ, (2000), The private finance initiative: spinning out the defence, *BMJ* (27th May) 320, p1460.

Boggan, S., (2001), 'We Blew It': Nike admits to mistakes over child labor, *The Independent*, 20th October.

Bolton, A., (1924), *Architectural Education a Century Ago: Being an account of the office of Sir John Soane, R.A., Architect of the Bank of England, with special reference to the career of George Basevi, his pupil, Architect of the Fitzwilliam Museum at Cambridge*, London, The Sir John Soane Museum.

Bonfield, D., (2015), *Disruptive Diversity: An external perspective on ways to increase diversity and inclusion within the Institution of Civil Engineers and beyond*, London, Institution of Civil Engineers, available from www.wes.org.uk/sites/default/files/Disruptive%20Diversity%20V9.pdf [Accessed 10th October 2017].

Bordass, B. and Leaman, A., (2013), A new professionalism: remedy or fantasy?, *Building Research & Information* 41(1), pp1–7.

Boswell, J., (1827), *The Life of Samuel Johnson LL.D.*, London, Jones & Co.

Boyle, B.M. (1977), Architectural practice in America 1865–1965 – ideal and reality, in Kostof, S. (Ed.), *The Architect: Chapters in the history of the profession*, Oxford, Oxford University Press, pp309–44.

Braudel, F., (1982), *The Wheels of Commerce: Civilization and capitalism 15th–18th Century*, Reynolds, S. (Trans.), Volume 2, London, Collins.

Briscoe, G., (2005), Women and minority groups in UK construction: Recent trends, *Construction Management and Economics* (December) 23, pp1001–5.

Broadbent, G., (1979), *Neo-Classicism*, Profile 23, Architectural Design, Volume 49, Numbers 8–9, London, Architectural Design.

Broadbent, J., Dietrich, M. and Roberts, J., (1997), *The End of the Professions: The restructuring of professional work*, London, Routledge.

Brown, J.A.C., (1954), *The Social Psychology of Industry*, Harmondsworth, Penguin Books.

Browne, J., (Ed.), (1780), *Browne's General Law List for the Year 1780*, London, John Browne.

BSRIA, (2017), Soft Landings, available from www.bsria.co.uk/services/design/soft-landings

Building (2006) Top 250 Consultants, available from www.building.co.uk/Journals/Builder_Group/Building/2006_issue_41/attachments/Top250Consultants.pdf [Accessed 27th July 2017].

Building (2016) Top 150 Consultants, available from www.building.co.uk/data/top-150-consultants-2016/5089708.article [Accessed 27th July 2017].

Building Futures, (2010), Practice Futures: Risk, Entrepreneurialism and the Professional Institute, available from www.buildingfutures.org.uk/assets/downloads/Practice_Futures.pdf

Building News, (1860), Editorial on a Surveyors' Institute, *Building News*, 8th June, VI.

Burg, D., (1976), *Chicago's White City of 1893*, Lexington, The University Press of Kentucky.

Burgess, P., (Ed.), (1983), *The Role of the Architect in Society*, Pittsburgh, PA, Carnegie-Mellon University Department of Architecture.

Buro Happold Engineering, (2016) *Quality Policy Statement*, Bath, Buro Happold Engineering.

Burrage, M., Jarausch, K. and Siegrist, H., (1990), An actor-based framework for the study of the professions, in Burrage, M. and Torstendahl, R. (Eds), *Professions in Theory and History: Rethinking the study of the professions*, London, Sage, pp203–25.

Busby, T. (1786) *The Age of Genius! A satire on the times: A poetical epistle to a friend.* London, Harrison and Co.

Business & Human Rights Resource Centre, (2016), Construction industry failing to tackle abuse of migrant workers in the Gulf, claims report / Qatar under scrutiny at ILO for forced labour, treatment of migrant workers / Labour rights and the Qatar World Cup, available from https://business-humanrights.org [Accessed 27th April 2017].

Byerley Thomson, H., (1857), *The Choice of Profession: A concise account and comparative review of the English professions*, London, Chapman and Hall.

CABE (2000) CABE's Objectives, available from www.cabe.org.uk/objectives.htm [Accessed 2nd May 2000].

CABE, (2001), *CABE – Commission for Architecture & the Built Environment: Improving people's lives through better buildings, spaces and places*, London, CABE.

CABE, (2011), Research: Practical findings, available from http://webarchive.nationalarchives.gov.uk/20110118095356/http://www.cabe.org.uk/research [Accessed 22nd March 2016].

CABE/Centre for Ethnic Minority Studies, (2005), *Black & Minority Ethnic Representation in the Built Environment Professions*, London, Royal Holloway, University of London / Commission for Architecture & the Built Environment.

California Architect and Building News, (1895), *California Architect and Building News*, 189. Available from https://archive.org/stream/californiaarchit16pacirich/californiaarchit16pacirich_djvu.txt [Accessed 26th August 2017].

Calvert M., (1967), *The Mechanical Engineer in America, 1830–1910: Professional cultures in conflict*, Baltimore, MD, Johns Hopkins University Press.

Cameron, D., (2008), Fixing our broken society, Speech in Glasgow 7th July 2008.

Cameron, E., (2012), *The European Reformation*, 2nd edition, Oxford, Oxford University Press.

Canadian Institute, (1851), Royal Charter, available from http://rciscience.ca/about-rci/history/ [Accessed 2nd August 2016].

Caplan, A. and Gilham, J., (2005), Included against the odds: Failure and success among minority ethnic built-environment professionals in Britain, *Construction Management and Economics* (December) 23, pp1007–15.

CarbonBuzz, (2017), CarbonBuzz, available from www.carbonbuzz.org

Carmona, M., de Margalhães, C., Natarajan, L, (2017), *Design Governance: The CABE experiment*, Abingdon, Routledge.

Carr-Saunders, A. and Wilson, P., (1933), *The Professions*, Oxford, Clarendon Press.

Cayley, G. (1837), *Prospectus of an Institution for the Advancement of the Arts and Practical Science, Especially in Connexion with Agriculture, Manufactures and other Branches of Industry*, London.

CBRE, (2017), CBRE Standards of Business Conduct, CBRE, available from www.cbre.com/-/media/files/corporate%20responsibility/ethics/cbre%20standards%20of%20business%20conduct%202017.pdf?la=en [Accessed 3rd February 2017].

Center for Ethical Business Cultures, (2005), *Corporate Social Responsibility: The shape of a history, 1945–2004*, Working paper No. 1, Washington, DC, Center for Ethical Business Cultures.

Chadwick, E., (1838), *Supplement No. 1 to the Fourth Annual Report of the Poor Law Commissioners for England and Wales*, Volume 4, London, HMSO.

Chadwick, E., (1842), *Report on an Inquiry into the Sanitary Condition of the Labouring Population of Great Britain*, London, HMSO.

Chaffee, R., (1977), The teaching of architecture at the Ecole des Beaux-Arts, in Drexler, A. (Ed.), *The Architecture of the Ecole des Beaux-Arts*, Cambridge, MA, MIT Press, pp61–109.

Chapman, C. and Levy, J., (2004), *A Chronicle of the Engineering Council*, London, The Engineering Council.

Chapman, J., (1858), Medical reform, *Westminster Review* (April) 8, pp478–530.

Chappell, D., (2016), *The Architect in Practice*, 11th edition, Oxford, Wiley Blackwell.

CIAT, (2014), Code of Conduct, effective 1st May 2014, Chartered Institute of Architectural Technologists, available from https://ciat.org.uk/membership/code-of-conduct.html [Accessed 2nd May 2018].

CIBSE, (1976), Royal Charter, The Chartered Institution of Building Services Engineers, available from www.cibse.org/about-cibse/governance/charter-and-by-laws-and-regulations [Accessed 2nd May 2018].

CIBSE, (2015), The Code of Professional Conduct, Chartered Institution of Building Services Engineers, available from www.cibse.org/about-cibse/governance/the-code-of-professional-conduct [Accessed 2nd May 2018].

CIBSE, (2017), PROBE – Post Occupancy Studies, available from www.cibse.org/building-services/building-services-case-studies/probe-post-occupancy-studies [Accessed 2nd May 2018].

CIC, (2006), *Survey of UK Construction Professional Services*, London, Construction Industry Council.

CIC, (2016), *A Blueprint for Change: Measuring success and sharing good practice*, London, Construction Industry Council.

CIOB, (1980), *Royal Charter of Incorporation*, Bracknell, Chartered Institute of Building.

CIOB, (2008), *International Policy Handbook*, Bracknell, Chartered Institute of Building.

CIOB, (2015a), *Rules and Regulations of Professional Competence and Conduct*, Bracknell, Chartered Institute of Building.

CIOB, (2015b), *Understanding the Value of Professionals and Professional Bodies*, Bracknell, Chartered Institute of Building.

Clark, G., (1965), History of the Royal College of Physicians of London, *British Medical Journal*, 1, pp79–82.

Clark, K., (1928), *The Gothic Revival*, London, Constable. 1964 edition, Harmondsworth, Penguin Books.

Clark, V., (1990), A struggle for existence: The professionalization of German architects, in Cocks, G. and Jarausch, K. (Eds), *German Professions 1800–1950*, Oxford, Oxford University Press, pp143–60.

Clarke, T., Ed. (1994) *International Privatisation: Strategies and practices*, Berlin, de Gruyter.

Clifton, G., (1989), Members and officers of the LCC, 1889–1965, in Saint, A. (ed.), *Politics and the People of London: The London County Council 1895–1965*, London, Hambledon Press, pp1–26.

Clifton, G., (1992), *Professionalism, Patronage and Public Service in Victorian London: The staff of the Metropolitan Board of Works 1856–1889*, London, The Athlone Press.

Clinton, B., (1996), Transcribed remarks by the President and the Vice President to the people of Knoxville on Internet for Schools, 10th October 1996, available from http://govinfo. library.unt.edu/npr/library/speeches/101096.html [Accessed 30th December 2016].

Coalwell, C., (2001), West Point: Jefferson's Military Academy, *Monticello Newsletter* (Winter) 12(2), available from www.monticello.org/sites/default/files/inline-pdfs/2001wjwstpnt.pdf.

Cole, R. and Lorch, R., (Eds), (2003), *Buildings, Culture and Environment: Informing local and global practices*, Oxford, Blackwell Publishing.

College of Physicians, (1688) *A Short Account of the Institution and Nature of the College of Physicians*, London.

Collins, S., Ghey, J. and Mills, G., (1989), *The Professional Engineer in Society: A textbook for engineering students*, London, Jessica Kingsley Publishers.

Colvin, H., (1978), *A Biographical Dictionary of British Architects: 1600 to 1840*, London, John Murray.

Commission des Titres d'Ingénieur (2018), Commission des Titres d'Ingénieur, available from www.cti-commission.fr/ [Accessed 15th January 2018].

Committee on Climate Change, (2008), *Building a Low-carbon Economy – the UK's contribution to tackling climate change*, London, TSO.

Committee on Climate Change, (2016), UK climate action following the Paris Agreement, Committee on Climate Change, available from www.theccc.org.uk/wp-content/uploads/2016/10/UK-climate-action-following-the-Paris-Agreement-Committee-on-Climate-Change-October-2016.pdf [Accessed 27th July 2017].

Committee on Standards in Public Life, (1995), The 7 principles of public life (the Nolan Principles), available from www.gov.uk/government/publications/the-7-principles-of-public-life/the-7-principles-of-public-life–2 [Accessed 22nd December 2015].

Connaughton, J. and Meikle, J., (2013), The changing nature of UK construction professional service firms, *Building Research & Information* 41(1), pp95–109.

Conrad, U., (1970), *Programs and Manifestoes on 20th-century Architecture*, Bullock, M. (Trans.), Cambridge, MA, MIT Press.

Cooley, M., (2016 [1980]), *Architect or Bee?: The human price of technology*, Nottingham, Spokesman.

Corfield, P., (1995), *Power and the Professions in Britain 1700–1850*, London, Routledge.

Corley, J., (Ed.), (2004), *The History of the National Council of Examiners for Engineering and Surveying: 1920–2004*, 3rd edition, Seneca, SC, NCEES.

Court, G. and Moralee, J., (1995), *Balancing the Building Team: Gender issues in the building professions*, Brighton, Institute for Employment Studies.

Cresswell, H.B., (1929), *The Honeywood File*, London, The Architectural Press.

Crinson, M. and Lubbock, J., (1994), *Architecture – Art or Profession?: Three hundred years of architectural education in Britain*, Manchester, Manchester University Press.

Crouch, C., (2002), *Design Culture in Liverpool 1880–1914: The origins of the Liverpool School of Architecture*, Liverpool, Liverpool University Press.

Crowther, G., (1959), *15 to 18: A report of the Central Advisory Council for Education (England), (The Crowther Report)*, London, HMSO.

Cunningham, A., (1999), Getting other, not better. The architectural profession was set on a false path in 1958. We must avoid this happening again, *The Architects' Journal* (23rd September). Available from www.architectsjournal.co.uk/home/getting-other-not-better-the-architectural-profession-was-set-on-a-false-path-in-1958-we-must-avoid-this-happening-again/773960.article [Accessed 7th May 2018].

Cunningham, T., (2011), *Professionalism and Ethics: A Quantity Surveying Perspective*, Dublin, Arrow, Dublin Institute of Technology.

Dainty, A. and Bagihole, B., (2005), Guest editorial, *Construction Management and Economics* (December) 23, pp995–1000.

Darby, H.C., (1933), The agrarian contribution to surveying in England, *The Geographical Journal* 82(6), pp529–35.

Darley, G., (1999) *John Soane, An accidental romantic*, New Haven, CT, Yale University Press.

Davies, W. and Knell, J., (2003), Conclusion, in Foxell, S. (Ed.), *The professionals' choice: the future of the built environment professions*, London, Building Futures, pp132–41. Available from www.buildingfutures.org.uk/assets/downloads/The_Professionals_ Choice2003.pdf

Davis Langdon Management Consulting, (2003), *Survey of UK Construction Professional Services 2001/2*, London, Construction Industry Council.

Davis Langdon Management Consulting & Experian BS, (2007), *Survey of UK Construction Professional Services 2005/6*, London, Construction Industry Council.

Day, N., (1988), *The Role of the Architect in Post-War State Housing: A case study of the housing work of the London County Council 1939–56*, University of Warwick. Available from http://wrap.warwick.ac.uk/34810/.

de Graft-Johnson, A., (1999) Gender, race and culture in the urban built environment, in Greed, C. (Ed.), *Social Town Planning*, London, Routledge, pp102–6.

de Graft-Johnson, A., Manley, S. and Greed, C., (2005), Diversity or the lack of it in the architectural profession, *Construction Management and Economics* (December) 23, pp1035–43.

de Graft-Johnson, A., Manley, S. and Greed, C., (2007), The gender gap in architectural practice: Can we afford it? In Dainty, A., Green, S. and Bagihole, B. (Eds), *People and Culture in Construction: A reader*, Abingdon, Taylor & Francis, pp159–83.

de Graft-Johnson, A. *et al.*, (2009), *Gathering and Reviewing Data on Diversity within the Construction Professions: Report for the CIC Diversity Panel*, Bristol, University of the West of England.

Deloitte, (2016), *Our Ethics Code*, Southampton, Deloitte LLP.

De Silva, C., (2017), First among equals, *Land Journal* (August/September), pp24–5.

Department for Communities and Local Government (2017), Table 241 House Building: Permanent dwellings completed, by tenure United Kingdom historical calendar year series2, 3, available from www.gov.uk/government/statistical-data-sets/live-tables-on-house-building [accessed 5th November 2017].

Department of the Built Environment (DoE), (1994), *Quality in Town and Country: a discussion document*, London, HMSO.

Design Commission for Wales, (2014), About the Design Commission for Wales, available from https://dcfw.org/about/ [Accessed 2nd May 2018].

Dickens, C., (1844), *The Life and Adventures of Martin Chuzzlewit*, London, Chapman & Hall.

Director General of Fair Trading, (2001), *Competition in Professions*, OFT328, London, Office of Fair Trading.

Dobson, A., (2014), *21 Things You Won't Learn in Architecture School*, London, RIBA Publishing.

Douglas, M., (1986), *How Institutions Think*, Syracuse, NY, Syracuse University Press.

Draper, J., (1977), The Ecole des Beaux Arts and the architectural profession in the United States: The case of John Galen Howard, in Kostof, S. (Ed.), *The Architect: Chapters in the history of the profession*, Oxford, Oxford University Press, pp209–35.

Drucker, P., (1950), *The New Society*, New York, Harper & Brothers.

Drucker, P., (1954), *The Practice of Management*, New York, Harper & Brothers.

Drucker, P., (1984), The meaning of Corporate Social Responsibility, *California Management Review* 26, pp53–63.

Duffy, F. and Hutton, L., (1998). *Architectural Knowledge, The Idea of a Profession*, London, E&FN Spon.

Duffy, F. and Rabeneck, A., (2013), Professionalism and architects in the 21st century, *Building Research & Information* 41(1), pp115–22.

Duncan, J., (Ed.), (2017), *Retropioneers: Architecture redefined*, London, RIBA Publishing.

Durkheim, E., (1893), *De la division du travail: Livre 1 / The Division of Labour in Society*, Book 1, 8th edition, Paris, Les Presses Universitaires de France.

Eccles, T., (2009), The English Building Industry in Late Modernity: An empirical investigation of the definition, construction and meaning of profession, PhD thesis for the London School of Economics and Political Science.

EEC, (1985), *Council Directive of 10 June 1985 on the Mutual Recognition of Diplomas, Certificates and Other Evidence of Formal Qualifications in Architecture, Including Measures to Facilitate the Effective Exercise of the Right of Establishment and Freedom to Provide Services* (85/384/EEC), Luxembourg, Official Journal of the European Communities.

Egan, J., (1998), *Rethinking Construction: The report of the Construction Task Force*, London, HMSO.

Eliot, G., (1991 [1872]), *Middlemarch*, London, Everyman's Library.

Ellison, L., (1999), *Surveying the Glass Ceiling: An investigation of the progress made by women in the surveying profession*, London, RICS.

Engineering Council, (2017a), Annual Report and Financial Statements for year ended 31st December 2016, Engineering Council, available from www.engc.org.uk/EngCDocuments/Internet/Website/Trustees%20Report%20and%20Accounts%20 2016.pdf [Accessed 3rd March 2017].

Engineering Council, (2017b), Professional Registration, available from www.engc.org.uk/professional-registration [Accessed 3rd March 2017].

Engineering Council/Royal Academy of Engineering, (2014), Statement of Ethical Principles for the Engineering Professions, Engineering Council, available from www.engc.org.uk/media/2334/ethical-statement-2017.pdf.

Enterprising Women, (2002) SET Women, available from www.enterprising-women.org/business-support/set-women [Accessed 13th December 2017].

EU, (2005), *Directive 2005/36/EC of the European Parliament and of the Council of 7 September 2005 on the Recognition of Professional Qualifications*, Strasbourg, Official Journal of the European Union.

EU, (2013), *Directive 2013/55/EC of the European Parliament and of the Council of 20 November 2013 on the Recognition of Professional Qualifications and Regulation (EU) No 1024/2012 on Administrative Cooperation through the Internal Market Information System ('the IMI Regulation')*, Strasbourg, Official Journal of the European Union.

Evans, E., (2013), *Thatcher and Thatcherism*, 3rd edition, Abingdon, Routledge.

Eve, H., (1967), *Report of Educational Policy Committee of the RICS (The Eve Report)*, London, RICS.

Evetts, J., (2003), The sociological analysis of professionalism: Occupational change in the modern world, *International Sociology* (June) 18(2), pp733–51.

Evetts, J., (2012a), Professionalism: value and ideology, Sociopedia.isa, DOI: 10.1177/205684601231. Available from www.sagepub.net/isa/resources/pdf/Professionalism.pdf [Accessed 2nd May 2018].

Evetts, J., (2012b), Professionalism in turbulent times: Changes, challenges and opportunities, Paper to Propel International Conference, Stirling, May 2012.

Evetts, J., (2014), The concept of professionalism: Professional work, professional practice and learning, in Billett, S. *et al.* (Eds), *International Handbook of Research in Professional and Practice-based Learning*, Dordrecht, Springer International, pp29–56.

Fairlie, H., (1955), Political commentary, The Spectator, 23rd September.

Fees Bureau, (2016), RIBA Business Benchmarking 2016, Executive Summary available from www.architecture.com/-/media/gathercontent/business-benchmarking/additional-documents/ribabenchmarking2016executivesummarypdf.pdf

Ferguson, H. and Chrimes, M., (2011), *The Civil Engineers: The story of the Institution of Civil Engineers and the people who made it*, London, ICE Publishing.

Ferrey, B., (1861), *Recollections of A. Welby Pugin and his father Augustus Pugin*, London, Edward Stanford.

Fewings, P., (2009), *Ethics for the Built Environment*, Abingdon, Taylor & Francis.

Financial Times, (2013), *Osborne wins the battle on austerity*, Editorial, 10th September.

Fisher, T., (2010), *Ethics for Architects: 50 dilemmas of professional practice*, New York, Princeton Architectural Press.

Fisher, T., (2016), Questions of Ethics – Ethics for Architects: Introductory comments, available from www.aiacc.org/2016/10/26/questions-of-ethics/ [Accessed 18th November 2017].

Foray, D., (2006), *The Economics of Knowledge*, Cambridge, MA, MIT Press.

Forty, A., (1980), The modern hospital in England and France: The social and medical uses of architecture, in King, A. (Ed.), *Buildings and Society: Essays on the social development of the built environment*, London, Routledge & Kegan Paul, pp61–93.

Forum for the Future, (2011), Futurescapes: The scenarios, Forum for the Future, available from www.forumforthefuture.org/sites/default/files/project/downloads/futurescapes-final-scenarios.pdf [Accessed 27th Decembr 2015].

Foster+Partners, (2015), *Corporate, Social & Environmental Responsibility Report: May 2014–April 2015*, London, Foster+Partners Ltd.

Fournier, V., (1999), The appeal to 'professionalism' as a disciplinary mechanism, *Social Review* 47(2), pp280–307.

Fox, W., (Ed.), (2000), *Ethics and the Built Environment*, London, Routledge.

Foxell, S., (Ed.), (2003), *The professionals' choice: the future of the built environment professions*, London, Building Futures. Available from www.buildingfutures.org.uk/assets/downloads/The_Professionals_Choice2003.pdf

Foxell, S., (2014), *A Carbon Primer for the Built Environment*, Abingdon, Routledge.

Foxell, S., (2017), *Professionalism 2.0*, in Duncan, J. (Ed.), *Retropioneers: Architecture redefined*, London, RIBA Publishing, pp36–8.

Foxell, S. and Cooper, I., (2015), Closing the policy gaps, *Building Research & Information* 43(4), pp339–406.

Foxhall, W., (1972), *Professional Construction Management and Project Administration*, New York, Architectural Record.

Franklin, G., (2009), *Inner-London Schools 1918–44: A thematic study*, English Heritage, Research Department Report Series 43–2009.

Frankopan, P., (2015), *The Silk Roads: A new history of the world*, London, Bloomsbury.

Frase, P., (2016), *Four Futures: Life after capitalism*, London, Verso.

Freidson, E., (2001), *Professionalism: The third logic*, Cambridge, Polity Press.

Frey, C. and Osborne, M., (2013), *The Future of Employment: How susceptible are jobs to computerisation?*, Oxford, Oxford Martin School, University of Oxford.

Frisch, T., (1987), *Gothic Art 1140–c1450: Sources and documents*, Toronto, University of Toronto Press.

Froud, D., (2014), Speaker's notes, Edge Commission of Inquiry on the Future of Professionalism, Session 4 – Future Value, 22nd May 2014.

Fulcher, M (2012) UK architects are 94 per cent white, *Architects' Journal*, 26th July.

Galbraith, J.K., (1970 [1958]), *The Affluent Society*, 3rd edition, Harmondsworth, Penguin.

Gardiner, J., (2014), Latham's report: Did it change us?, *Building*, 27th June available from www.building.co.uk/focus/lathams-report-did-it-change-us/5069333.article

Gardner & Theobold, (2017), Culture and Values, Gardner & Theobold, available from www.gardiner.com/about-us [Accessed 3rd February 2017].

Gerber, A., (2015), Independent of bureaucrat? The early career choice of an architect at the turn of the twentieth century in Germany, France and England, *Footprint, Delft Architecture Theory Journal* (Autumn) 9(2), pp47–68.

Gerou, P., (2008), *Ethics and Professional Rules of Conduct: Distinction and clarification*, NCARB Monograph Series 2008 Volume 11, Issue 1, Washington DC, National Council of Architectural Registration Boards.

Gershon, P., (2004), *Releasing Resources to the Front Line: Independent review of public sector efficiency*, London, HM Treasury.

GfK Verein, (2015), Trust in Professions, available from www.gfk-verein.org/en/compact/focustopics/worldwide-ranking-trust-professions

Gibbs, S., (2016), Chatbot lawyer overturns 160,000 parking tickets in London and New York, *The Guardian*, 28th June. Available from www.theguardian.com/technology/2016/jun/28/chatbot-ai-lawyer-donotpay-parking-tickets-london-new-york

Gibson, D. *et al.*, (1952a), Public architecture: The programme described, *The Architects' Journal* (31st January), pp145–6.

Gibson, D. *et al.*, (1952b), The scope of the work (I), *The Architects' Journal* (14th February), pp206–8.

Gibson, D. *et al.*, (1952c), Group working and the large office, *The Architects' Journal* (15th May), pp597–600.

Gibson, D. *et al.*, (1952d), An example of group working, *The Architects' Journal* (26th June), pp785–8.

Gibson, D. *et al.*, (1952e), Building controls and public architecture, *The Architects' Journal* (9th October), pp428–9.

Gilbert, C., (1909), Convention of the American Institute of Architects: Address of Mr. Cass Gilbert, the retiring President, *The American Architect* 96(1774), pp272–4.

Gispen, K., (1989), *New Profession, Old Order: Engineers and German society, 1815–1914*, Cambridge, Cambridge University Press.

Gispen, K., (1996), The long quest for professional identity: German engineers in historical perspective, 1850–1990, in Meiksins, P. and Smith, C., *Engineering Labour: Technical workers in comparative perspective*, London, Verso, p143.

Gleeds, (2017), *Social Value Policy*, Southampton, Gleeds.

Glendinning, M., (2008), *Modern Architect: The life and times of Robert Matthew*, London, RIBA.

Glenigan, CITB *et al.*, (2016), UK Industry Performance Report 2016, available from www.glenigan.com/sites/default/files/UK_Industry_Performance_Report_2016_LR.pdf [Accessed 2nd May 2018].

Global Alliance for Building Sustainability, (2002), *The Charter for Action*, London, RICS Foundation.

Goode, W., (1957), Community within a community: The professions, *American Social Review* 22, pp194–200.

Goodhead, T., (2010), *Technician and Associate Routes to Professional Membership – An opportunity for FIG?*, FIG Commission 2, London, UEL.

Gordimer, N., (1997), In Nigeria, the price for oil is blood, *New York Times*, 25th May.

Gotch, J.A., (Ed.), (1934), *The Growth and Work of the Royal Institute of British Architects 1834–1934*, London, Simson and Co.

GOV.UK, (2017), Whistleblowing for employees, available from www.gov.uk/whistleblowing [Accessed 14th December 2017].

Grant Thornton (2014), Corporate Social Responsibility: Beyond financials, Grant Thornton International Business Report 2014, available from www.grantthornton.global/en/insights/articles/Corporate-social-responsibility/ [Accessed 3rd May 2018].

Gratton, L., (2011), *The Shirt: The future of work is already here*, London, Collins.

Greed, C., (Ed.), (1999), *Social Town Planning*, London, Routledge.

Greed, C., (2000), Women in the construction professions: Achieving critical mass, *Gender, Work & Organization* (July) 7(3), pp181–96.

Griggs, F., (2003), 1852–2002: 150 Years of Civil Engineering in the United States of America, in Russell, J. (Ed.), *Perspectives in Civil Engineering Commemorating the 150th Anniversary of the American Society of Civil Engineers*, Reston, VA, American Society of Civil Engineers, pp111–22.

Gromov, R., (1995–2011), The Roads and Crossroads of the Internet's History, available from www.netvalley.com/intvalweb.html [Accessed 30th December 2016].

Guadet, J., (1885), *Les devoirs professionnels de l'architecte*, *L'Architecture*, 10th August, Paris, La Société Centrale des Architectes.

Guest, D., (1991), The hunt is on for the Renaissance man of computing, *The Independent*, 17th September.

Gunther, R.T., (Ed.), (1928), *The Architecture of Sir Roger Pratt*, Oxford, Oxford University Press.

Gutman, R., (1988), *Architectural Practice: A critical view*, New York, Princeton Architectural Press.

GVA, (2017), Governance and compliance, GVA, available from www.gva.co.uk/about-us/governance-and-compliance/ [Accessed 3rd February 2017].

GVA Grimley, (2017), Our Approach to Sustainability, GVA Grimley, available from www.gva.co.uk/environmental-services/our-approach-to-sustainability/ [Accessed 3rd May 2018].

H.M. Government, (1855), *Metropolis Management Act 1855*, London, George Edward Eyre and William Spottiswood.

Halliday, S., (1999), *The Great Stink of London: Sir Joseph Bazalgette and the cleansing of the Victorian capital*, Stroud, Sutton Publishing.

Handy, C., (1984), *The Future of Work: A guide to a changing society*, London, John Wiley & Sons.

Handy, C., (1994), *The Empty Raincoat*, London, Hutchinson.

Handy, C., (2001), *The Elephant and the Flea*, London, Hutchinson.

Hanlon, G., (1999), *Lawyers, the State and the Market, Professionalism Revisited*, Basingstoke, Palgrave Macmillan.

Hansard, (1858), House of Commons Second Reading: Medical Practitioners' Bill, 02 June 1858, vol 150 cc1406–21.

Hansard, (1967), House of Lords Debate, 30 January 1967, vol 279 cc779–87.

Harford, J., (1840), *Life of Thomas Burgess DD*, London, Longman, Orme, Brown, Green & Longmans.

Harrison, P., (1983), *Inside the Inner City: Life under the cutting edge*, Harmondsworth, Penguin Books.

Hart, J.M., (1874), *German Universities: A narrative of personal experience together with recent statistical information, practical suggestions and a comparison of the German, English and American systems of higher education*, New York, G.P. Putnam's Sons.

Hart P., (2001), *History Notes: A background history of the CSCE*, Montreal, Canadian Civil Engineer.

Harvey, L., (1870), The French mind, *The Builder*, 9th April, p280.

Harwood, E., (2014a), Women Architects, English Heritage, available from https://content.historicengland.org.uk/content/docs/listing/women-architects.pdf

Harwood, E., (2014b), London County Council architects (act. C.1940–1965), *Oxford National Dictionary of Biography*, Oxford, Oxford University Press.

Harwood, E., (2015), *Space, Hope and Brutalism: English architecture 1945–1975*, New Haven, CT, Yale University Press.

Hatfield, D., (2005), The Women's Engineering Society: A little bit of history, Presentation to WES Conference 2005, available from www.wes.org.uk/files/WESHistory.ppt [accessed 12th October 2017].

Haug, M. and Sussman, M., (1971), Professionalization and unionism: A jurisdictional dispute?, Professions in Contemporary Society, special issue of *American Behavioural Scientist* 14, pp525–50.

Hawkes, D. (1978), The architectural partnership of Barry Parker and Raymond Unwin, *Architectural Review* (June) 163, pp327–32.

Hawney, W., (1717), *The Compleat Measurer or The Whole Art of Measuring*, London (2nd edition, 1721). Available from https://babel.hathitrust.org/cgi/pt?id=mdp.39015065314059;view=1up;seq=7

Hayes, J., (2016), The journey to beauty, available from www.gov.uk/government/speeches/the-journey-to-beauty

Hibbert, C., (1988), *George IV: Regent and king 1811–1830*, London, Penguin.

Hill, S. and Lorenz, D., (2011), Rethinking professionalism: Guardianship of land and resources, *Building Research & Information* 39(3), pp314–19.

Hill, S. *et al.*, (2013), Professionalism and ethics in a changing economy, *Building Research & Information* 41(1), pp8–27.

Historic England, (2017), Leverkus, Gertrude, available from https://historicengland.org.uk/research/inclusive-heritage/womens-history/visible-in-stone/biographies/l-z/leverkus-gertrude/ [Accessed: 9th October 2017].

HM Government (1919), *Sex Disqualification (Removal) Act, 1919*, London, King's Printer of Acts of Parliament. Available from www.legislation.gov.uk/ukpga/1919/71/pdfs/ukpga_19190071_en.pdf [Accessed: 7th October 2017].

HM Government, (1975), *Sex Discrimination Act 1975*, London, HMSO, available from www.legislation.gov.uk/ukpga/1975/65/pdfs/ukpga_19750065_en.pdf

HM Government, (1994), *The Construction (Design and Management) Regulations 1994*, London, HMSO.

HM Government, (2006), *Companies Act 2006 (c46)*, London, HMSO.

HM Government, (2006, corrected 2010), *Companies Act 2006*. London, HMSO. Available from www.legislation.gov.uk/ukpga/2006/46/section/172

HM Government Cabinet Office, (2011), Government Construction Strategy, available from assets.publishing.service.gov.uk/government/uploads/system/uploads/attachment_data/file/61152/Government-Construction-Strategy_0.pdf

Hobsbawm, E., (1977a [1962]), *The Age of Revolution 1789–1848*, London, Abacus.

Hobsbawm, E., (1977b [1975]), *The Age of Capital 1848–1875*, London, Abacus.

Hobsbawm, E., (1994 [1987]), *The Age of Empire 1875–1914*, London, Abacus.

Hobsbawm, E., (1995 [1994]), *The Age of Extremes: The short twentieth century 1914–1991*, London, Abacus.

Hobsbawm, E. and Ranger, T., (Eds), (1983), *The Invention of Tradition*, Cambridge, Cambridge University Press.

Holmes, G., (1986), *Politics, Religion and Society in England, 1679–1742*, London, Hambledon Press.

Home Secretary and Secretary of State for Health, (2007), *Learning from Tragedy, Keeping Patients Safe: Overview of the Government's action programme in response to the recommendations of the Shipman Inquiry*, Cm 7014, London, TSO.

House of Commons Official Report, (1937), Parliamentary Debates, available from api.parliament.uk/historic-hansard/commons/1937/dec/17/architects-registration-bill

House of Commons, (1977), Orders of the Day – Property Services Agency, 11th March 1977, available from www.theyworkforyou.com/debates/?id=1977-03-11a.1873.16 [Accessed 4th November 2016].

Howard, A., (2016), Women in Surveying, Royal Institution of Chartered Surveyors, available from www.rics.org/uk/news/rics150/women-in-surveying [Accessed 6th October 2017].

Howard, A., (2017), Celebrating the pioneer spirit of trail blazing women making history in surveying, Womanthology. Available from www.womanthology.co.uk/celebrating-pioneer-spirit-trail-blazing-women-making-history-surveying-annette-howard-information-officer-rics/ [Accessed 12th October 2017].

Howell-Ardila, D., (2010), Writing our own Program: The USC experiment in modern architectural pedagogy, 1930 to 1960, Thesis, Faculty of the USC School of Architecture, University of Southern California.

Hoxley, M., (2001), Purchasing UK public sector property and construction professional services: Competition v quality, *European Journal of Purchasing & Supply Management* (June) 7(2), pp133–9.

HSE, (2015a), The Construction (Design and Management) Regulations 2015, Health and Safety Executive, available from www.hse.gov.uk/construction/cdm/2015/index.htm [Accessed 27th April 2017].

HSE, (2015b), *Managing Health and Safety in Construction*, L153, Liverpool, Health and Safety Executive.

HSE, (2015c), *Regulation 8*, Liverpool, Health and Safety Executive.

Hughes, E., (1994 [1948]), Good people and dirty work. Originally delivered as a lecture at McGill University in 1948. Republished in Coser, L. (Ed.), *On Work, Race and the Sociological Imagination*, Chicago, University of Chicago Press, pp180–91.

Hughes, W., (2003), De-professionalised, automated construction, in Foxell, S. (Ed.), *The professionals' choice: the future of the built environment professions*, London, Building Futures, pp82–97.

Hughes, W. and Hughes C., (2013), Professionalism and professional institutions in times of change, *Building Research & Information* 41(1), pp28–38.

Hyett, P., (2000), *In Practice*, London, Emap Construct.

Ibsen, H., (1961), *The Master Builder*, Meyer, M. (Trans.), London, Rupert Hart-Davis.

ICE, (1828), Charter, in *Transactions of the Institution of Civil Engineers*, Volume 1, 2nd edition, London, John Weale.

ICE, (1842), *Transactions of the Institution of Civil Engineers*, Volume 1, 2nd edition, London, John Weale.

ICE, (1853), Minutes of the Proceedings of the Institution of Civil Engineers, Volume 12, Issue 1853, Institution of Civil Engineers

ICE, (1873), *Proceedings of the Institute of Civil Engineers*, Vol 36, London, The Institution of Civil Engineers.

ICE, (1975), *Royal Charter*, London, The Institution of Civil Engineers.

ICE, (2004), *ICE Code of Professional Conduct*, London, The Institution of Civil Engineers.

ICE, (2013), *ICE Strategy 2013–2025: Civil Engineers: Shaping the World*, London, The Institution of Civil Engineers.

ICE, (2014), *ICE Code Of Professional Conduct*, London, The Institution of Civil Engineers.

ICE, (2015), *ICE Diversity and Inclusivity Action Plan*, London, The Institution of Civil Engineers.

ICE, (2016), *Continuing Professional Development Guidance, M018 Version 1, Revision 1*, London, The Institution of Civil Engineers.

ICE, (2017a), ICE Thinks, available from www.ice.org.uk/ice-thinks [Accessed on 2nd March 2017].

ICE, (2017b), *Civil Engineering through the Ages*, London, Institution of Civil Engineers. Available from www.ice.org.uk/what-is-civil-engineering#historical-engineers [Accessed: 6th October 2017].

IEI, (2000), *An Exhibition of Engineering in the West of Ireland*, Galway, The Institution of Engineers in Ireland. Available from www.realizedvision.com/ap.php [Accessed 6th October 2017].

IESC, (2016), *International Ethics Standards: An ethical framework for the global property market*, International Ethics Standards Coalition.

IET, (2015), The Role of the Professional Engineering Institutions in 2025 (Futures paper), pdf, Institution of Engineering and Technology, available from www.theiet.org/about/people/trustees/op-bot/t15100.cfm [Accessed 28th June 2016].

IET, (2017), Women and engineering: An online exhibition, The Institution of Engineering and Technology, available from www.theiet.org/resources/library/archives/exhibition/women/index.cfm [Accessed: 6th October 2017].

Immigrant Entrepreneurship, (2010–17), German–American business biographies: 1720 to the present, available from www.immigrantentrepreneurship.org [Accessed 3rd November 2017].

Institute for Fiscal Studies, (2017), Institute for Fiscal Studies, available from www.ifs.org.uk

Institute of British Architects, (1837), Royal Charter 1837, available from www.architecture.com [Accessed 22nd July 2016].

Institute of British Architects of London, (1835), Laws and regulations of the Architectural Society, *Transactions of the Institute of British Architects of London (Sessions 1835–36)* 1(1), p.x.

Institution of Engineers, (1920), *Transactions of the Institution of Engineers, Australia*, Vol. 1, Sydney, Institution of Engineers.

Institution of Structural Engineers, (2005), *Supplemental Charter*, London, The Institution of Structural Engineers.

Institution of Structural Engineers, (2014), *Code of Conduct and Guidance Notes*, London, The Institution of Structural Engineers.

Institution of Structural Engineers, (2015), *History of the Institution of Structural Engineers*, London, The Institution of Structural Engineers. Available from www.istructe.org/downloads/about-us/who-we-are/history-of-the-institution-of-structural-engineers [Accessed 4th November 2015].

Institution of Surveyors, (1869), *Royal Charter*, in Transactions Vol. II, Session 1869–70, London, Institution of Surveyors.

IPPR (2001), *Building Better Partnerships: The final report of the Commission on Public Private Partnerships*, London, IPPR.

Ipsos MORI, (2016), Veracity Index 2015: Trust in Professions, available from www.ipsos-mori.com

ISO, (2005), *ISO 9000:2005: Quality management systems – fundamentals and vocabulary*, Geneva, International Organization for Standards.

ISO, (2010a), *ISO 26000, Introduction*, Geneva, International Organization for Standards.

ISO, (2010b), *ISO 26000, Guidance on social responsibility*, Geneva, International Organization for Standards.

Jackson, A., (1970), *The Politics of Architecture*, London, The Architectural Press.

Jackson, B., (Ed.), (1950), *Recollections of Thomas Graham Jackson 1834–1924*, Oxford, Oxford University Press.

Jacobs, J., (1961), *The Death and Life of Great American Cities: The failure of town planning*, New York, Random House.

Jarausch, K., (1990a), The German professions in history and theory, in Cocks, G. and Jarausch, K. (Eds), *German Professions 1800–1950*, Oxford, Oxford University Press, pp9–24.

Jarausch, K., (1990b), *The Unfree Professions: German lawyers, teachers, and engineers, 1900–1950*, Oxford, Oxford University Press.

Jardine, L., (2002), *On a Grander Scale: The outstanding career of Sir Christopher Wren*, London, Harper Collins.

Jenkins, F., (1961), *Architect and Patron*. Oxford, Oxford University Press.

Jennings, J., (2005), *Cheap and Tasteful Dwellings: Design competitions and the convenient interior, 1879–1909*, Knoxville, University of Tennessee Press.

Jephson, H., (1907), *The Sanitary Evolution of London*, London, Fisher Unwin.

JLL, (2016), *Jones Lang LaSalle Incorporated and LaSalle Investment Management Code of Business Ethics: Our code of ethics*, Southampton, Jones Lang LaSalle IP.

Jonson, B., (1631), 'An Expostulacon with Inigo Jones', in *The Works of Ben Jonson, Vol 8: The Poems; The Prose Works*, Herford, C. *et al.* (eds), Oxford, Oxford University Press, p34.

Journal of the RIBA, (1935), *Journal of the RIBA*, 3rd Series, XLII (8th June).

Journal of the Royal Institute of British Architects, (1898), The admission of lady associates, Chronicle, *Journal of the Royal Institute of British Architects* (10th December), pp77–8.

Judt, T., (2005), *Postwar: A history of Europe since 1945*, London, William Heinemann.

Judt, T., (2010), *Ill Fares the Land*, London, Penguin Books.

Kant, I., (1993 [1785]), *Grounding of the Metaphysics of Morals*, Ellington, J. (Trans.), Cambridge, MA, Hackett.

Kaye, B., (1960), *The Development of the Architectural Profession in Britain: A sociological study*. London, George Allen & Unwin.

Keynes, J.M., (1930), *Economic Possibilities for our Grandchildren: Essays in persuasion*, New York, W.W. Norton & Co.

Killick, A.H., (1958), British Surveyors: Their training, activities and emoluments, Delft, Compte Rendu du IXème Congres.

Kitchen, R. and Hill, M., (2007), *The Story of the Original CMK*, Cirencester, Lacey Books.

Kiwana, L., Kumar, A. and Randerson, N., (2011), *An Investigation into Why the UK has the Lowest Proportion of Female Engineers in the EU: A summary of the key issues*, London, EngineeringUK.

Knapton, S., (2016), Robots will take over most jobs within 30 years, experts warn, *The Telegraph*, 13th February. Available from www.telegraph.co.uk/news/science/science-news/12155808/Robots-will-take-over-most-jobs-within-30-years-experts-warn.html [Accessed 27th July 2017].

Krause, E., (1996), *Death of the Guilds: Professions, states, and the advance of capitalism, 1930 to the present*, New Haven, CT, Yale.

Kynaston, D., (2007), *Austerity Britain 1945–48: A world to build*, London, Bloomsbury.

Kynaston, D., (2009), *Family Britain 1951–57*, London, Bloomsbury.

La Vopa, A., (1990), Specialists against specialization: Hellenism as professional ideology in German classical studies, in Cocks, G. and Jarausch, K. (Eds), *German Professions 1800–1950*, Oxford, Oxford University Press, pp27–45.

Lamond, E., (Ed. and Trans.), (1890), *Walter of Henley's Husbandry together with An Anonymous Husbandry, Seneschaucie and Robert Grosseteste's Rules*, New York, Longmans, Green & Co.

Landau, R., (1968), *New Directions in British Architecture*, New York, Brazilier.

Landscape Institute, (2008), Royal Charter, London, Landscape Institute.

Landscape Institute, (2012), Code of Standards of Conduct and Practice, London, Landscape Institute.

Lang, R., (2014), Architects take command: The LCC architects department, in *How to Build a Nation, Volume #41*, Amsterdam, Archis, pp32–9.

Latham, M., (1994), *Constructing the Team: Final report of the government/industry review of procurement and contractual arrangements in the UK construction industry*, London, HMSO.

Latrobe, B.J., (1984), Correspondence and Miscellaneous Papers of Benjamin Henry Latrobe, Vols 1 and 2, van Horne, J. and Formwalt, Lee (Eds), New Haven, CT, Yale University Press.

Lawrence, E., (1934), President's Address, Twenty-first Annual Meeting of the Association of Collegiate Schools of Architecture, 13th–14th May, Washington, DC.

Laxton, W., (1840), College for civil engineers, *The Civil Engineer and Architect's Journal, Scientific and Railway Gazette*, 3 (February), pp57–9.

Layton, E., (1971), *The Revolt of the Engineers: Social responsibility and the American engineering profession*, Cleveland, OH, The Press of Case Western Reserve University.

Leaman, A., Stevenson, F. and Bordass, B., (2010), Building evaluation: practice and principles, *Building Research & Information* 38(5), pp564–77.

Le Seur, T., Jacquier, F. and Boscovitch, R., (1743), *Parere di tre Matematici sopra i danni che si sono trovati nella Cupola di S. Pietro* (On the opinion of three mathematicians concerning the damage to the dome of St Peter's), Rome.

Leick, R., Schreiber, M. and Stoldt, H-U., (2010), Out of the ashes: A new look at Germany's postwar reconstruction, *Der Speigel*. Available from www.spiegel.de/international/germany/out-of-the-ashes-a-new-look-at-germany-s-postwar-reconstruction-a-702856-2.html [Accessed 28th June 2017].

Leske, N., (2013), Doctors seek help on cancer treatment from IBM supercomputer, Reuters, available from http://in.reuters.com/article/ibm-watson-cancer-idINDEE9170G120130208

Linnean Society of London, (1848), *Charter and Bye-laws of the Linnean Society of London*, London, Richard and John E. Taylor.

Littman, W., (1995), Battling the Beaux-Arts: The student campaign for modernism at the University of California, 83rd ACSA Annual Meeting, History/Theory/Criticism.

Llewelyn-Davies, R., (1958), Deeper knowledge: Better design, *Architects' Journal*, (23rd May).

Llewelyn-Davies, R., (1960), Education of an architect, *Architects' Journal* (17th December).

Lloyd, B., (2008), *Engineering in Australia: A professional ethos*, Melbourne, Histec Publications.

Loudon, J.C., (Ed.), (1834), Domestic notices: – England, *The Architectural Magazine and Journal of improvement in architecture, building and furnishing, and in the various arts and trades connected therewith*, Volume 1, London, Longman, Rees, Orme, Green & Longman.

Lukes, S. and Urbinati, N., (2012), *Condorcet: Political writings*, Cambridge, Cambridge University Press.

Lupton, D., (1632), *London and the Countrey Carbonadoed and Quartred into Severall Characters*, London.

MacCormac, R., (2005), Architecture, art and accountability, in Ray, N. (Ed.), *Architecture and its Ethical Dilemmas*, London, Routledge, pp49–54.

Macdonald, J. and Mills, J., (2011), The potential of BIM to facilitate collaborative AEC education, Proceedings of the 118th ASEE Annual Conference, 2011.

Macdonald, K., (1995), *The Sociology of the Professions*, London, Sage Publications.

Mace, (2017), Our Commitments, Mace Group, available from www.macegroup.com/about-us/our-commitments [Accessed 3rd February 2017].

Mace, A., (1986), *The Royal Institute of British Architects: A guide to its archive and history*. London, Mansell.

Mackay, T., (1900), *The Life of Sir John Fowler, Engineer, Bart., K.C.M.G, Etc.*, London, John Murray.

Mainstone, R., (1975), *Developments in Structural Form*, London, Allen Lane.

Malatesta, M., (2011), *Professional Men, Professional Women: The European professions from the nineteenth century to today*, Belton, A. (Trans.), Sage Studies in International Sociology 58. London, Sage Publications.

Malik, C., (2014), Speaker's notes, Edge Commission of Inquiry on the Future of Professionalism, Session 4 – Future Value, 22nd May 2014.

Malik, K., (2014), *The Quest for a Moral Compass*, London, Atlantic Books.

Malleson, A., (2016), BIM Survey: Summary of findings, *National BIM Report 2016*, NBS, London, RIBA Enterprises, pp28–42.

Mallgrave, H.F., (1996) *Gottfried Semper: Architect of the nineteenth century*, New Haven, CT, Yale University Press.

Mance, H., (2016), Britain has had enough of experts, says Gove, *Financial Times*, 3rd June.

Mandey, V., (1682), *Mellificium mensionis or The Marrow of Measuring*, London, (4th edition, 1727). Available from https://babel.hathitrust.org/cgi/pt?id=gri.ark:/13960/t77t2vc82;view=1up;seq=9

Manley, S., de Graft-Johnson, A. and Greed, C., (2003), *Why Do Women Leave Architecture?*, Technical Report, London, RIBA.

Manley, S., de Graft-Johnson, A. and Lucking K., (2011), Disabled architects: unlocking the potential for practice, Project Report, University of West of England / Royal Institute of British Architects, available from http://eprints.uwe.ac.uk/16961/

Martin, L., (1958), RIBA Conference on Architectural Education: Report by the Chairman, Sir Leslie Martin.

Mason, P., (2016), *Postcapitalism: A guide to our future*, London, Penguin Books.

Mawer, P., (2010), Public Trust in the Professions, Keynote address to International Actuarial Association Conference, 25th February 2010. Available from www.actuaries.org.uk/documents/ethics-trust-and-integrity [Accessed 19th April 2017].

McArthur Butler, C., (1931), The registration of architects in Great Britain, Journal of the Royal Institute of British Architects (1st August), pp687–9.

McBride, E., (2009), The Development of Architectural Office Specialization as Evidenced by Professional Journals, 1890–1920, Dissertation, Paper 454, Washington University in St. Louis.

McKean, C., (1994), *Value or Cost: Scottish architectural practice in the 1990s, RIAS insurance services/Royal Incorporation of Architects in Scotland (RIAS)*, London, Rutland Press.

McWilliams, A. and Siegel, D., (2001), Corporate Social Responsibility: A theory of the firm perspective, *Academy of Management Review* 26, pp117–27.

Meiksins, P. and Smith, C., (1996), *Engineering Labour: Technical workers in comparative perspective*, London, Verso.

Mellors-Bourne, R. *et al.*, (2017), *Engineering UK 2017: The state of engineering*, London, EngineeringUK.

Merritt, R., (1969), *Engineering in American Society: 1850–1875*, Lexington, University Press of Kentucky.

Mill, J.S., (1867), *Inaugural Address Delivered to the University of St. Andrews, Feb.1st 1867*, London, Longmans, Green, Reader, and Dyer.

Millar, J., (2010), The first woman architect, *The Architects' Journal* (11th November), available from www.architectsjournal.co.uk/the-first-woman-architect/8608009.article [Accessed 7th May 2018].

Millerson, G., (1964), *The Qualifying Associations: A study in professionalization*, Abingdon, Routledge.

Ministry of Housing and Local Government, (1961), *Homes for Today and Tomorrow (The Parker–Morris Report)*, London, HMSO.

Ministry of Housing, (1963), *Design Bulletin 6 – Space in the Home*, London, HMSO.

Mitcham, C. and Nan, W. (2015), From engineering ethics to engineering politics, in Christensen, S. *et al.* (Eds), *Engineering Identities, Epistemologies and Values: Engineering, education and practice in context*, Volume 2, Berlin, Springer, pp307–24.

Monopolies and Mergers Commission (MMC), (1977), *Surveyors' Services: A report on the supply of surveyors' services with reference to scale fees*, London, HMSO.

Monopolies Commission, (1970), *A Report on the General Effect on the Public Interest of Certain Restrictive Practices so far as they Prevail in Relation to the Supply of Professional Services*, Cmnd 4463, London, HMSO.

Mordaunt Crook, J., (1972), *The Greek Revival*, London, John Murray.

Morrell, P., (2015), *Collaboration for Change: The Edge Commission Report on the Future of Professionalism*, available from www.edgedebate.com/wp-content/uploads/2015/05/150415_collaborationforchange_book.pdf

Morrill Act, (1862), Chap. CXXX, AN ACT Donating Public Lands to the several States and Territories which may provide Colleges for the Benefit of Agriculture and Mechanic Arts, available from www.ourdocuments.gov/print_friendly.php?flash=true&page=transcript&doc=33&title=Transcript+of+Morrill+Act+(1862) [Accessed 3rd November 2017].

Mott MacDonald, (2016), Mott MacDonald Group Policy Statement: Ethics, Mott MacDonald Group, available from www.mottmac.com/our-policies [Accessed 3rd February 2017].

Mudie, R,. (1841), *The Surveyor, Engineer and Architect for the Year 1841*, London, Wm. S Orr and Co.

Mulgan, G., (2005), *Public Value: Physical capital and the potential of value maps in physical capital: How great places boost public value*, London, CABE.

Mulgan, G., (2009), *The Art of Public Strategy: Mobilizing power and knowledge for the common good*, Oxford, Oxford University Press.

Mumford, L., (1955), *Sticks & Stones: A study of American architecture and civilization*, 2nd edition. New York, Dover Publications.

Mundy, J., (1991 [1973]), *Europe in the High Middle Ages: 1150–1309*, 2nd edition, London, Longman.

NAAB, (2014), *2014 Conditions for Accreditation*, Washington, DC, The National Architectural Accrediting Board.

Najemy, J., (2006), *A History of Florence 1200–1575*, Oxford, Blackwell.

NCARB, (2014), *Rules of Conduct*, Washington, DC, National Council of Architectural Registration Boards.

NCARB, (2016), *Architectural Registration Examination 5.0 Guidelines*, Washington, DC, National Council of Architectural Registration Boards.

NCARB, (2017), *Architectural Experience Program Guidelines*, Washington, DC, National Council of Architectural Registration Boards.

NCEES (2015a), *Model Rules*, Seneca, SC, National Council of Examiners for Engineering and Surveying.

NCEES, (2015b), *The History of NCEES: 1920–2015*, Seneca, SC, National Council of Examiners for Engineering and Surveying.

Neve, R. (T.N. Philomath), (1703), *The City and Countrey Purchaser & Builder's Dictionary*, London, J. Sprint, G. Conyers & T. Ballard.

Newton, R. and Ormerod, M., (2005), Do disabled people have a place in the UK construction industry?, *Construction Management and Economics* (December) 23, pp1071–81.

North, J., (2010), *Building a Name: The history of the Royal Institution of Chartered Surveyors*, Bury St Edmunds, Arima Publishing.

O'Neill, O., (2002a), *A Question of Trust, The BBC Reith Lectures 2002*, Cambridge, Cambridge University Press.

O'Neill, O., (2002b), A Question of Trust, Lecture 3: Called to Account, BBC Radio 4, April 2002.

O'Neill, O., (2002c), A Question of Trust, Lecture 5: Licence to Deceive, BBC Radio 4, April 2002.

O'Neill, O., (2005), Accountability, trust and professional practice: The end of professionalism?, in Ray, N. (Ed.), *Architecture and its Ethical Dilemmas*, London, Routledge, pp77–88.

Oldfield, P., (2015), UK scraps zero carbon homes plan, *The Guardian*, 10th July. Available from www.theguardian.com/environment/2015/jul/10/uk-scraps-zero-carbon-home-target

ONS, (2015), *International Trade in Services, UK: 2015*, London, Office for National Statistics. Available from www.ons.gov.uk

ONS, (2016), *Output and Labour Force Statistics*, London, Office for National Statistics. Available from www.ons.gov.uk

ONS, (2017a), *Overview of the UK Population: July 2017*, London, Office for National Statistic. Aavailable from www.ons.gov.uk/peoplepopulationandcommunity/populationandmigration/populationestimates/articles/overviewoftheukpopulation/july2017#the-european-context [Accessed 27th July 2017].

ONS, (2017b), *Construction Output in Great Britain: Feb 2017*, London, Office for National Statistics.

Order of the Engineer, (2017), Obligation, available from www.order-of-the-engineer.org/?page_id=6 [Accessed 5th November 2017].

Osborne, G., (2009), Speech to the Conservative Party Conference, 6th October 2009.

Osubor, O., (2017), Headship of professional bodies in the built environment: The challenges facing the female gender, *IOSR Journal of Engineering (IOSRJEN)* 7(3) (March), pp81–7.

Owen, D., (1982), *The Government of Victorian London, 1855–1889: The Metropolitan Board of Works, the vestries and the City Corporation*, Cambridge, MA, Harvard University Press.

Palmer, J., (1972), Introduction to the British Edition, in Goodman, R. (1972), *After the Planners*, Harmondsworth, Penguin, pp9–56.

Park, K., (1985), *Doctors and Medicine in Early Renaissance Florence*, Princeton, NJ, Princeton University Press.

Parliamentary Papers (1800), Enclosing the Land, Report from the Select Committee of the House of Commons on the expense of and Mode of obtaining Bills of Enclosure, London, available from www.parliament.uk [Accessed 27 May 2016].

Parliamentary Papers, (1828), Report from the Select Committee on the Office of Works and Public Buildings, Sess. 1828, (446).

Parliamentary Papers, (1836), Report of the Commissioners appointed to consider the Plans for Building the Houses of Parliament, (1836) XXXVL, (488).

Parsons, T., (1939), The professions and social structure, *Social Forces* 17, pp457–67.

Parsons, T., (1951), *The Social System*, London, Collier-Macmillan.

Pawley, M., (1984), Architects in crisis, *Marxism Today* (May), pp39–40.

pba, (2017), Anti-Slavery and Human Trafficking Statement, Peter Brett Associates LLP, available from www.peterbrett.com/anti-slavery-and-human-trafficking-statement [Accessed 3rd February 2017].

Penrose, F., (1985), Session 1895–96: The opening address, *Journal of The Royal Institute of British Architects*, pp1–14.

Pepys, S., (1669), *Diary of Samuel Pepys*, Everyman Edition Volume 2 (1906), London, J.M. Dent.

Percy, E., (1945), *The Percy Report (1945), Higher technological education: Report of a Special Committee appointed in April 1944*, London, HMSO.

Perkin, H., (1989), *The Rise of Professional Society: England since 1880*, London, Routledge.

Peterson, M.J., (1978), *The Medical Profession in Mid-Victorian London*, Berkeley, University of California Press.

Pevsner, N., (1942), The term architect in the Middle Ages, *Speculum* 17, pp549–62.

Pevsner, N., (1960), *Pioneers of Modern Design*, Harmondsworth, Penguin Books.

Piketty, T., (2014), *Capital in the Twenty-First Century*, Goldhammer, A. (Trans.), Cambridge, MA, Harvard University Press.

Plimmer, F., (2003), Education for Surveyors: An RICS Perspective, Paper to 2nd FIG Regional Conference, Marrakech.

Plimmer, F. and Sayce, S., (2003), Ethics and Professional Standards for Surveyors towards a Global Standard?, Paper to FIG Working Week 2003, Paris.

Polanyi, K., (1957 [1944]), *The Great Transformation: The political and economic origins of our time*, Boston, MA, Beacon Press.

Polanyi, M., (1962), Tacit knowing: its bearing on some problems of philosophy, *Reviews of Modern Physics* 34(4) (October), pp601–16.

Polanyi, M., (1966), *The Tacit Dimension*, Chicago, The University of Chicago Press.

Pontremoli, E., (1959), *Propos d'un solitaire*, Vanves, Imprimerie Kapp.

Porter, D. and Clifton, G., (1988), Patronage, professional values, and Victorian public works: Engineering and contracting the Thames embankment, *Victorian Studies* 31(3), pp319–49.

Porter, D. and Porter R., (1989), *Patient's Progress: Doctors and doctoring in eighteenth-century England*, Redwood City, CA, Stanford University Press.

Porter, M., (2008), The five competitive forces that shape strategy, *Harvard Business Review* (January), pp25–40.

Porter, R., (1997), *The Greatest Benefit to Mankind*, London, Harper Collins.

Pound, R., (1944), What is a profession? The rise of the legal profession in antiquity, *Notre Dame Lawyer* 19(3) (March), pp203–28.

Powell, J.M., (2008), *The New Competitiveness in Design & Construction*, Oxford, Wiley.

Powers, A., (1979), Goodhart-Rendel: The appropriateness of style, *Architectural Design* 49(10–11), pp41–54.

Prasuhn, A. and FitzSimons, N., (2003), ASCE history and heritage programs, in Russell, J. (Ed.), *Perspectives in Civil Engineering Commemorating the 150th Anniversary of the American Society of Civil Engineers*, Reston, VA, American Society of Civil Engineers, pp123–29.

Pringle, M., (2000), Editorial, *British Journal of General Practice*, Royal College of General Practitioners, May, pp355–6.

Prior, E. Shaw, R.N., Sedding, J., Micklethwaite, J., Jackson, T., Macartney, M., Blomfield, A. and Blomfield R., (1899), Letter to the RIBA, reprinted in *RIBA Proceedings*, New series VII, pp220–1.

Privy Council, (2016), Royal Charters, available from https://privycouncil.independent.gov. uk/royal-charters/ [Accessed 12th July 2016].

Privy Council, (2017), Chartered bodies, available from https://privycouncil.independent.gov. uk/royal-charters/chartered-bodies [Accessed 30th June 2017].

PwC, (2017), *UK Economic Outlook March 2017*, Southampton, PricewaterhouseCoopers LLP.

Pye, M., (2014), *The Edge of the World: How the North Sea made us who we are*, London, Viking.

RAIC, (2016), Architecture Canada: History, available from www.raic.org/raic/history [Accessed 2nd August 2016].

Ramboll, (2012), *Code of Practice: Our code of conduct in practical terms*, Copenhagen, Ramboll Group A/S.

Rao, N., (2005), *Composition of the Register, Report to the Board, 31st October 2005*, London, Architects Registration Board.

Ratcliffe, J., (2011), *Just Imagine!, RICS Strategic Foresight 2030*, London, Royal Institution of Chartered Surveyors.

Read, J. (2014), Cartographers John Walker and his son, John Walker, available from www. essexlifemag.co.uk/home/cartographers-john-walker-and-his-son-john-walker-1-3754070

Reilly, C.H., (1905), *The Training of Architects*, London, Sherratt & Hughes.

Rennie, J., (1875), *The Autobiography of Sir John Rennie FRS, Past President of the Institution of Civil Engineers: Comprising the history of his professional life together with reminiscences dating from the commencement of the century to this present time*, London, E & FN Spon.

Rensselaer, (2011), Rensselaer, available from https://rpi.edu/

Rhodes, C., (2015), Construction industry: statistics and policy, Briefing paper 01432, 6th October 2015, House of Commons Library.

Rhys Jones, T., (2014), Obituary: Bryan Jefferson (1928–2014), *The Architects' Journal* (28th October). Available from www.architectsjournal.co.uk/news/obituary-bryan-jefferson-1928-2014/8671657.article [Accessed 7th May 2018].

Riach, J., (2014), Zaha Hadid defends Qatar World Cup role following migrant worker deaths, *The Guardian*, 25th February.

RIBA, (1863), *Regulations and Course of Examination, with forms of Declaration and Recommendation, for the Voluntary Architectural Examination*, London, Royal Institute of British Architects.

RIBA, (1890), *RIBA Proceedings*, New series VI, London, Royal Institute of British Architects.

RIBA, (1925), International Congress on Architectural Education, July 28–Aug. 2 1924, Proceedings.

RIBA, (1962), *The Architect and his Office*, London, Royal Institute of British Architects.

RIBA, (1971), *Supplemental Charter 1971*, London, Royal Institute of British Architects.

RIBA, (1976, 1981, 1997), *Codes of Conduct*, London, Royal Institute of British Architects.

RIBA, (1992), *Strategic Study of the Profession*, London, Royal Institute of British Architects.

RIBA, (2004), *RIBA Employment Policy*, London, Royal Institute of British Architects.

RIBA, (2005a), *Code of Professional Conduct*, London, Royal Institute of British Architects.

RIBA, (2005b), *RIBA Constructive Change: A strategic industry study into the future of the architects' profession*, London, Royal Institute of British Architects.

RIBA, (2012a), *RIBA Chartered Practice Accreditation Criteria and Standards: Standards and enforcement procedures*, London, Royal Institute of British Architects.

RIBA, (2012b), *RIBA Chartered Practice Membership and Student Employment*, London, Royal Institute of British Architects.

RIBA, (2015a), *Trustees' Report and Financial Statements*, London, Royal Institute of British Architects.

RIBA, (2015b), *Post Occupancy Evaluation and Building Performance Evaluation Primer*, London, Royal Institute of British Architects. Available from www.architecture.com/ knowledge-and-resources/resources-landing-page/post-occupancy-evaluation.

RIBA, (2017), Corporate statement, available from www.architecture.com [Accessed 3rd March 2017].

Richards, J.M., (1940), *An Introduction to Modern Architecture*, Harmondsworth, Penguin.

Richards, J.M., (1962), *An Introduction to Modern Architecture*, revised edition, Harmondsworth, Penguin.

Richmond, P., (2001), *Marketing Modernisms: The architecture and influence of Charles Reilly*, Liverpool, Liverpool University Press.

RICS, (1974), *Supplemental Charter of the Royal Institution of Chartered Surveyors*, London, Royal Institution of Chartered Surveyors.

RICS, (2007a), *Regulation: The global professional and ethic standards*, London, Royal Institution of Chartered Surveyors. Available from www.rics.org/Global/The%20 Global%20Professional%20and%20Ethical%20Standards%20UK.pdf [Accessed 30th October 2006].

RICS, (2007b), *Rules of Conduct for Members*, London, Royal Institution of Chartered Surveyors.

RICS, (2008), *Policy and Guidance on University Partnerships*, London, Royal Institution of Chartered Surveyors.

RICS, (2010), *Practice Management Guidelines: The management of surveying businesses*, 3rd edition, RICS Practice standards, London, Royal Institution of Chartered Surveyors.

RICS, (2011), *Rules of Conduct for Members*, Version 4, London, Royal Institution of Chartered Surveyors.

RICS, (2012a), *RICS Valuation Professional Standards: Incorporating the International Valuation Standards*, Global and UK edition (the Red Book), London, Royal Institution of Chartered Surveyors.

RICS, (2012b), *Global Professional and Ethical Standards*, London, Royal Institution of Chartered Surveyors.

RICS, (2013), *Rules of Conduct for Members*, Version 6, London, Royal Institution of Chartered Surveyors.

RICS, (2014), *UK Membership Infographic*, London, Royal Institution of Chartered Surveyors.

RICS, (2015a), *Diversity and Inclusion*, London, Royal Institution of Chartered Surveyors, available from https://rics.turtl.co/story/552836456c80ce09009ab72b/ [Accessed 10th October 2017].

RICS, (2015b), *Fellowship Applicant Guide*, London, Royal Institution of Chartered Surveyors.

RICS, (2015c), *Our Changing World: Let's be ready*, London, Royal Institution of Chartered Surveyors.

RICS, (2016a), *CPD FAQ 2016*, London, Royal Institution of Chartered Surveyors.

RICS, (2016b), *Disciplinary Panel Hearing, 17th August 2016*, London, Royal Institution of Chartered Surveyors, available from www.rics.org/Global/2016 08 17 Sweett Group.pdf

RICS, (2016c), *Diversity in Practice: The 6 principles*, London, Royal Institution of Chartered Surveyors, available from www.rics.org/uk/about-rics/responsible-business/inclusive-employer/the-six-principles

RICS, (2016d), *Building Inclusivity: Laying the foundations for the future*, London, Royal Institution of Chartered Surveyors and Ernst & Young LLP.

RICS, (2017a), *Assessment of Professional Competence Quantity Surveying and Construction*, London, Royal Institution of Chartered Surveyors.

RICS, (2017b), *Assessment of Professional Competence Candidate Guide*, London, Royal Institution of Chartered Surveyors.

RICS (2017c), *Professional Statements*, London, Royal Institution of Chartered Surveyors, available from www.rics.org/uk/knowledge/professional-guidance/professional-statements [Accessed on 25th October 2017].

RICS, (2017d) *Irene Barclay FRICS: Social housing pioneer*, London, Royal Institution of Chartered Surveyors, available from www.rics.org/uk/news/rics150/irene-barclay-social-housing-pioneer/ [Accessed: 6th October 2017].

RICS, (2017e), *5 Benefits of RICS Status*, London, Royal Institution of Chartered Surveyors, available from www.rics.org/uk/the-profession/benefits-of-membership [Accessed 3rd March 2017].

Rifkin, J., (1995), *The End of Work: The decline of the global labor force and the dawn of the post-market era*, New York, Putnam.

Rifkin, J., (2011), *The Third Industrial Revolution: How lateral power is transforming energy, the economy and the world*, Palgrave Macmillan, New York.

Roaf, S. and Bairstow, A., (Eds), (2008), *The Oxford Conference: A re-evaluation of education in architecture*, Southampton, WIT Press.

Robbins, C., (1963), *The Robbins Report: Committee on Higher Education: Report of the Committee appointed by the Prime Minister under the Chairmanship of Lord Robbins 1961–63*. London, HMSO.

Roberts, M.J.D., (2009), The politics of professionalization: MPs, medical men, and the 1858 Medical Act, *Medical History* (January) 53(1), pp37–56.

Robinson, D., *et al.*, (2011), *The Future for Architects?*, Building Futures. Available from www.buildingfutures.org.uk/assets/downloads/The_Future_for_Architects_Full_Report_2.pdf

Robson, R., (1959), *The Attorney in Eighteenth Century England*, Cambridge, Cambridge University Press.

Rodriguez Tomé, D., (2006), L'organisation des architects sous la IIIe République, *Le Mouvement Social* 214, pp55–76.

Rogers, J., (1883), Statement as to the origin, arrangement, and results of the system of professional examinations, *Transactions of the Institution*, XV (30th April), The Surveyors Institution.

Rolt, L., (1970), *Victorian Engineering*. Harmondsworth, Penguin Books.

Romba, K., (2006), Aesthetics and the Professional Identity of the Modern German Engineer, Proceedings of the Second International Congress on Construction History, Cambridge University, Vol 3, Construction History Society.

Root, J., (1994), Thomas Baldwin: His public career in Bath, 1775–1779, in Fawcett, T. (Ed.), *Bath History: Volume V*, Bath, Millstream Books, pp80–103.

Roscoe Pound, N., (1944), What is a profession? The rise of the legal profession in antiquity, *Notre Dame Lawyer* 19(3), pp203–28.

Rosenberg, C., (1962), *The Cholera Years*, Chicago, University of Chicago Press.

Rotblat, J., (1995) Nobel lecture: Remember your humanity, available from www.nobelprize.org/nobel_prizes/peace/laureates/1995/rotblat-lecture.html

Rothery, H., (1880), *Tay Bridge Disaster: Report of the Court of Inquiry and report of Mr. Rothery, upon the circumstances attending the fall of a portion of the Tay Bridge on the 28th December 1879*, London, HMSO.

Rowlands, A., (2001), The conditions of life for the masses, in Cameron, E. (Ed.), *Early Modern Europe: An Oxford history*, Oxford, Oxford University Press, pp31–62.

Royal Academy of Engineering, (2013), *Engineering Diversity Concordat: Resource guide (updated 27th July 2016)*, London, Royal Academy of Engineering. Available from www.raeng.org.uk/publications/other/concordat-resource-guide [Accessed 12th October 2017].

Royal College of Surgeons of England, (1900), *Souvenir of the Centenary of the Royal College of Surgeons of England*, London, Ballantyne, Hanson & Co.

Royal Commission, (1863), *Report of the Commissioners Appointed to Inquire into the Present Position of the Royal Academy in Relation to the Fine Arts, together with the Minutes of Evidence*, London, Eyre & Spottiswod.

RSL, (1823), Constitution and Regulations of the Royal Society of Literature, Volumes 2–3. Available online.

RTPI, (2015a), *RTPI 2020: Corporate Strategy 2015–2020*, London, Royal Town Planning Institute.

RTPI, (2015b), *The Royal Town Planning Institute Code of Professional Conduct*, London, Royal Town Planning Institute.

Ruskin, J., (1865), An Inquiry into some of the Conditions at present affecting the Study of Architecture in our Schools, A paper read by Ruskin at the ordinary meeting of the Royal Institute of British Architects on May 15th 1865, in Cook, E. and Wedderburn, A. (Eds), *The Works of John Ruskin*, London, George Allen, pp15–40.

Rykwert, J., (2014), RIBA Royal Gold Medal Lecture, available from www.youtube.com/watch?v=Jgflqy2VxAM

Ryle, G., (1949), *The Concept of Mind*, Chicago, University of Chicago Press.

Sadri, H., (2017), Profession vs ethics, *Contemporary Urban Affairs* 1(2), pp76–82.

Sailer, K. *et al.*, (2007), Changing the architectural profession – evidence-based design, the new role of the user and a process-based approach, Paper submission, Conference 'Ethics and the Professional Culture', UCL, available from http://discovery.ucl.ac.uk/4828/1/4828.pdf

Saint, A., (1976), *Richard Norman Shaw*. New Haven, CT, Yale University Press.

Saint, A., (1983), *The Image of the Architect*, New Haven, CT, Yale University Press.

Saint, A., (1987), *Towards a Social Architecture: The role of school building in post-war England*, New Haven, CT, Yale University Press.

Sampson, A., (1962), *Anatomy of Britain*, London, Hodder and Stoughton.

Samuel, F., (2018), *Why Architects Matter: Evidencing and communicating the value of architects*, Abingdon, Routledge.

Samuelson, B. (1884), *Report of the Royal Commission on Technical Instruction (The Samuelson Report)*, London, HMSO.

Sandel, M., (2012), *What Money Can't Buy: The moral limits of markets*, London, Penguin.

Sarfatti Larson, M., (2013 [1977]), *The Rise of Professionalism: Monopolies of competence and sheltered markets*, New Brunswick, NJ, Transaction Publishers.

Scheeler, J. and Smith, A., (2009), *Building the Culture of the Architects' Professional Society – 150 years of the American Institute of Architects*, AIA Issues Database, New York, American Institute of Architects.

Schmitter, P., (1974), Still the century of corporatism?, *Review of Politics* 36(1), pp85–131.

Schön, D. (1983), *The Reflective Practitioner: How professionals think in action*, London, Temple Smith.

Schultz, C., (2010), The first modern systematic inquiry into the Institution's ..., available from https://chaseschultz.wordpress.com/2010/07/17/the-first-modern-systematic-inquiry-into-the-institutions/ [Accessed 1st October 2017].

Scott, W.R., (1995), *Institutions and Organizations*, Thousand Oaks, CA, Sage Publications.

Scottish Executive, (1999), *The Development of a Policy on Architecture for Scotland*, Edinburgh, Scottish Executive.

Scruton, R., (2009), The modern cult of ugliness: With desolate city centres and sordid 'art' like Tracey Emin's, *Daily Mail*, 2nd December.

Sennett, R. (1998), *The Corrosion of Character: Personal consequences of work in the new capitalism*, New York, W.W. Norton & Co.

Sennett, R., (2012), *Together: The rituals, pleasures and politics of cooperation*, London, Penguin.

Shaw, G.B., (1906), *The Doctor's Dilemma: Preface on doctors*, London, Constable.

Shaw, R.N. and Jackson, T.G. (Eds) (1892), *Architecture: A profession or an art: Thirteen short essays on the qualifications and training of architects*, London, John Murray.

Silim, A. and Crosse, C., (2014), *Women in Engineering: Fixing the talent pipeline*, London, Institute for Public Policy Research.

Simon, M., (1996), The Beaux-Arts Atelier in America, 84th ACSA Annual Meeting – Practice.

Simpson, F., (1895), *The Scheme of Architectural Education*, Liverpool, Marples.

Simpson, F., (1896), Architectural education. A school of architecture: a paper read before the Birmingham Architectural Association on Friday, 21st February, 1896, *The Builders' Journal* (3rd March).

Sinha, S., (2002), Diversity in architectural education: Teaching and learning in the context of diversity, available from http://women-in-architecture.com/fileadmin/wia/pdfs/WIA24092010/DIVERSITY_IN_ARCHITECTURAL_EDUCATION.pdf [Accessed: 6th October 2017].

Skaife, T., (1774), *A Key to Civil Architecture or the Universal British Builder*, London, I. Moore & Co.

Smith, A., (1776), *An Inquiry into the Nature and Causes of the Wealth of Nations*, Book 1, (1970 edition), Books I–III, London, Penguin Books.

Smith, D., (2001), *Civil Engineering Heritage: London and the Thames Valley*, London, Thomas Telford.

Smith, J., (2002), *The Shipman Inquiry: First report; death disguised*, Norwich, TSO.

Smith, J., (2003), *The Shipman Inquiry: Second report; the police investigation of March 1998*, Norwich, TSO.

Smith, R., (2004), The GMC: Expediency before principle, *British Medical Journal* 330, 253.

Smollett, T., (1753), *The Adventures of Ferdinand Count Fathom*, London, W. Johnston.

SNC-Lavalin, (2017), SNC-Lavalin completes transformative acquisition of WS Atkins, 3rd July 2017, Montreal, available from www.snclavalin.com/en/media/press-releases/2017/snc-lavalin-completes-transformative-acquisition-ws-atkins.aspx [Accessed 8th December 2107].

Soane, J., (1835), *Memoirs of the Professional Life of an Architect, between the years 1768 and 1835*, London, privately printed.

Soane, J., (1788), *Plans Elevations and Sections of Buildings*, (1971 reprint), London, Gregg.

Spector, T., (2001), *The Ethical Architect: The dilemma of ethical practice*, New York, Princeton Architectural Press.

Spencer, A., (Ed.), (1913–25), *Memoirs of William Hickey, Vol 1 (1749–1775)*, London, Hurst & Blackett.

Staley, J.E., (1905), *The Guilds of Florence*, London, Methuen & Co.

Startup, H.M., (1984), Institutional Control of Architectural Education and Registration 1834–1960, MPhil Thesis at the University of Greenwich.

Steffen, W. *et al.*, (2015), Planetary boundaries: Guiding human development on a changing planet, *Science* 347(6223), 1259855.

Stephen, L., (Ed.), (1888), *The Dictionary of National Biography, 1885–1900*, Vol 13 Craik–Damer. London, Smith, Elder & Co.

Stewart, W., (1989), *Higher Education in Postwar Britain*, London, Macmillan.

Sturge, W., (1868), The Education of a Surveyor, Transactions Volumes 1–2 (1869). London, Institution of Surveyors, pp60–61.

Sullivan, L., (1922), *The Autobiography of an Idea*, New York, Press of the American Institute of Architects.

Summerson, J., (1942), Bread & butter and architecture, *Horizon* (October), pp233–43.

Summerson, J., (1993 [1953]), *Architecture in Britain 1530 to 1610*, 9th edition, New Haven, CT, Yale University Press.

Summerson, J., (1957), The case for a theory of 'modern' architecture, in *The Unromantic Castle and Other Essays*, London, Thames and Hudson, 1990, pp257–66. Originally published in the *Royal Institute of British Architects Journal*, June 1957, pp307–10.

Summerson, J., (1966), *Inigo Jones*, Harmondsworth, Penguin.

Susskind, R. and Susskind, D., (2015), *The Future of the Professions: How technology will transform the work of human experts*, Oxford, Oxford University Press.

Svensson, L. and Evetts, J., (Eds), (2003), Conceptual and Comparative Studies of Continental and Anglo-American Professions, Goteborg Studies in Sociology, No129, Goteborg, Goteborg University.

Taleb, N.N., (2007), *The Black Swan: The impact of the highly improbable*, Harmondsworth, Penguin.

Tawney, R.H., (1920), *The Acquisitive Society*, New York, Harcourt, Brace and Company.

Taylor, M., (Ed.), (1837), Preface to the first edition, *Reports of the Late John Smeaton FRS*, London, M. Taylor.

Tett, G., (2015), *The Silo Effect: The peril of expertise and the promise of breaking down barriers*, London, Little, Brown.

The American Architect, (1909), Convention of the American Institute of Architects held at Washington, D.C., Dec. 14,15 and 16, *The American Architect* 96(1774), pp272–4.

The Architect, (1869), *The Architect* 1.

The Builder, (1845), *The Builder* 146 (22nd November).

The Builder, (1846), *The Builder* 4.

The Builder, (1847), *The Builder* (18th September).

The Builder, (1866), *The Builder* 24 (23rd June and 7th July).

The Builder, (1890), *The Builder* 48 (25th January).

The Edge, (2014a), Edge Commission on Future Professionalism: Note 7, March 2014, www.edgedebate.com/wp-content/uploads/2014/03/edge-professionalism-coi-2014-7-4.pdf

The Edge, (2014b), Edge Debate 61 – Edge Commission on Future Professionalism: Session 2 – The Economy, available from www.edgedebate.com/wp-content/uploads/2014/05/edge-debate-61-coi-future-professionalism-session-2.pdf

The Geological Society (1825) Governance Charter, available from www.geolsoc.org.uk/About/Governance/Charter [Accessed 29th May 2017].

The Royal College of Surgeons in London, (1800), *Charter of The Royal College of Surgeons in London*, available from https://archive.org/details/b28036517 [Accessed 7th May 2018].

The Society of Attorneys, Solicitors, Proctors, and others not being Barristers, practising in the Courts of Law and Equity of the United Kingdom, (1845), The Charter of the Society, available from www.lawsociety.org.uk/About-us/documents/royal-charters/.

The Society of Construction Law, (2017), Statement of Ethical Principles, available from www.scl.org.uk/resources/ethics [Accessed 28th April 2017].

The Surveyors' Institution, (1891), *Transactions*, Vol 21, London, The Surveyors' Institution.

Thompson, F.M.L., (1968), *Chartered Surveyors: The growth of a profession*, London, Routledge & Kegan Paul.

Till, J., (2009), *Architecture Depends*, Cambridge, MA, MIT Press.

Titmuss, R., (1958), *Essays on 'The Welfare State'*, London, Allen & Unwin.

Tomasi di Lampedusa, G., (1958), *The Leopard*, Colquhoun, A. (Trans.) (1960), London, Collins.

Tompkins, S., (2014), Written evidence to the Edge Commission of Inquiry on the Future of the Professions, available from www.edgedebate.com.

Tubbs, R., (1945), *The Englishman Builds*, London, Penguin.

Tudsbery-Turner, S., (2003), *William, Earl Lovelace, 1805–1893*, West Horsley, Surrey, The Horsley Countryside Preservation Society (rpt. from Surrey Archaeological Collections, LXX, 1974).

Turner & Townsend, (n.d.), Industry integrity, Turner & Townsend, www.turnerandtownsend. com/ [Accessed 3rd February 2017].

Turner, H., Clack, G. and Roberts, G., (1967), *Labour Relations in the Motor Industry*, London, Allen & Unwin.

Turner, J.B., (1853), *Industrial Universities for the People: Published in compliance with resolutions of the Chicago and Springfield Conventions and under the Industrial League of Illinois*, Jacksonville.

Turrell, P. *et al.*, (2002), A Gender for Change: The future for women in surveying, FIG XXII International Congress, 19th–26th April.

Turrell, P., Wilkinson, S. and Berry, M., (2005), The business skills of graduate surveyors, Proceedings of the Queensland University of Technology Research week international Conference, Queensland University of Technology, Brisbane.

Twinn, C., (2013), Professionalism, sustainability and the public interest: what next? *BRI* 41(1), pp123–8.

Uglow, J., (2012), *The Pinecone: The story of Sarah Losh, forgotten romantic heroine – antiquarian, architect, and visionary*, London, Faber and Faber.

United Nations, (2015), Paris Agreement, New York, United Nations.

United Nations, (2017), *World Population Prospects: The 2017 revision, key findings and advance tables*, Working Paper No. ESA/PWP/248, New York, United Nations Department of Economic and Social Affairs, Population Division.

University of Leicester, (2003), Income, inequalities and industrialisation, Unit 7, Occupational Classifications: The Census returns for England and Wales. Available from www.le.ac. uk/eh/teach/ug/modules/eh3107/occupations.pdf [Accessed 27th July 2017].

Urban Task Force, (1999), *Towards an Urban Renaissance, Mission Statement*, London, E&FN Spon.

van Schaik, L., (2005), *Mastering Architecture: Becoming a creative innovator in practice*, Chichester, Wiley-Academy.

Vesilind, P.A. and Gunn, A., (1998), *Engineering, Ethics and the Environment*, Cambridge, Cambridge University Press.

Volkwein, J.F. *et al.*, (2004), Engineering change: A study of the impact of EC2000, *International Journal of Engineering Education* 2(3), pp318–28.

Walford, S., (2009), Architecture in Tension: an examination of the position of the architect in the private and public sectors, focusing on the training and careers of Sir Basil Spence (1907–1976) and Sir Donald Gibson (1908–1991), University of Warwick, available from http://wrap.warwick.ac.uk/id/eprint/3216

Walker, L., (n.d.) *Golden Age or False Dawn? Women architects in the early 20th century*, London, English Heritage. Available from https://content.historicengland.org.uk/.../ women-architects-early-20th-century.pdf [Accessed: 29th August 2017].

Waterman, (2015), *Standing Out from the Crowd: Annual report and financial statement 2015*, Leeds, Waterman Group Plc.

Watkin, D., (Ed.), (2000), *Sir John Soane: The Royal Academy lectures*, Cambridge, Cambridge University Press.

Watson, G., (1988), *The Civils: The story of the Institution of Civil Engineers*, London, Thomas Telford.

Way, M. and Bordass, B., (2005), Making feedback and post-occupancy evaluation routine 2: Soft landings – involving design and building teams in improving performance, *Building Research & Information*, 33(4), pp353–60.

Weatherhead, A., (1941), *The History of Collegiate Architectural Education in the United States*, New York, Columbia University Press.

Webb, M., (1924), The first international congress on architectural education, *Journal of the Royal Institute of British Architects* (16th August) 31(18), p585.

Weisberg, D., (2006), The Engineering Design Revolution, available from www.cadhistory. net [Accessed 7th May 2018].

Welbourn, D., (2001), The Cambridge Engineering Tradition, available from www-g.eng. cam.ac.uk/125/achievements/tradition/index.htm [Accessed 18th August 2017].

Wells, H.W. (1960), *Wells Committee Report on the Education Policy of the RICS*, London, Royal Institution of Chartered Surveyors.

Welter, V., (2010), The limits of community – the possibilities of society: On modern architecture in Weimar Germany, *Oxford Art Journal* 33(1), pp63–80.

Whaley, J., (2012), *Germany and the Holy Roman Empire: Volume II: The peace of Westphalia to the dissolution of the Reich 1648–1806*, Oxford, Oxford University Press.

Wheeler, K., (2014), *Victorian Perceptions of Renaissance Architecture*, Farnham, Ashgate.

White, A., (2014), The Bartlett, architectural pedagogy and Wates House – an historical study, *Opticon1826* 16(26), pp1–19.

White, W., (1885), *The Past, Present and Future of the Architectural Profession: Being a paper read to the Leeds and Yorkshire Architectural Society*, London, Spottiswood.

Whittock, N. *et al.*, (1842), *The Complete Book of Trades or the Parents' Guide and Youth's Instructor: Forming a popular encyclopædia of trades, manufactures, and commerce as at present pursued in England with a particular regard to its state in and near the Metropolis: including a very copious table of every trade, profession, occupation, and calling, however divided and subdivided: together with the apprentice fee usually given with each, and an estimate of the sums required for commencing business*, London, Thomas Tegg.

Whyte, J. and Hartmann, T., (2017), How digitizing building information transforms the built environment, *Building Research & Information*, 45(6), pp591–5.

Wilkinson, S. and Hoxley, M., (2003), The Impact of RICS education reform on building surveying, *Structural Survey* 23(5), pp359–70.

Williams, D., (2009), Morrill Act's contribution to engineering's foundation, *The Bent of Tau Beta Pi* (Spring), pp15–20.

Wisely, W., (1974), *The American Civil Engineer: 1852–1974*, New York, American Society of Civil Engineers.

Withers, I., (2016), Qatar: The scrutiny will be on them, *Building* (19th May), available from www.building.co.uk/focus/qatar-the-scrutiny-will-be-on-them/5081742.article

Wolf, I., (2008), AC 2008–205: Engineering Technology and the 75th Anniversary Retrospective of ABET, Washington, DC, American Society for Engineering Education.

Wolfenden, J., (1958), The architect's role in society, *RIBA Journal* 65(3), pp186–90.

women in architecture, (2012), Ethel Mary Charles Bessie Ada Charles, available from http://women-in-architecture.com/fileadmin/wia/pdfs/pdfs_creating_change_profiles/Ethel_Mary_Charles_Bessie_Ada_Charles.pdf [Accessed: 6th October 2017].

Woodley, R., (1999), Professionals: Early episodes among architects and engineers, *Construction History* 15, pp15–22.

Woods, M., (1999), *From Craft to Profession: The practice of architecture in nineteenth-century America*, Berkeley, University of California Press.

Woollard, M., (1999), The Classification of Occupations in the 1881 Census of England and Wales, Historical Censuses and social surveys research group: Occasional paper No. 1, Colchester, University of Essex.

Worsop, E., (1582), *Discoverie of Sundrie Errors and Faults Daily Committed by Landemeaters*, London. British Library.

Wren, S., (1750), *Parentalia or Memoir of the Family of the Wrens*, London, T. Osborne and R. Dodsley.

WSP Parsons Brinkerhoff, (2015), *Code of Conduct*, WSP Parsons Brinkerhoff.

Yale College, (1828), *Reports on the Course of Instruction in Yale College by a Committee of the Corporation and the Academical Faculty*, New Haven, CT, Yale College. Available from http://collegiateway.org/reading/yale-report-1828/

Yoskuhl, A., (2014), Baumeister, bildung and civil service. Social and intellectual disputes in architecture and civil engineering in Germany during the second industrial revolution. *Wolkenkuckucksheim, Internationale Zeitschrift zur Theorie der Architektur* 19(33), pp111–26.

Young, E., (2015), Inclusion zone, *RIBA Journal* (2nd June), available from www.ribaj.com/intelligence/riba-role-models

Young, M. and Muller, J., (Eds), (2014), *Knowledge, Expertise and the Professions*, Abingdon, Routledge.

INDEX

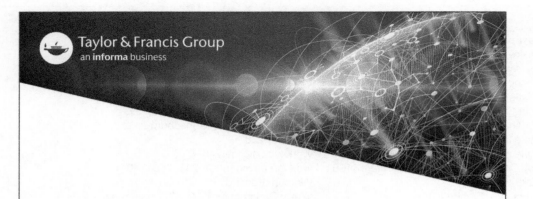